BIOZONE

Biology Modular Workbook Series

Environmental Science

The Biozone Writing Team

Tracey Greenwood

Lissa Bainbridge-Smith

Richard Allan

Daniel Butler

Published by:

Biozone International Ltd

109 Cambridge Road, Hamilton 2034, New Zealand

Printed by REPLIKA PRESS PVT LTD using paper
produced from renewable and waste materials

Distribution Offices:

United Kingdom & Europe	**Biozone Learning Media (UK) Ltd**, Scotland	
	Telephone:	+44 131-557-5060
	Fax:	+44 131-557-5030
	Email:	sales@biozone.co.uk
	Website:	www.biozone.co.uk
USA, Canada, South America, Africa	**Biozone International Ltd**, New Zealand	
	Telephone:	+64 7-856-8104
	Freefax:	1-800717-8751 (USA-Canada only)
	Fax:	+64 7-856-9243
	Email:	sales@biozone.co.nz
	Website:	www.biozone.co.nz
Asia & Australia	**Biozone Learning Media Australia**, Australia	
	Telephone:	+61 7-5575-4615
	Fax:	+61 7-5572-0161
	Email:	sales@biozone.com.au
	Website:	www.biozone.com.au

ISBN: 978-1-877462-15-3

Front cover photographs:

City in smog. Image courtesy of CDC ©2006

Iceberg, Antarctica. Image ©2007 istockphotos (www.istockphoto.com)

Biology Modular Workbook Series

The Biozone *Biology Modular Workbook Series* has been developed to meet the demands of customers with the requirement for a modular resource which can be used in a flexible way. Like Biozone's popular Student Course Workbooks, these resources provide a collection of visually interesting and accessible activities, which cater for students with a wide range of abilities and background. The workbooks are divided into a series of chapters, each comprising an introductory section with detailed learning objectives and useful resources, and a series of write-on activities ranging from paper practicals and data handling exercises, to questions requiring short essay style answers. Page tabs identifying "**Related activities**" in the workbook and "**Web links**" help students to locate related material within the workbook and provide access to web links and activities (including video clips and animations) that will enhance their understanding of the topic. Material for these workbooks has been drawn from Biozone's popular, widely used course books, but the workbooks have been structured with greater ease of use and flexibility in mind. During the development of this series, we have taken the opportunity to improve the design and content, while retaining the basic philosophy of a student-friendly resource which spans the gulf between textbook and study guide. With its unique, highly visual presentation, it is possible to engage and challenge students, increase their motivation and empower them to take control of their learning.

Environmental Science

This title in the *Biology Modular Workbook Series* provides students with a set of comprehensive guidelines and highly visual worksheets through which to explore aspects of environmental science theory and practice. *Environmental Science* is the ideal companion for students of environmental biology, encompassing the basic principles of geology, ecology and field biology, as well as the impact of humans on the natural environment. This workbook comprises six chapters each focusing on one particular area within this broad topic. These areas are explained through a series of activities, usually of one or two pages, each of which explores a specific concept (e.g. food chains or quadrat sampling). Model answers (on CD-ROM) accompany each order free of charge. *Environmental Science* is a student-centred resource and is part of a larger package, which also includes the ***Environmental Science Presentation Media CD-ROM*** (to be released in the near future). Students completing the activities, in concert with their other classroom and practical work, will consolidate existing knowledge and develop and practise skills that they will use throughout their course. This workbook may be used in the classroom or at home as a supplement to a standard textbook. Some activities are introductory in nature, while others may be used to consolidate and test concepts already covered by other means. Biozone has a commitment to produce a cost-effective, high quality resource, which acts as a student's companion throughout their biology study. Please do not photocopy from this workbook; we cannot afford to provide single copies to schools and continue to develop, update, and improve the material they contain.

Acknowledgements & Photo Credits

Royalty free images, purchased by Biozone International Ltd, are used throughout this manual and have been obtained from the following sources: istockphotos (www.istockphoto.com) • Corel Corporation from various titles in their Professional Photos CD-ROM collection; ©Hemera Technologies Inc, 1997-2001; © 2005 JupiterImages Corporation www.clipart.com; PhotoDisc®, Inc. USA, www.photodisc.com; ©Digital Vision; Wikimedia Commons. 3D models created using Poser IV, Curious Labs, 3D landscapes, Bryce 5.5. Biozone's authors also acknowledge the generosity of those who have kindly provided photographs for this edition: • Campus Photography at the University of Waikato for photographs monitoring instruments • Kurchatov Institute for the photograph of Chornobyl • Exxon Valdez Oil Spill Trustee Council for their photograph of dead seabirds • Stephen Moore for his photos of aquatic invertebrates • PASCO for their photographs of probeware • **EPA**: US Environmental Protection Agency • Jane Ussher for her photograph of the Albatross • Sam Banks for the Wombat scot photo • Coded credits are: **BH**: Brendan Hicks (Uni. of Waikato), **COD**: Colin O'Donnell, **DEQ**: Dept. of Environment QL, **DoC**: Dept. of Conservation (NZ), **DRNI**: Dept. of Natural Resources, Illinois, **EII**: Education Interactive Imaging, **EW**: Environment Waikato, **IF**: I. Flux (DoC), **JB-BU**: Jason Biggerstaff, Brandeis University, **JDG**: John Green (University of Waikato), **NASA**: National Aeronautics and Space Administration, **NASA-GSFC**: National Aeronautics and Administration- Goddard Space Flight Centre, **NOAA**: National Oceanic & Atmospheric Administration, www.photolib.noaa.gov, **RA**: Richard Allan, **RCN**: Ralph Cocklin, **TG**: Tracey Greenwood, • **USDA**: United States Department of Agriculture, • **USGS**: United States Geological Survey.

Also in this series:

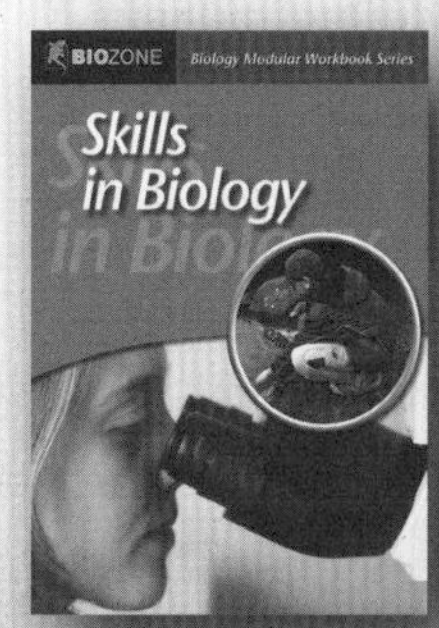

Skills in Biology

Ecology

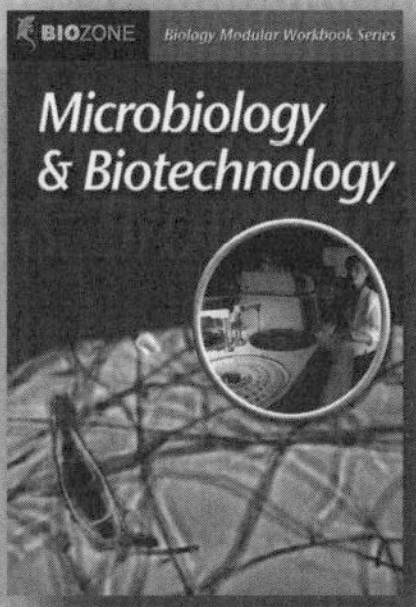

Microbiology & Biotechnology

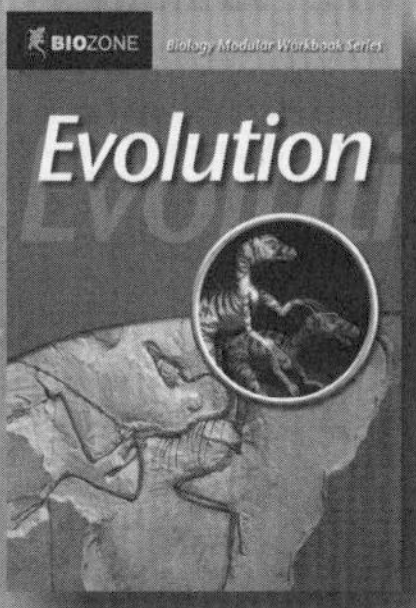

Evolution

For other titles in this series go to: www.thebiozone.com/modular.html

Contents

Activity is marked: • to be done; ✓ when completed

How to Use this Workbook

Environmental Science is designed to provide students with a resource that will make the acquisition of knowledge and skills in this area easier and more enjoyable. An understanding of geological and ecological theory is important in most biology curricula. Moreover, this subject is of high interest, with many opportunities to combine theory and practical work in the field.

This workbook is suitable for all students of the biological sciences, and will reinforce and extend the ideas developed by teachers. It is **not a textbook**; its aim is to complement the texts written for your particular course. *Environmental Science* provides the following resources in each chapter. You should refer back to them as you work through each set of worksheets.

Guidance Provided for Each Topic

Learning objectives:
These provide you with a map of the chapter content. Completing the learning objectives relevant to your course will help you to satisfy the knowledge requirements of your syllabus. Your teacher may decide to leave out points or add to this list.

Chapter content:
The upper panel of the header identifies the general content of the chapter. The lower panel provides a brief summary of the chapter content.

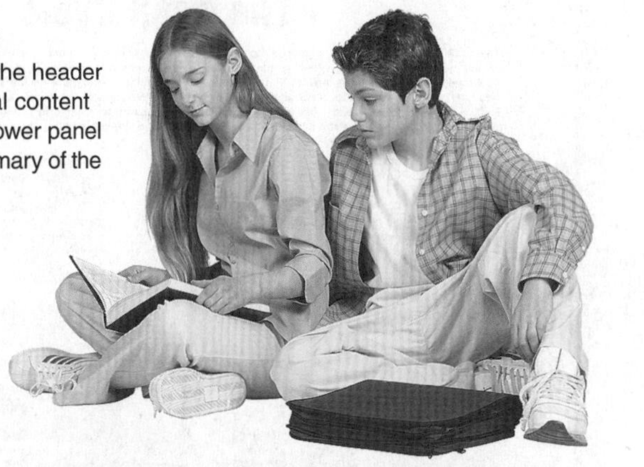

Key words:
Key words are displayed in **bold** type in the learning objectives and should be used to create a glossary as you study each topic. From your teacher's descriptions and your own reading, write your own definition for each word.

Note: Only the terms relevant to your selected learning objectives should be used to create your glossary. Free glossary worksheets are also available from our web site.

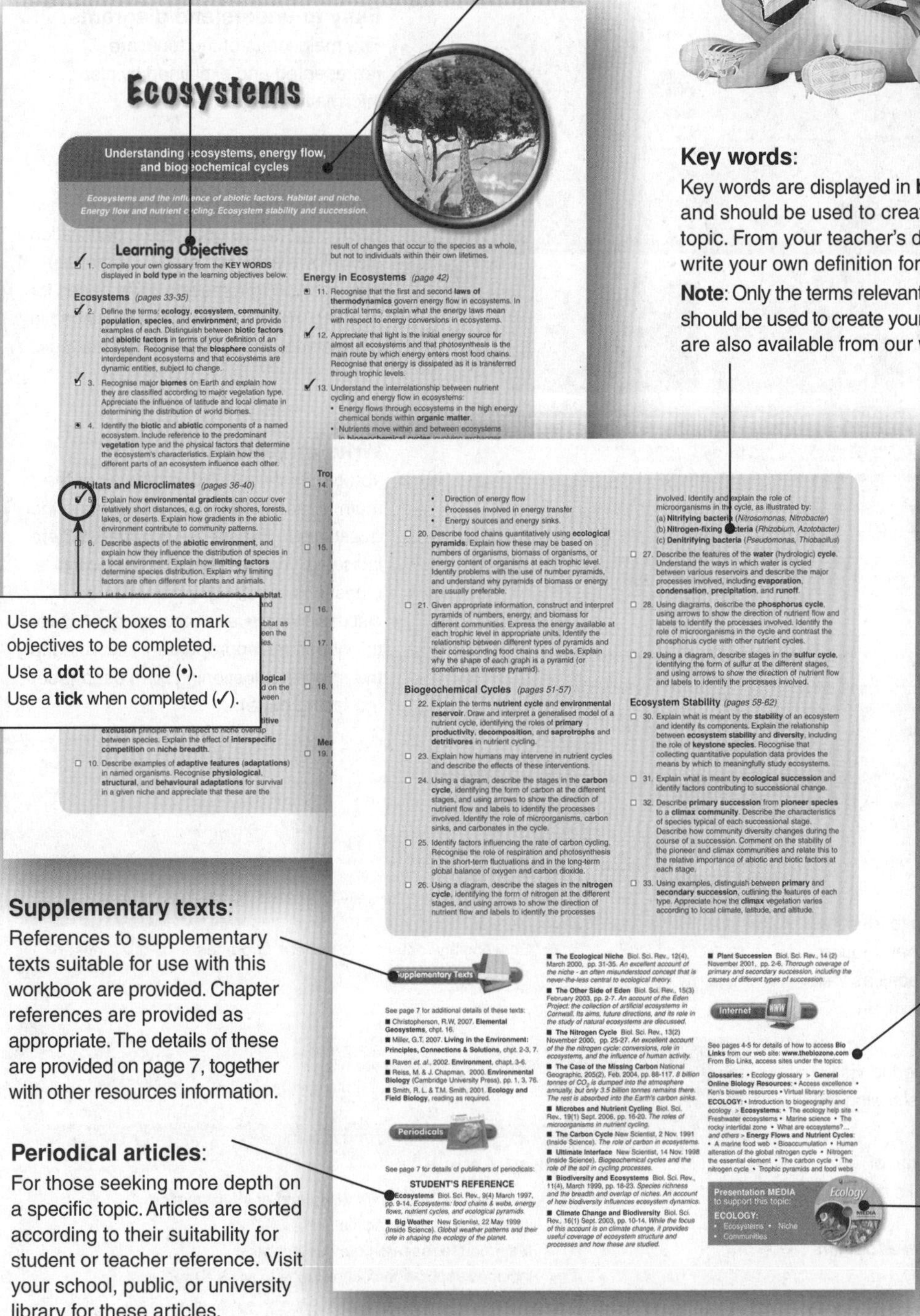

Use the check boxes to mark objectives to be completed.
Use a **dot** to be done (•).
Use a **tick** when completed (✓).

Supplementary texts:
References to supplementary texts suitable for use with this workbook are provided. Chapter references are provided as appropriate. The details of these are provided on page 7, together with other resources information.

Periodical articles:
For those seeking more depth on a specific topic. Articles are sorted according to their suitability for student or teacher reference. Visit your school, public, or university library for these articles.

Internet addresses:
Access our database of links to more than **800** web sites (updated regularly) relevant to the topics covered. Go to Biozone's own web site: **www.thebiozone.com** and link directly to listed sites using the *BioLinks* button.

Supplementary resources
Biozone's Presentation MEDIA are noted where appropriate.

Activity Pages

The activities and exercises make up most of the content of this workbook. They are designed to reinforce the concepts you have learned about in the topic. Your teacher may use the activity pages to introduce a topic for the first time, or you may use them to revise ideas already covered. They are excellent for use in the classroom, and as homework exercises and revision. In most cases, the activities should not be attempted until you have carried out the necessary background reading from your textbook. As a self-check, model answers for each activity are provided on CD-ROM with each order of workbooks.

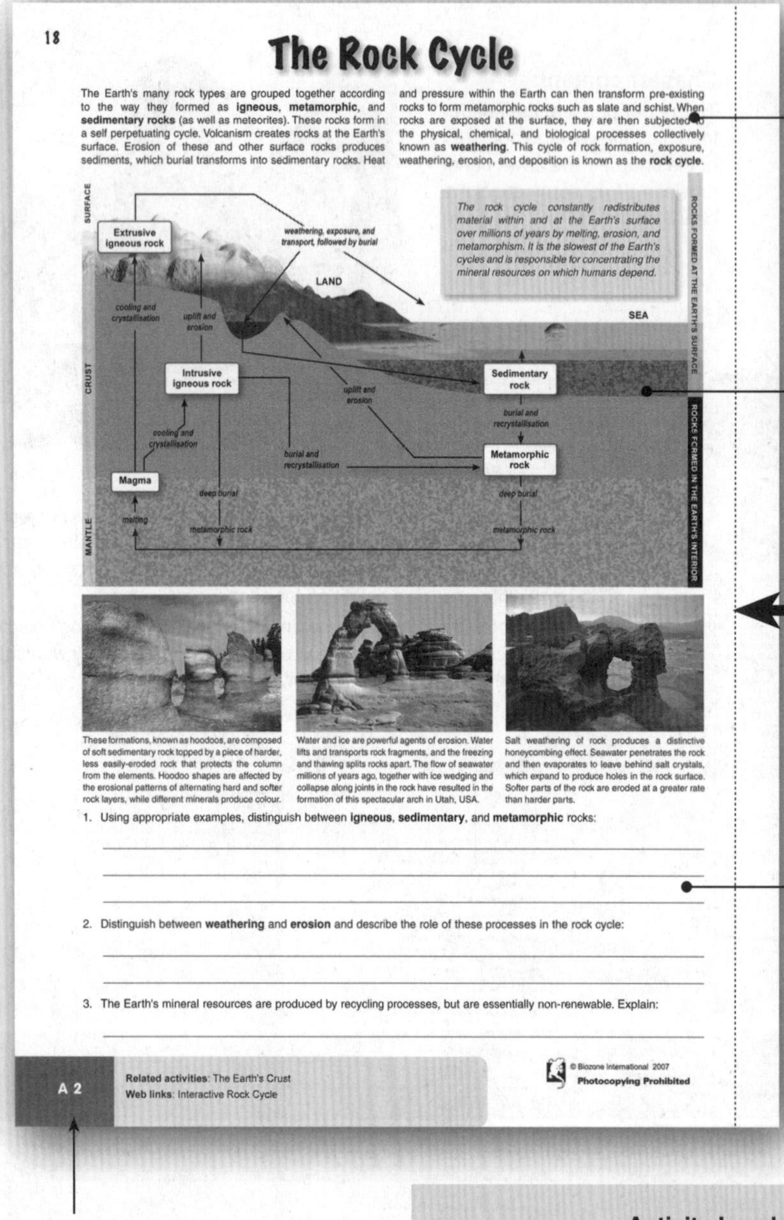
18

The Rock Cycle

The Earth's many rock types are grouped together according to the way they formed as **igneous**, **metamorphic**, and **sedimentary rocks** (as well as meteorites). These rocks form in a self perpetuating cycle. Volcanism creates rocks at the Earth's surface. Erosion of these and other surface rocks produces sediments, which burial transforms into sedimentary rocks. Heat and pressure within the Earth can then transform pre-existing rocks to form metamorphic rocks such as slate and schist. When rocks are exposed at the surface, they are then subjected to the physical, chemical, and biological processes collectively known as **weathering**. This cycle of rock formation, exposure, weathering, erosion, and deposition is known as the **rock cycle**.

The rock cycle constantly redistributes material within and at the Earth's surface over millions of years by melting, erosion, and metamorphism. It is the slowest of the Earth's cycles and is responsible for concentrating the mineral resources on which humans depend.

These formations, known as hoodoos, are composed of soft sedimentary rock topped by a piece of harder, less easily-eroded rock that protects the column from the elements. Hoodoo shapes are affected by the erosional patterns of alternating hard and softer rock layers, while different minerals produce colour.

Water and ice are powerful agents of erosion. Water lifts and transports rock fragments, and the freezing and thawing splits rocks apart. The flow of seawater millions of years ago, together with ice wedging and collapse along joints in the rock have resulted in the formation of this spectacular arch in Utah, USA.

Salt weathering of rock produces a distinctive honeycombing effect. Seawater penetrates the rock and then evaporates to leave behind salt crystals, which expand to produce holes in the rock surface. Softer parts of the rock are eroded at a greater rate than harder parts.

1. Using appropriate examples, distinguish between **igneous**, **sedimentary**, and **metamorphic** rocks:

2. Distinguish between **weathering** and **erosion** and describe the role of these processes in the rock cycle:

3. The Earth's mineral resources are produced by recycling processes, but are essentially non-renewable. Explain:

A 2

Related activities: The Earth's Crust
Web links: Interactive Rock Cycle

© Biozone International 2007
Photocopying Prohibited

Introductory paragraph:

The introductory paragraph sets the 'scene' for the focus of the page and provides important background information. Note any words appearing in **bold**; these are 'key words' which could be included in a glossary of biological terms for the topic.

Easy to understand diagrams:

The main ideas of the topic are represented and explained by clear, informative diagrams.

Tear-out pages:

Each page of the book has a perforation that allows easy removal. Your teacher may ask you to remove activity pages for marking, or so that they can be placed in a ringbinder with other work on the topic.

Write-on format:

You can test your understanding of the main ideas of the topic by answering the questions in the spaces provided. Where indicated, your answers should be concise. Questions requiring explanation or discussion are spaced accordingly. Answer the questions appropriately according to the specific questioning term used (see the facing page).

Activity code and links:

Activity codes (explained right) help to identify the type of activities and the skills they require. Most activities require knowledge recall as well as the application of knowledge to explain observations or predict outcomes.

Use the **'Related activities'** indicated to visit pages that may help you with understanding the material or answering the questions.

Web links indicate additional material of assistance or interest available via links to helpful web pages). Access these from: *www.thebiozone.com/weblink/EnvSci-2153.html*

Activity Level

1 = Generally a simpler activity with mostly "describe" questions
2 = More challenging material (including "explain" questions)
3 = Challenging content or questions (more "discuss" questions)

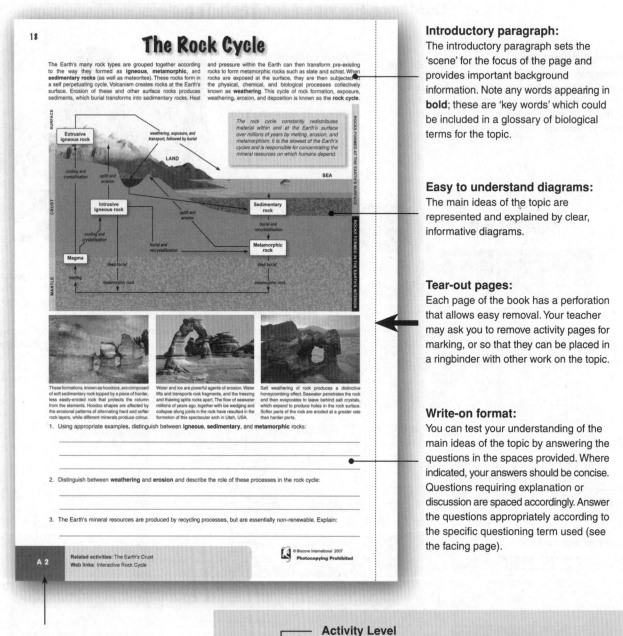

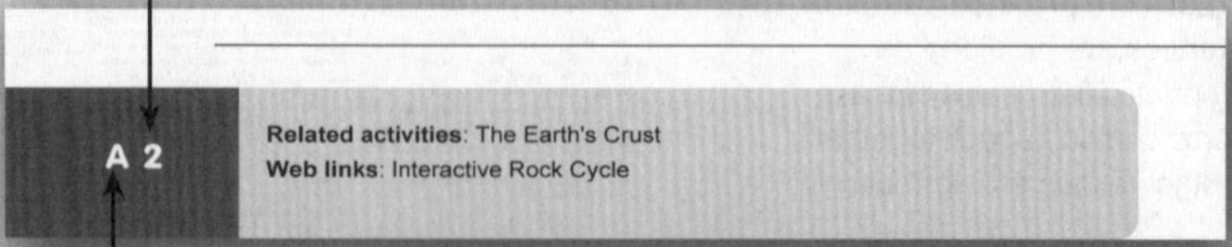

Type of Activity

D = Includes some data handling and/or interpretation
P = includes a paper practical
R = May require research outside the page
A = Includes application of knowledge to solve a problem
E = Extension material

Explanation of Terms

Questions come in a variety of forms. Whether you are studying for an exam or writing an essay, it is important to understand exactly what the question is asking. A question has two parts to it: one part of the question will provide you with information, the second part of the question will provide you with instructions as to how to answer the question. Following these instructions is most important. Often students in examinations know the material but fail to follow instructions and do not answer the question appropriately. Examiners often use certain key words to introduce questions. Look out for them and be clear as to what they mean. Below is a description of terms commonly used when asking questions in biology.

Commonly used Terms in Biology

The following terms are frequently used when asking questions in examinations and assessments. Students should have a clear understanding of each of the following terms and use this understanding to answer questions appropriately.

Account for: Provide a satisfactory explanation or reason for an observation.

Analyse: Interpret data to reach stated conclusions.

Annotate: Add **brief** notes to a diagram, drawing or graph.

Apply: Use an idea, equation, principle, theory, or law in a new situation.

Appreciate: To understand the meaning or relevance of a particular situation.

Calculate: Find an answer using mathematical methods. Show the working unless instructed not to.

Compare: Give an account of similarities and differences between two or more items, referring to both (or all) of them throughout. Comparisons can be given using a table. Comparisons generally ask for similarities more than differences (see contrast).

Construct: Represent or develop in graphical form.

Contrast: Show differences. Set in opposition.

Deduce: Reach a conclusion from information given.

Define: Give the precise meaning of a word or phrase as concisely as possible.

Derive: Manipulate a mathematical equation to give a new equation or result.

Describe: Give a detailed account, including all the relevant information.

Design: Produce a plan, object, simulation or model.

Determine: Find the only possible answer.

Discuss: Give an account including, where possible, a range of arguments, assessments of the relative importance of various factors, or comparison of alternative hypotheses.

Distinguish: Give the difference(s) between two or more different items.

Draw: Represent by means of pencil lines. Add labels unless told not to do so.

Estimate: Find an approximate value for an unknown quantity, based on the information provided and application of scientific knowledge.

Evaluate: Assess the implications and limitations.

Explain: Give a clear account including causes, reasons, or mechanisms.

Identify: Find an answer from a number of possibilities.

Illustrate: Give concrete examples. Explain clearly by using comparisons or examples.

Interpret: Comment upon, give examples, describe relationships. Describe, then evaluate.

List: Give a sequence of names or other brief answers with no elaboration. Each one should be clearly distinguishable from the others.

Measure: Find a value for a quantity.

Outline: Give a brief account or summary. Include essential information only.

Predict: Give an expected result.

Solve: Obtain an answer using algebraic and/or numerical methods.

State: Give a specific name, value, or other answer. No supporting argument or calculation is necessary.

Suggest: Propose a hypothesis or other possible explanation.

Summarise: Give a brief, condensed account. Include conclusions and avoid unnecessary details.

In Conclusion

Students should familiarise themselves with this list of terms and, where necessary throughout the course, they should refer back to them when answering questions. The list of terms mentioned above is not exhaustive and students should compare this list with past examination papers / essays etc. and add any new terms (and their meaning) to the list above. The aim is to become familiar with interpreting the question and answering it appropriately.

Using the Internet

The internet is a powerful resource for locating information. There are several key areas of Biozone's web site that may be of interest to you. Go to the **BioLinks** area to browse through the hundreds of web sites hosted by other organisations. These sites provide a supplement to the activities provided in our workbooks and have been selected on the basis of their accurate, current, and relevant content. We have also provided links to biology-related **podcasts** and **RSS newsfeeds**. These provide regularly updated information about new discoveries in biology; perfect for those wanting to keep abreast of changes in this dynamic field.

The BIOZONE website: www.thebiozone.com

The current internet address (URL) for the web site is displayed here. You can type a new address directly into this space.

Use Google to search for web sites of interest. The more precise your search words are, the better the list of results. EXAMPLE: If you type in "biotechnology", your search will return an overwhelmingly large number of sites, many of which will not be useful to you. Be more specific, e.g. "biotechnology medicine DNA uses".

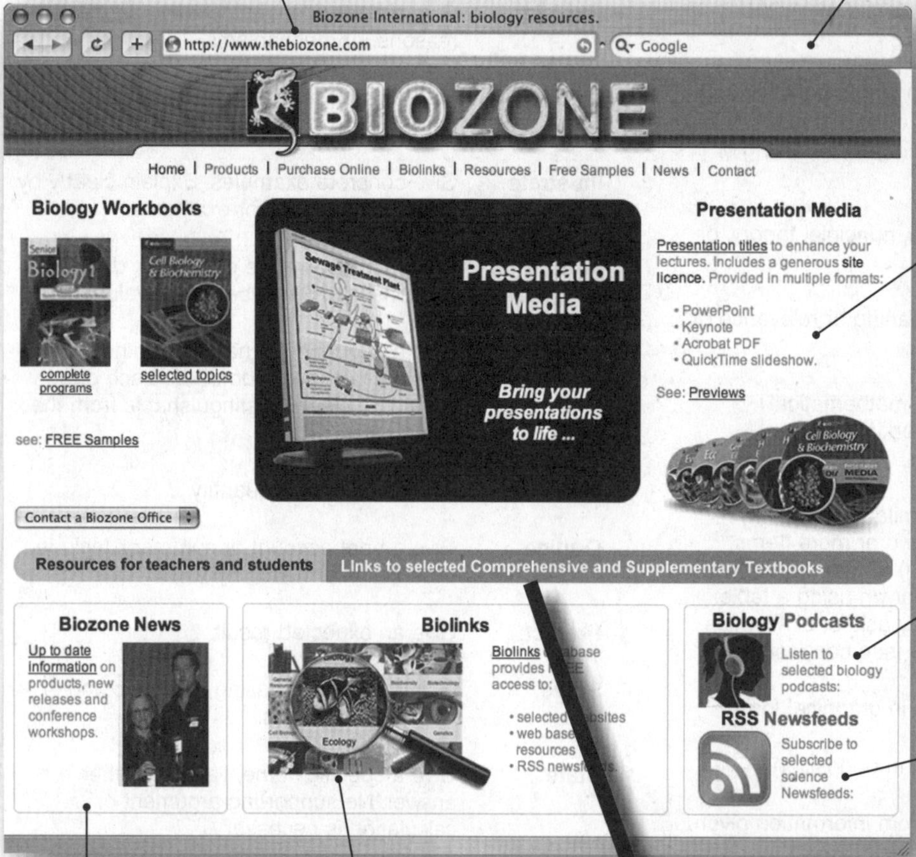

Find out about our superb **Presentation Media**. These slide shows are designed to provide in-depth, highly accessible illustrative material and notes on specific areas of biology.

Podcasts: Access the latest news as audio files (mp3) that may be downloaded to your ipod (mp3 player) or played directly off your computer.

RSS Newsfeeds: See breaking news and major new discoveries in biology directly from our web site.

Access the **BioLinks** database of web sites related to each major area of biology.

The **Resource Hub** provides links to the supporting resources referenced in the workbook. These resources include comprehensive and supplementary texts, biology dictionaries, computer software, videos, and science supplies.

News: Find out about product announcements, shipping dates, and workshops and trade displays by Biozone at teachers' conferences around the world.

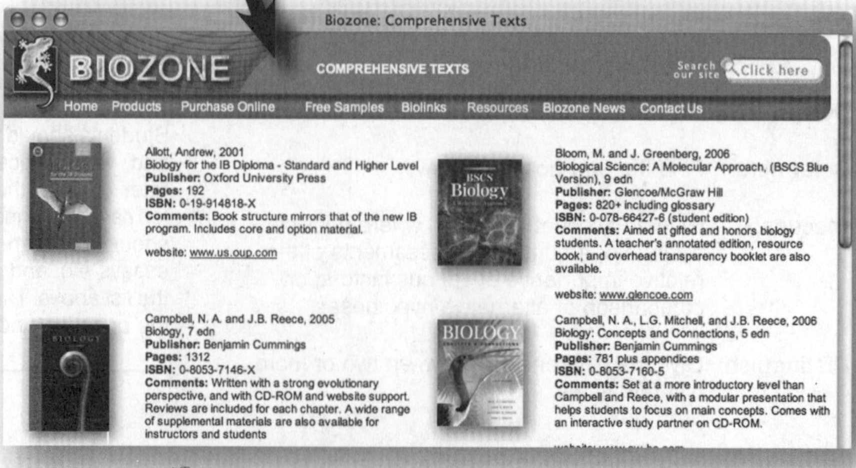

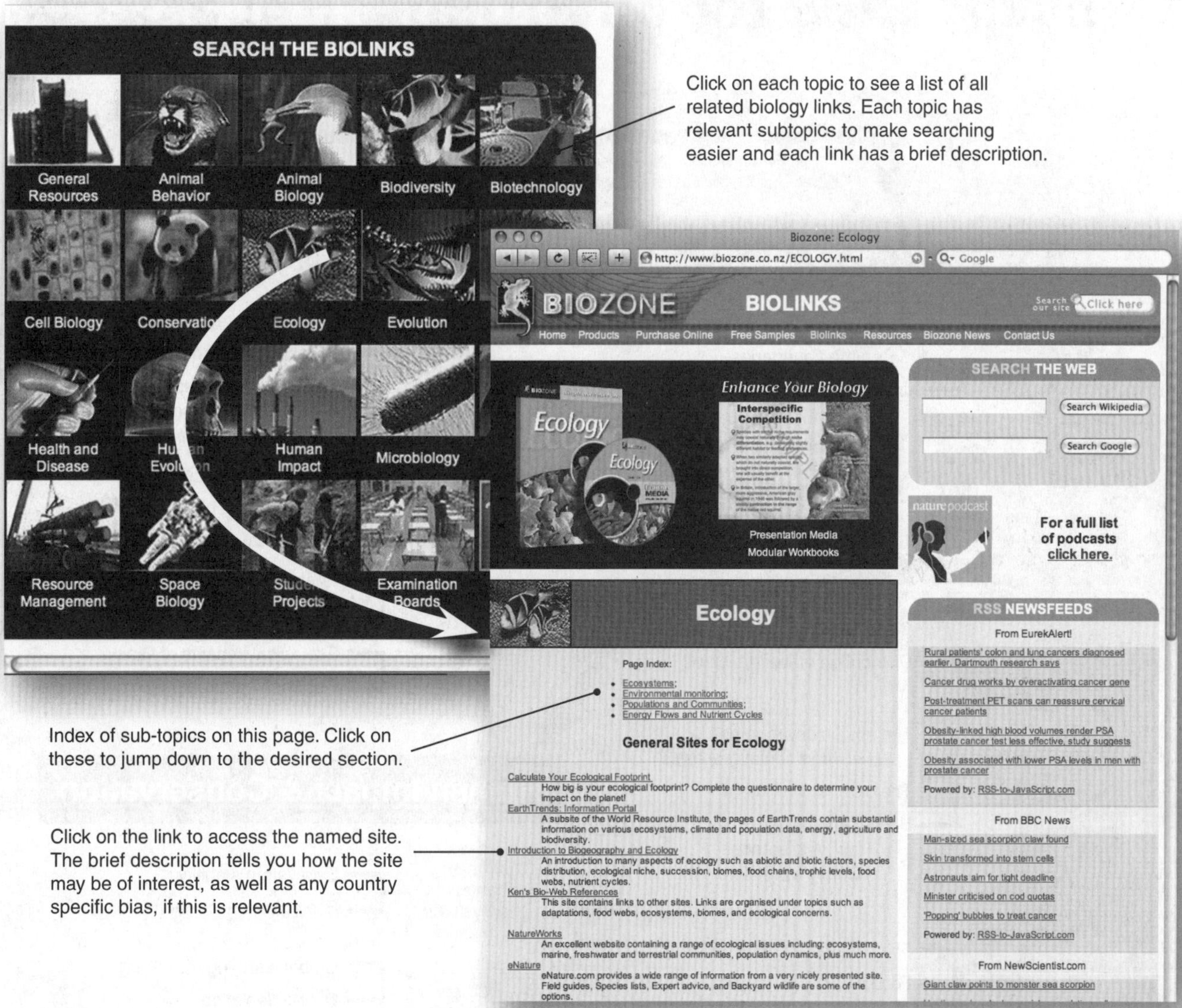

Click on each topic to see a list of all related biology links. Each topic has relevant subtopics to make searching easier and each link has a brief description.

Index of sub-topics on this page. Click on these to jump down to the desired section.

Click on the link to access the named site. The brief description tells you how the site may be of interest, as well as any country specific bias, if this is relevant.

Weblinks:

Go to: **www.thebiozone.com/weblink/EnvSci-2153.html**

Throughout this workbook, some pages make reference to additional or alternative activities, as well as web sites that have particular relevance to the activity. See example of page reference below:

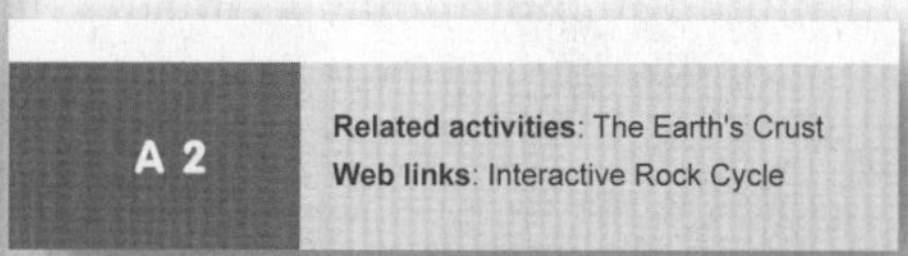

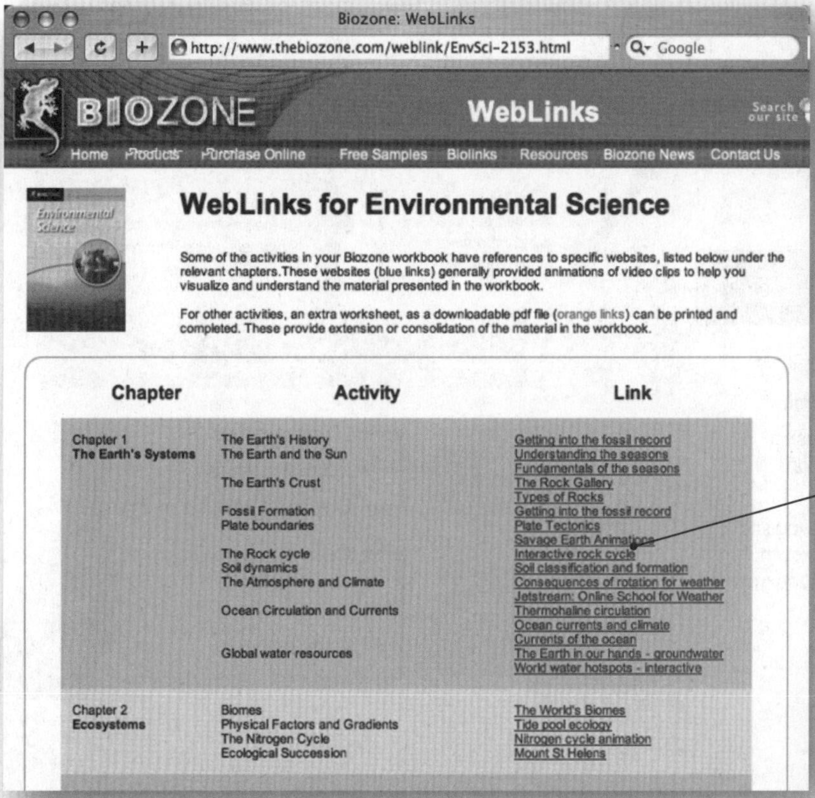

Web Link: provides a link to an **external web site** with supporting information for the activity. These sites have been specifically chosen for their clear presentation, accuracy, and appeal. They often include helpful explanatory animations.

Concept Map for Environmental Science

Human exploitation of the environment is dependent on, and influenced by the Earth's systems: the soil, the water, the climate

The Earth's Systems

Earth's systems
- Earth's crust
  - plate boundaries
- Atmosphere and climate

Ecosystems encompass communities and their environments on Earth

Ecosystems

Biomes
- Classification of world biomes
- Distribution of world biomes

Components of ecosystems
- The abiotic environment
  - environmental gradients
- The biotic environment
- Habitat and niche
  - law of tolerances
  - Gause's principle

Energy flow in ecosystems
- Food chains & trophic levels
- Food webs
- Ecological pyramids

Biogeochemical cycles
- Nutrient cycles
  - carbon cycle
  - nitrogen cycle
  - phosphorus cycle
- Hydrologic cycle

Populations make up communities; the biotic component of ecosystems.

Populations

Population dynamics
- Features of populations
- Density and distribution
- Life tables & survivorship
  - *r* and K selection
- Population growth
  - exponential vs logistic
  - regulation of growth
  - human population growth
  - urbanisation

Interspecific interactions
- Predator-prey
- Symbiosis

Humans can change ecosystems

Food and Agriculture

Farming & nutrition
- The green revolution
- World food crops
- Farming practices
  - industrialised vs traditional
  - sustainable agriculture

Pest control
- Integrated pest mangagement
- Pesticides & bioaccumulation
- Biological control

Energy
- Energy resources and use
- Energy conservation

Fisheries
- Ecological impacts of fishing
- Fisheries management

Human Impact & Conservation

Natural changes in ecosystems
- Ecosystem stability
- Ecological succession

Human induced changes
- Global warming
- Ozone depletion
- Pollution
- Deflected successions

Resource management
- Energy, waste, sewage
- Grasslands, forestry, fisheries

Biodiversity & conservation issues
- Loss of biodiversity
- Management practices

We can sample the components of ecosystems

Monitoring ecosystems allows biologists to detect change

Investigating Ecosystems

Sampling populations
- Direct methods
- Indirect methods
- Statistical analysis of data

Sampling physical factors
- Collecting samples
  - water
  - soil
- Dataloggers
- Meters & secchi discs

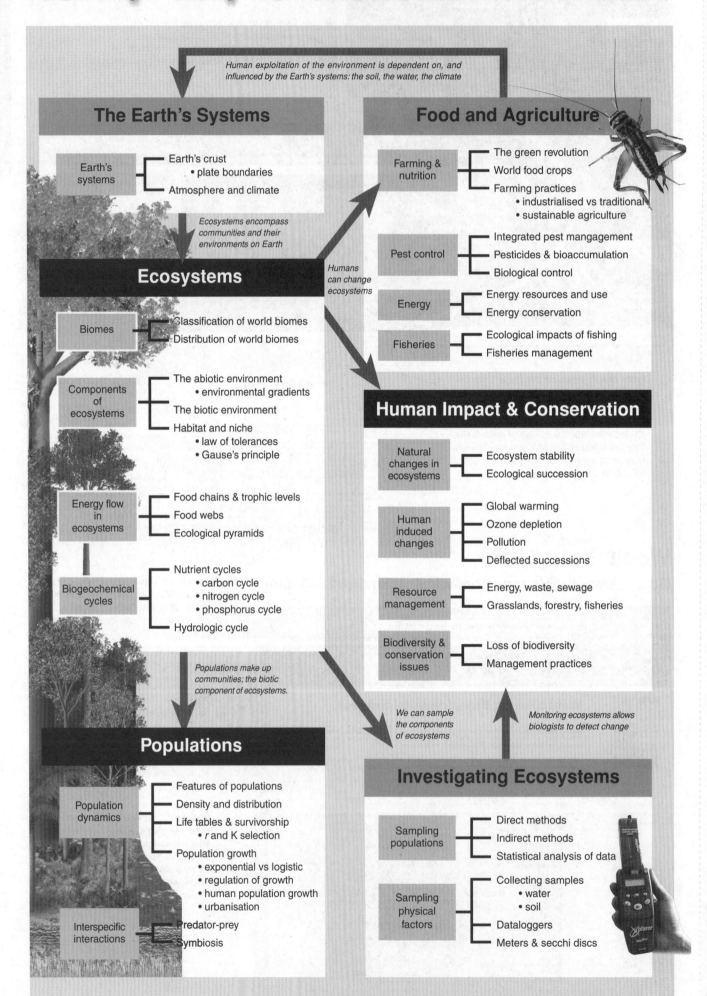

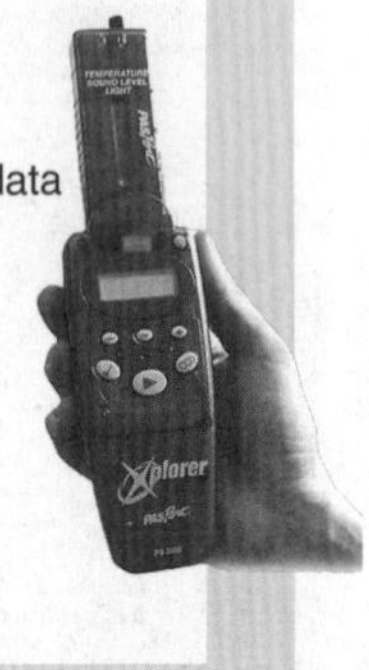

Resources Information

Your set textbook should always be a starting point for information, but there are also many other resources available. A list of readily available resources is provided below. Access to the publishers of these resources can be made directly from Biozone's web site through our resources hub: **www.thebiozone.com/resource-hub.html**. Please note that our listing of any product in this workbook does not denote Biozone's endorsement of it.

Supplementary Texts

Brower, J.E, J.H. Zar, & C.N. von Ende, 1997.
Field and Laboratory Methods for General Ecology, 288 pp. (spiral bound)
Publisher: McGraw-Hill
ISBN: 0697243583
Comments: *An introductory manual for ecology, focussing on the collection, recording, and analysis of data. Provides balanced coverage of plants and animals, and physical elements.*

Christopherson, R.W, 2007. (5 edn).
Elemental Geosystems, 640 pp.
ISBN: 9780131497023
Although written as a geography textbook, much of the material in this book is relevant to environmental sciences courses. Relevant material includes chapters on the atmosphere, weather, climate, and landscape and biogeographical systems.

Miller, G.T. 2007. (15 edn).
Living in the Environment: Principles, Connections and Solutions, 627 pp.
ISBN: 9780495015987
A comprehensive textbook covering aspects of ecology, biodiversity, natural resources, human impact on the environment and sustainability issues. A large number of appendices topics expand the scope of the material further.

Raven, P.H, Berg, L.R, & Hassenzahi, D.M. 2008. (6 edn).
ISBN: 9780470119266
A comprehensive textbook providing an overview of ecosystems, populations, the world's resources, and the impact of human activity on the environment.

Reiss, M. & J. Chapman, 2000.
Environmental Biology, 104 pp.
ISBN: 0521787270
An introduction to environmental biology covering agriculture, pollution, resource conservation and conservation issues, and practical work in ecology. Questions and exercises are provided and each chapter includes an introduction and summary.

Smith, R.L. and T.M. Smith, R. 2001 (6 edn).
Ecology and Field Biology, 700 pp.
ISBN: 0321042905
A comprehensive overview of all aspects of ecology, including evolution, ecosystems theory, practical application, biogeochemical cycles, and global change. A field package, comprising a student "Ecology Action Guide" and a subscription to the web based "The Ecology Place" is also available.

Biology Dictionaries

Access to a good biology dictionary is useful when dealing with biological terms. Some of the titles available are listed below. Link to the relevant publisher via Biozone's resources hub or by typing: **www.thebiozone.com/resources/dictionaries-pg1.html**

Clamp, A. **AS/A-Level Biology. Essential Word Dictionary**, 2000, 161 pp. Philip Allan Updates.
ISBN: 0-86003-372-4.
Carefully selected essential words for AS and A2. Concise definitions are supported by further explanation and illustrations where required.

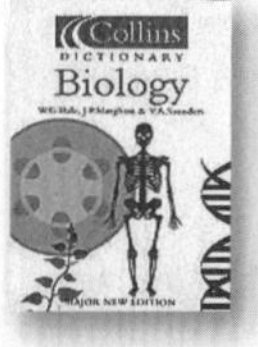

Hale, W.G., J.P. Margham, & V.A. Saunders.
Collins: Dictionary of Biology 3 ed. 2003, 672 pp.
HarperCollins. **ISBN**: 0-00-714709-0.
Updated to take in the latest developments in biology from the Human Genome Project to advancements in cloning (new edition pending).

Henderson, I.F, W.D. Henderson, and E. Lawrence.
Henderson's Dictionary of Biological Terms, 1999, 736 pp. Prentice Hall. **ISBN**: 0582414989
This edition has been updated, rewritten for clarity, and reorganised for ease of use. An essential reference and the dictionary of choice for many.

McGraw-Hill (ed). **McGraw-Hill Dictionary of Bioscience**, 2 ed., 2002, 662 pp. McGraw-Hill.
ISBN: 0-07-141043-0
22 000 entries encompassing more than 20 areas of the life sciences. It includes synonyms, acronyms, abbreviations, and pronunciations for all terms.

Periodicals, Magazines, & Journals

Biological Sciences Review: *An informative quarterly publication for biology students.* Enquiries: **UK**: Philip Allan Publishers **Tel**: 01869 338652 **Fax**: 01869 338803 **E-mail**: sales@philipallan.co.uk **Australasia**: **Tel**: 08 8278 5916, **E-mail**: rjmorton@adelaide.on.net

New Scientist: *Widely available weekly magazine with research summaries and features.* Enquiries: Reed Business Information Ltd, 51 Wardour St. London WIV 4BN **Tel**: (UK and intl):+44 (0) 1444 475636 **E-mail**: ns.subs@qss-uk.com *or subscribe from their web site.*

Scientific American: *A monthly magazine containing specialist features. Articles range in level of reading difficulty and assumed knowledge.* Subscription enquiries: 415 Madison Ave. New York. NY10017-1111 **Tel**: (outside North America): 515-247-7631 **Tel**: (US& Canada): 800-333-1199

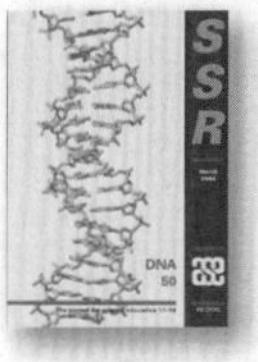

School Science Review: *A quarterly journal which includes articles, reviews, and news on current research and curriculum development. Free to Ordinary Members of the ASE or available on subscription.* Enquiries: **Tel**: 01707 28300 **Email**: info@ase.org.uk *or visit their web site.*

The American Biology Teacher: *The peer-reviewed journal of the NABT. Published nine times a year and containing information and activities relevant to biology teachers.* Contact: NABT, 12030 Sunrise Valley Drive, #110, Reston, VA 20191-3409 **Web**: www.nabt.org

Advanced Placement Course

The following guide provides an outline of the major topics within the AP Environmental Science course, matched to their corresponding chapters in this workbook. Support material is provided by weblinks and references to suggested reading.

Topic		See workbook
Topic I:	**Earth's Systems and Resources**	
A	**Earth Science Concepts**	
	Geologic time scale, plate tectonics, earthquakes, volcanism, seasons, solar intensity and latitude.	The Earth's Systems
B	**The Atmosphere**	
	Composition, structure, weather, climate, atmospheric circulation, the Coriolis Effect, atmosphere-ocean interactions.	The Earth's Systems
C	**Global Water Resources and Use**	
	Freshwater, saltwater, ocean circulation, water use, problems & conservation.	The Earth's Systems Pollution & Global Change
D	**Soil and Soil Dynamics**	
	Rock cycle, formation, composition, properties, soil types and issues, erosion.	The Earth's Systems Land, Water & Energy
Topic II:	**The Living World**	
A	**Ecosystem Structure**	
	Biological populations and communities, niches, species interactions, keystone species, species diversity, biomes.	Ecosystems Investigating Ecosystems
B	**Energy Flow**	
	Photosynthesis, cellular respiration, food webs, trophic levels, ecological pyramids.	Ecosystems Populations
C	**Ecosystem Diversity**	
	Biodiversity, natural selection, evolution, ecosystem services.	Ecosystems
D	**Natural Ecosystem Change**	
	Climate shifts, species movement, ecological succession.	Ecosystems Pollution & Global Change
E	**Natural Biogeochemical Cycles**	
	Carbon, nitrogen, phosphorus, sulfur, and water cycles. Conservation of matter.	Ecosystems
Topic III:	**Population**	
A	**Population Biology Concepts**	
	Population ecology, carrying capacity, reproductive strategies, survivorship.	Populations
B	**Human Population**	
1.	Human population dynamics; historical information, distribution, fertility rates, growth rates and doubling times, demographic transition, age-structure.	Populations
2.	Population size; sustainability strategies, case studies, national policies.	Populations
3.	Impacts of population growth; hunger, disease, economic effects, resource use and habitat destruction.	Populations Land, Water & Energy
Topic IV:	**Land and Water Use**	
A	**Agriculture**	
1.	Human nutritional requirements, Green Revolution, agriculture types and sustainability, genetic engineering and crop production, deforestation, irrigation.	The Earth's Systems Land, Water & Energy Pollution & Global Change
2.	Pest control; pesticides and their use, integrated pest management.	Land, Water & Energy
B	**Forestry**	
	Tree plantations, old growth forests, forest fire, forest management, national forests.	Pollution & Global Change
C	**Rangelands**	
	Overgrazing, deforestation, desertification, rangeland management, federal rangelands.	Land, Water & Energy Pollution & Global Change
D	**Other Land Use**	
1.	Urban land development (urbanisation).	Populations
2.	Impact of transport and infrastructure.	Populations
3.	Public and Federal lands; management, types of land areas.	Pollution & Global Change
4 - 5.	Land conservation and sustainable use.	Land, Water & Energy
E	**Mining**	
	Mineral formation, extraction, global reserves, relevant laws and treaties.	
F	**Fishing**	
	Fishing techniques, overfishing, aquaculture, relevant laws and treaties.	Land, Water & Energy

Topic		See workbook
G	**Global Economics**	
	Globalisation, World Bank, Tragedy of the Commons, relevant laws and treaties.	
Topic V:	**Energy Resources and Consumption**	
A	**Energy Concepts**	
	Energy forms, power, units, conversion, Laws of Thermodynamics.	Ecosystems Land, Water & Energy
B	**Energy Consumption**	
1.	Historical energy use.	
2.	Present global energy use.	
3.	Future energy requirements	
C	**Fossil Fuel Resources and Use**	
	Coal, oil and natural gas formation, extraction/purification methods, reserves and demands, synfuels, environmental considerations.	Land, Water & Energy Ecosystems
D	**Nuclear Energy**	
	Fission process, nuclear fuel, electricity production, reactor types, environmental considerations, safety issues.	Land, Water & Energy
E	**Hydroelectric Power**	
	Dams, flood control, impact.	Land, Water & Energy
F	**Energy Conservation**	
	Energy efficiency, CAFE standards, hybrid electric vehicles, mass transit.	Land, Water, & Energy
G	**Renewable Energy**	
	Types of renewable energy and their environmental impact.	
Topic VI:	**Pollution**	
A	**Pollution Types**	
1.	Air pollution; sources, major air pollutants, measurement, smog, acid deposition, heat islands and temperature inversions, indoor air pollution, remediation and reduction strategies, relevant laws.	Pollution & Global Change
2.	Noise pollution; sources, effects and controls.	
3.	Water pollution; types and sources, cause and effects, eutrophication, groundwater pollution, water quality, purification, sewage, relevant acts.	The Earth's Systems Pollution & Global Change
4.	Solid waste; types, disposal and reduction.	Pollution & Global Change
B	**Impacts on the Environment and Human Health**	
1.	Hazards to human health, risk analysis, acute and chronic effects, dose response, air pollutants, smoking and other risks.	Pollution & Global Change
2.	Hazardous chemicals; types of hazardous waste, treatment and disposal, cleaning contaminated sites, biomagnification, relevant laws.	
C	**Economic Impacts**	
	Cost-benefit analysis, marginal cost, sustainability, externalities.	
Topic VII:	**Global Change**	
A	**Stratospheric Ozone**	
	Formation of stratospheric ozone and its depletion, UV radiation, laws and treaties.	Pollution & Global Change
B	**Global Warming**	
	Greenhouse gases, greenhouse effect, global warming, reducing climate change, relevant laws and treaties.	Pollution & Global Change
C	**Loss of Biodiversity**	
1.	Habitat loss, overuse, pollution, introduced species, endangered and extinct species.	Pollution & Global Change
2.	Maintenance through conservation	Pollution & Global Change
3.	Relevant laws and treaties	
Laboratory and Field Investigation		
	This course comprises diverse laboratory and field investigations such as: collecting and analysing samples, long term studies of a local system or environmental feature, analysis of a real data set, or a visit to a public facility (e.g. water treatment plant).	Investigating Ecosystems

International Baccalaureate Course

The following guide provides an outline of the major topics within the IB Environmental Systems course, matched to their corresponding chapters in this workbook. Support material is provided by weblinks and references to suggested reading.

Topic		See workbook
CORE: *(All students)*		
1	**Systems and Models**	
1.1.1- 1.1.2	The concept and characteristics of a system (open, closed and isolated systems).	
1.1.3 - 1.1.4	Describe how the laws of thermodynamics are relevant to environmental systems. Equilibria.	Ecosystems
1.1.5	Positive and negative feedback regulation.	
1.1.6	Transfer and transformation processes.	Ecosystems
1.1.7 - 1.1.8	Energy flows (inputs and outputs) and storages (stock) in a system. Quantitative flow models.	Ecosystems
2	**The Ecosystem**	
2.1	**Structure**	
2.1.1	Distinguish between biotic and abiotic.	Ecosystems
2.1.2- 2.1.3	Define trophic level. Identify and explain trophic levels in local food chains and food webs.	Ecosystems
2.1.4- 2.1.5	Ecosystem pyramids (numbers, biomass, productivity). Affects on ecosystem functions.	Ecosystems
2.1.6	Define species, population, community, niche and habitat. Give local examples.	Ecosystems
2.1.7	Define the term biome.	Ecosystems
2.1.8- 2.1.9	Distribution, relative productivity and comparison of several different ecosystems.	Ecosystems
2.1.10	Describe and explain populations interactions.	Populations
2.2	**Function**	
2.2.1	Role of producers, consumers & decomposers.	Ecosystems
2.2.2 - 2.2.3	Describe photosynthesis and respiration. Energy transfer through an ecosystem.	Ecosystems
2.2.4 - 2.2.5	Productivity terms and definitions. Gross and net productivity calculations.	Ecosystems
2.2.6	Negative and positive feedback mechanisms.	
2.3	**Changes**	
2.3.1- 2.3.2	Population limiting factors and carrying capacity. S and J population growth curves.	Populations
2.3.3 - 2.3.4	Effects of density, internal and external factors on population regulation. *K*- and *r*- strategies.	Populations
2.3.5	Describe succession in a named habitat.	Ecosystems
2.3.6	Describe photosynthesis and respiration.	Ecosystems
2.3.7	Describe factors affecting climax communities.	Ecosystems
3	**Global Cycles and Physical Systems**	
3.1	**The Atmosphere**	
3.1.1- 3.1.2	The structure & composition of the atmosphere. Describe the global atmospheric energy budget.	The Earth's systems
3.1.3 - 3.1.5	Atmospheric circulation & heat distribution, circulation patterns. Affect on climate & biomes.	The Earth's systems
3.2	**Depletion of Stratospheric Ozone**	
3.2.1- 3.2.2	The role of ozone in UV radiation absorption and interaction with halogenated organic gases.	Pollution & Global Change
3.2.3	Effects of UV on biological activity & productivity.	Pollution & Global Change
3.2.4 - 3.2.5	Methods to reduce ozone depleting substances, and the role of organisations to achieve this.	Pollution & Global Change
3.3	**Tropospheric Ozone**	
3.3.1- 3.3.2	Source and effect of tropospheric ozone and the formation of photochemical smog.	Pollution & Global Change
3.4	**The Issue of Global Warming**	
3.4.1- 3.4.3	The role of greenhouse gases in global temperature and the impact of human activities. Reducing greenhouse gases.	Pollution & Global Change
3.4.4	Effect of global temperature increases.	Pollution & Global Change
3.5	**Acid Deposition**	
3.5.1- 3.5.2	The formation of acidified precipitations, effects on soil, water and living organisms.	Pollution & Global Change
3.5.3	Why acid deposition effects are regional.	
3.5.4 - 3.5.5	Methods to reduced causal agents of acid deposition, and restoration of affected sites.	
3.6	**The Hydrosphere**	
3.6.1 - 3.6.2	Earth's water budget, freshwater resources and sustainability.	The Earth's Systems
3.6.3 - 3.6.4	Role of ocean currents in global energy transfer, and climate regulation.	The Earth's Systems
3.6.5	The El Niño Southern Oscillation phenomenon.	The Earth's Systems
3.7	**The Lithosphere**	
3.7.1- 3.7.2	Earth's internal zones, plate tectonic theory, and its effect on evolution and biodiversity.	The Earth's Systems
3.8	**The Soil System**	
3.8.1- 3.8.5	Soil systems and living systems, soil formation, structure, degradation, conservation.	The Earth's Systems
4	**Human Population & Carrying Capacity**	
4.1	**Population Dynamics**	
4.1.1 - 4.1.4	Growth of human populations, population calculations and analysis, growth models.	Populations
4.2	**Resources - Natural Capital**	
4.2.1 4.2.3	Resources as natural capital, types of natural capital, natural income.	Land, Water & Energy
4.2.4 - 4.2.6	Sustainability, appraisal and use of natural resources.	Land, Water & Energy
4.3	**Limits to Growth**	
4.3.1 - 4.3.4	Carrying capacity. Effects of energy & resources, developmental policies & culture.	Populations Land, Water & Energy

OPTIONS:

All students must complete option A, plus one of the remaining options.

A:	**Analysing Ecosystems**	**(All students)**
A1	Physical factors and their measurement.	Ecosystems Investigating Ecosystems
A2	Measuring biotic components of the system	Investigating Ecosystems
A3	Measuring productivity of the system	Ecosystems
A4	Measuring changes in the system.	Investigating Ecosystems
B:	**Impacts of Resource Exploitation**	
B2	Exploitation of energy resources.	Land, Water & Energy
B2	Exploitation of food resources.	Land, Water & Energy
B3	Environmental demands of human populations.	Populations Land, Water & Energy
C:	**Conservation and Biodiversity**	
C1	Biodiversity and ecosystems.	Ecosystems Pollution & Global Change
C2	Evaluating biodiversity and vulnerability.	Pollution & Global Change
C3	Conservation of biodiversity.	Pollution & Global Change
D:	**Pollution Management**	
D1	Nature of pollution.	Pollution & Global Change
D2	Detection and monitoring of pollution.	Investigating Ecosystems Pollution & Global Change
D3	Impacts of pollution.	Pollution & Global Change
D4	Approaches to pollution management.	Pollution & Global Change

The Earth's Systems

Investigating the Earth's systems: hydrosphere, lithosphere, and atmosphere

Concepts in Earth science: the Earth's crust, plate tectonics, soil dynamics and the rock cycle, ocean and air, global water resources.

Learning Objectives

- ☐ 1. Compile your own glossary from the **KEY WORDS** displayed in **bold type** in the learning objectives below.

Earth Systems and Resources *(pages 11-17)*

- ☐ 2. Describe how the history of the Earth is divided into a hierarchical scheme based on information from many sources. Include reference to the time divisions and identify the role of **radiometric dating** in providing a reliable method of dating rocks, minerals, and fossils.
- ☐ 3. Describe the basic structure of the Earth including the composition and properties of the **core**, **mantle**, and **crust**. Describe the composition of the Earth's crust, distinguishing between **igneous**, **metamorphic**, and **sedimentary rocks**.
- ☐ 4. Describe the theory of **plate tectonics**, including reference to the evidence for past plate movement. Describe the three main types of **plate boundaries**: **convergent**, **divergent**, and **transform**, and relate these to any associated tectonic activity such as sea-floor spreading, earthquakes, volcanism, and faulting.
- ☐ 5. Explain the role of the sun and the Earth's rotation and tilt in determining the Earth's day-night cycle and seasons. Explain how the Earth's curvature influences the amount of radiant heat received at different latitudes and relate this to atmospheric circulation and global weather patterns.

The Atmosphere *(pages 21-26, 34-35)*

- ☐ 6. Describe the composition of the Earth's **atmosphere**, including the important features associated with the **troposphere** and **stratosphere**. Explain how the **Coriolis Effect** influences the movement of air (winds) across the Earth's surface, including the direction of cyclonic movements in each hemisphere.
- ☐ 7. Describe features of **atmospheric circulation** and its interaction with the ocean including:
  (a) Features of latitudinal circulation (**tricellular model**)
  (b) The **El Niño-Southern Oscillation** (ENSO) cell
  (c) Tropical cyclones and depressions
  (d) The **intertropical convergence zone** (the doldrums)
- ☐ 8. Explain how atmospheric circulation gives rise to broad climatic regions and consequently, **biomes**.

Global Water Resources and Use *(pages 27-30)*

- ☐ 9. Describe the nature and extent of the Earth's freshwater and saltwater resources and explain how this resource is used by humans for agricultural, industrial, and domestic use.
- ☐ 10. Discuss local and global problems with use of the water resource, including:
  (a) Pollution of surface and **groundwater**.
  (b) Damming and flood control schemes.
  (c) Conservation of water and supply of potable water.

Soil and Soil Dynamics *(pages 18-20, 103-104)*

- ☐ 11. Describe the **rock cycle** and recognise the relationship between the **regolith** and soil development. Explain the basic composition of **soil** and its physical and chemical properties, identifying how soil profiles may vary regionally according to latitude and rock type.
- ☐ 12. Discuss problems with conserving the soil resource, including soil loss (erosion), salinisation, and pollution as a result of excessive fertiliser and pesticide use.

Textbooks

See page 7 for additional details of these texts:

- Christopherson, R.W, 2007. **Elemental Geosystems**, chpt. 1-10, 15.
- Miller, G.T, 2007. **Living in the Environment: Principles, Connections & Solutions**, Chpt. 13-14.
- Raven *et. al.*, 2002. **Environment**, chpt. 1, 14-15.
- Reiss, M. & J. Chapman, 2000. **Environmental Biology** (Cambridge University Press), pp. 1, 3, 76.
- Smith, R. L. & T.M. Smith, 2001. **Ecology and Field Biology**, reading as required.

Periodicals

See page 7 for details of publishers of periodicals:

STUDENT'S REFERENCE

- **Big Weather** New Scientist, 22 May 1999 (Inside Science). *Global weather patterns and their role in shaping the ecology of the planet.*
- **How Old is...** National Geographic, 200(3) September 2001, pp. 79-101. *A comprehensive discussion of dating methods and their application.*
- **The Quick and the Dead** New Scientist, 5 June 1999, pp. 44-48. *The formation of fossils: fossil types and preservation in different environments.*
- **Earth, Fire and Fury** New Scientist, 27 May 2006, pp. 32-36. *How global warming might affect the Earth's crust and what this might mean for volcanism and seismic activity.*
- **Climate and the Evolution of Mountains** Scientific American, August 2006, pp. 54-61. *Although climate can sculpt the Earth's surface, new research shows that climate may also play a role in deformational history of mountain systems.*
- **Water Pressure** National Geographic, Sept. 2002, pp. 2-33. *The demand for freshwater for human consumption and hygiene and the problems associated with increased pressure on supplies.*

Internet

See pages 4-5 for details of how to access **Bio Links** from our web site: **www.thebiozone.com** From Bio Links, access sites under the topics:

ENVIRONMENTAL SCIENCE: • Earth science for schools • Atmospheric and ocean circulation • The thermohaline ocean circulation • The rock cycle • Rocks and soils • Interactive rock cycle animation • Consequences of rotation for weather • US geological Survey • USGS science education • How volcanoes work

The Earth's History

Geologists gather information from many sources to reconstruct the Earth's history. Analysis of extra-terrestrial material, such as meteorites, can provide information about the early history of the Earth, while the Earth's rocks, minerals, and fossils provide information about the crust and the nature of the Earth's deeper layers. The history of the Earth (bottom figure) spans the last 4600 million years and scientists have made sense of this enormous time span by dividing it into hierarchical scheme of time periods. The boundaries of these time periods are based on the worldwide correlation of distinctive fossils and rock types (a process made more reliable with the advent of radiometric dating). This **stratigraphic record** documents the nature of the ancient Earth and the increasing diversification of life from its origins in the seas approximately 3800 mya to the present day.

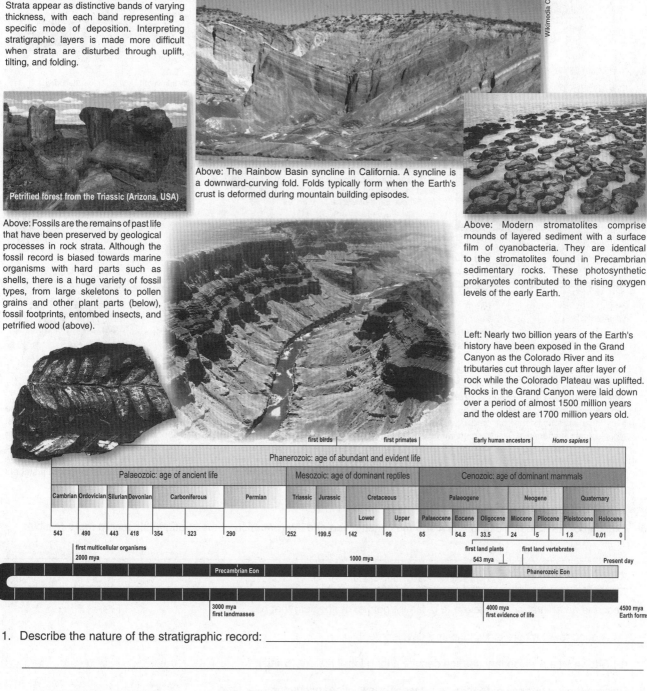

Rock **strata** are built up through the deposition of material by natural processes, with younger layers overlying older ones. Strata appear as distinctive bands of varying thickness, with each band representing a specific mode of deposition. Interpreting stratigraphic layers is made more difficult when strata are disturbed through uplift, tilting, and folding.

Above: The Rainbow Basin syncline in California. A syncline is a downward-curving fold. Folds typically form when the Earth's crust is deformed during mountain building episodes.

Petrified forest from the Triassic (Arizona, USA)

Above: Fossils are the remains of past life that have been preserved by geological processes in rock strata. Although the fossil record is biased towards marine organisms with hard parts such as shells, there is a huge variety of fossil types, from large skeletons to pollen grains and other plant parts (below), fossil footprints, entombed insects, and petrified wood (above).

Above: Modern stromatolites comprise mounds of layered sediment with a surface film of cyanobacteria. They are identical to the stromatolites found in Precambrian sedimentary rocks. These photosynthetic prokaryotes contributed to the rising oxygen levels of the early Earth.

Left: Nearly two billion years of the Earth's history have been exposed in the Grand Canyon as the Colorado River and its tributaries cut through layer after layer of rock while the Colorado Plateau was uplifted. Rocks in the Grand Canyon were laid down over a period of almost 1500 million years and the oldest are 1700 million years old.

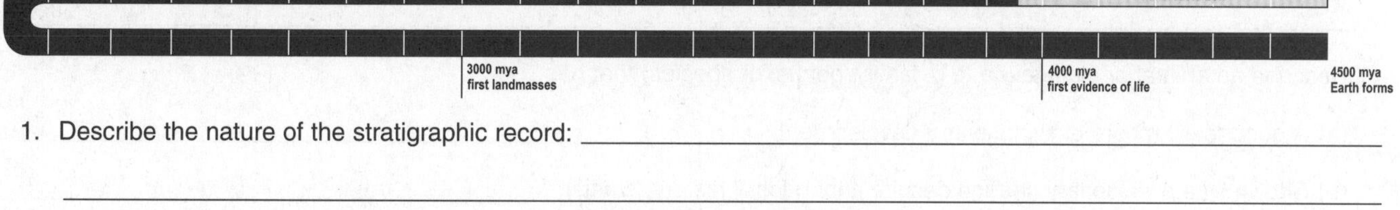

1. Describe the nature of the stratigraphic record: ______________________________

____

____

2. (a) Describe the fundamental basis on which the record of the Earth's history is compiled: ______________________________

____

(b) Explain why the interpretation of strata on this basis is sometimes difficult: ______________________________

____

____

Related activities: The Earth's Crust
Web links: Getting into the Fossil Record, Fossil Formation

Sedimentary Rock Profile

This diagram represents a cutting through layers of sedimentary rock in which fossils are exposed. Fossils are the remains or impressions of plants or animals that become trapped in the sediments after death. Layers of sedimentary rock are arranged in the order that they were deposited, with the most recent layers near the surface (unless they have been disturbed).

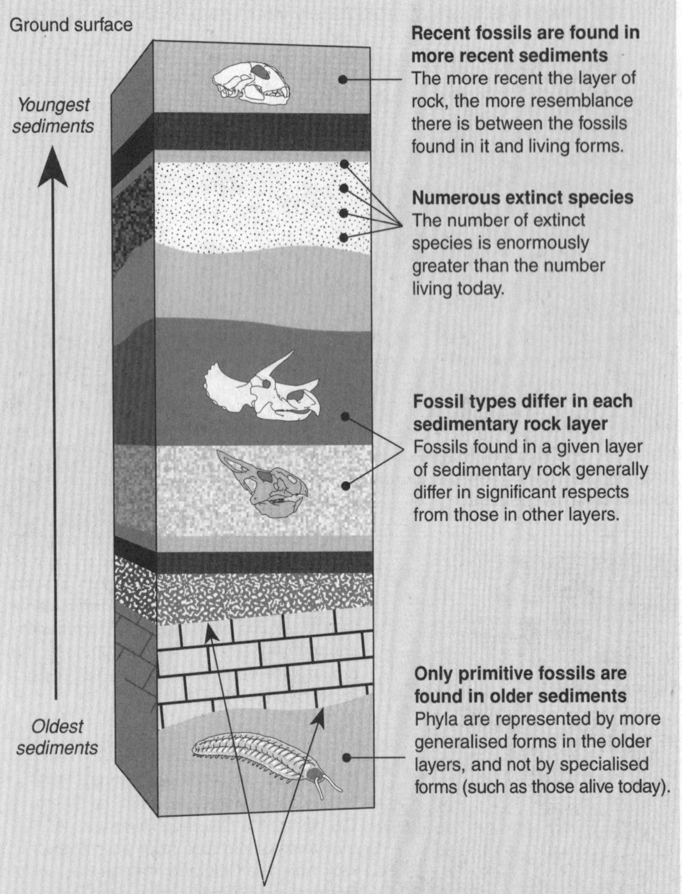
Ground surface
Youngest sediments
Oldest sediments
Recent fossils are found in more recent sediments
The more recent the layer of rock, the more resemblance there is between the fossils found in it and living forms.
Numerous extinct species
The number of extinct species is enormously greater than the number living today.
Fossil types differ in each sedimentary rock layer
Fossils found in a given layer of sedimentary rock generally differ in significant respects from those in other layers.
Only primitive fossils are found in older sediments
Phyla are represented by more generalised forms in the older layers, and not by specialised forms (such as those alive today).

New fossil types mark changes in environment
In the rocks marking the end of one geological period, it is common to find new fossils that become dominant in the next. Each geological time period had an environment very different from those before and after. The boundaries of these coincided with considerable environmental change and the creation of new niches. These produced new selection pressures and resulted in diversification of surviving genera.

A Case Study in the Fossil Record

The history of modern day species can be traced
The evolution of many modern species can be well reconstructed. For instance, the evolutionary history of modern elephants is well documented for the last 40 million years. and the modern horse has a well understood fossil record spanning 50 million years.

Fossil species are similar to but different from today's species
Most fossil animals and plants belong to the same major taxonomic groups as organisms living today. However, they do differ from the living species in many features.

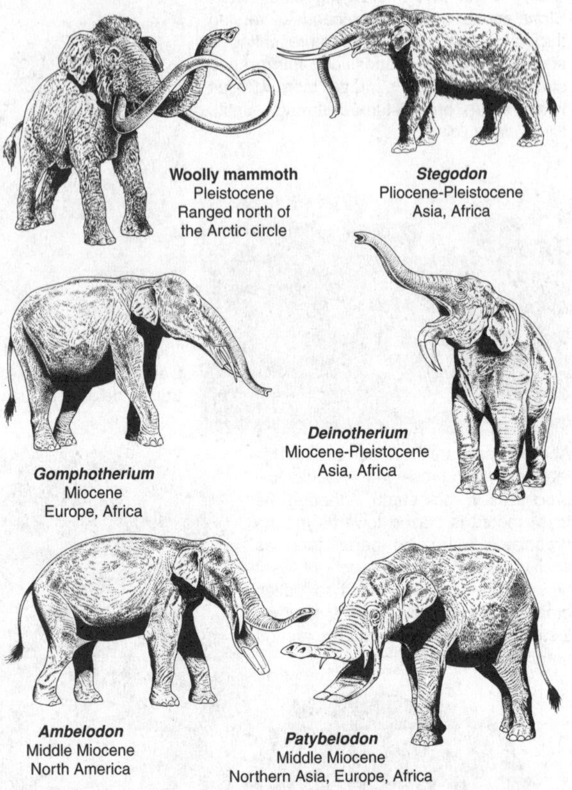
Woolly mammoth
Pleistocene
Ranged north of the Arctic circle
Stegodon
Pliocene-Pleistocene
Asia, Africa
Gomphotherium
Miocene
Europe, Africa
Deinotherium
Miocene-Pleistocene
Asia, Africa
Ambelodon
Middle Miocene
North America
Patybelodon
Middle Miocene
Northern Asia, Europe, Africa

African and Indian elephants have descended from a diverse group of **proboscideans** (named for their long trunks). The first pig-sized, trunkless members of this group lived in Africa 40 million years ago. From Africa, their descendants invaded all continents except Antarctica and Australia. As the group evolved in response to predation, they became larger. Examples of extinct members of this group are illustrated above.

3. Explain how radiometric dating has made the construction and interpretation of the stratigraphic record more reliable:

4. Describe an animal or plant taxon (e.g. family, genus, or species) that has:

   (a) A good fossil record of evolutionary development:

   (b) Shown little evolutionary change despite a long fossil history (stasis):

5. Discuss the importance of **fossils** as a record of evolutionary change over time:

Fossil Formation

Fossils are the remains of organisms that have escaped decay and have become preserved in rock strata. A fossil may be the preserved remains of the organism itself, a mould or cast, or the marks made by it during its lifetime (trace fossils). Fossilisation requires the normal processes of decay to be permanently arrested. This can occur if the remains are buried rapidly and isolated from air, water, or decomposing microbes. Fossils provide a record of the appearance and extinction of organisms. Once this record is calibrated against a time scale, it is possible to build up an evolutionary history of life on Earth.

Some Fossil Types and Modes of Preservation

Brachiopod (lamp shell), Jurassic (New Zealand)

Mould: An organism-shaped impression left after the original remains were dissolved or otherwise destroyed.

Ammonite cast, late Cretaceous (Charmouth, England).

Cast: The original materials of the organism have been replaced by new unrelated ones, in this case, iron pyrite.

This fossil fern frond Carboniferous (USA).

This **compression fossil** of a fern frond shows traces of carbon and wax from the original plant. Compression fossils are preserved in sedimentary rock that has been compressed.

Ants in amber about 25 mya (Madagascar).

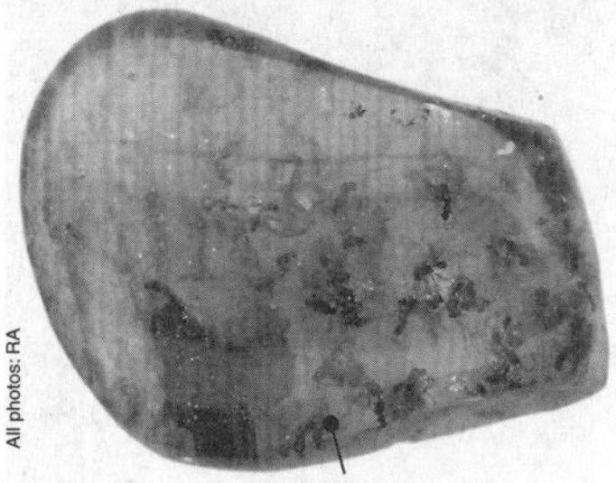

All photos: RA

Fossilised resin (or amber) produced by some ancient conifers trapped organisms, such as these ants, before it hardened.

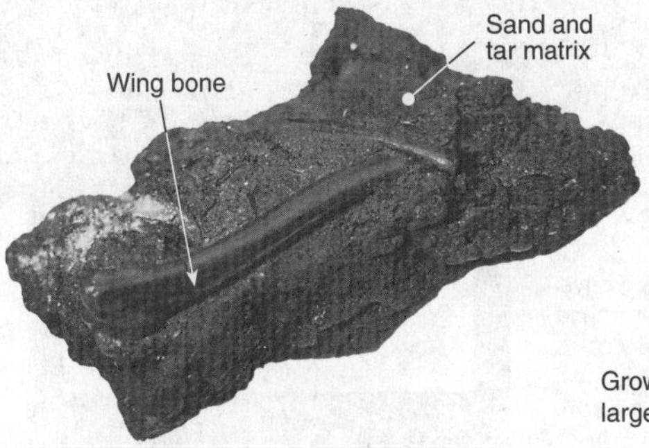

Fossilised bones of a bird that lived about 5 mya and became stuck in the **tar pits** at la Brea, Los Angeles, USA.

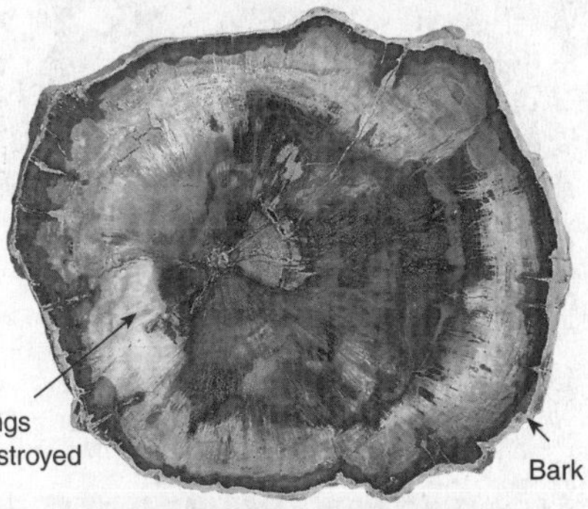

Permineralisation: In some fossils, the organic material is replaced by minerals, as in this **petrified wood** from Madagascar.

Teeth and bones (being hard) are often well preserved. This shark tooth is from Eocene phosphate beds in Morocco.

Ammonite, Jurassic (Madagascar).

The fossil record is biased toward organisms with hard parts. This ammonite still has a layer of the original shell covering the stone interior.

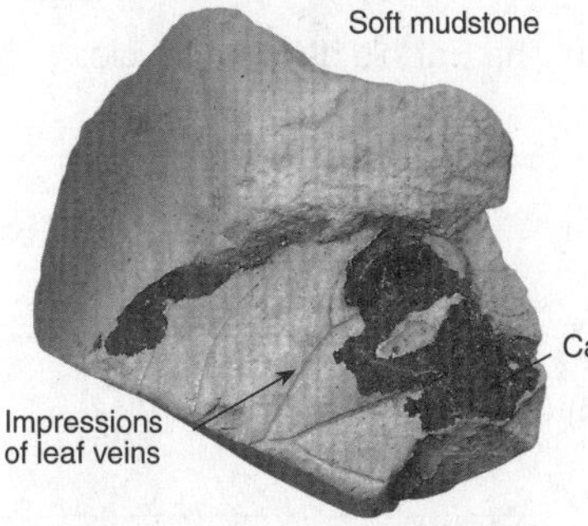

In a **sub-fossil**, the fossilisation process is incomplete. In this leaf impression in soft mudstone, some of the leaf remains are still intact (a few thousand years old, New Zealand).

1. Describe the natural process that must be arrested in order for fossilisation to take place: ______________________

2. Explain why the fossil record is biased towards marine organisms with hard parts: ______________________

____

____

3. Fossils tell us much about the organisms that lived in the past. Suggest what other information they might provide:

____

____

The Earth and the Sun

Of all celestial bodies, the Sun has the greatest influence on the Earth, affecting its movements, determining the day-night and seasonal cycles, driving climatic systems and longer term climate cycles, and providing the energy for most life on the planet. The Sun also lays a part in tidal movement on Earth, modifying the effect of the Moon to produce monthly variations in the tidal range. The Sun emits various types of radiation, but most is absorbed high in the Earth's atmosphere. Only visible light, some infra-red radiation, and some ultraviolet light reach the surface in significant amounts. Visible light is pivotal to the producer base of Earth's biological systems, but infra-red is also important because it heats the atmosphere, oceans, and land. The intensity of solar radiation is not uniform around the Earth and this uneven heating effect, together with the Earth's rotation, produce the global patterns of wind and ocean circulation that profoundly influence the Earth's climate.

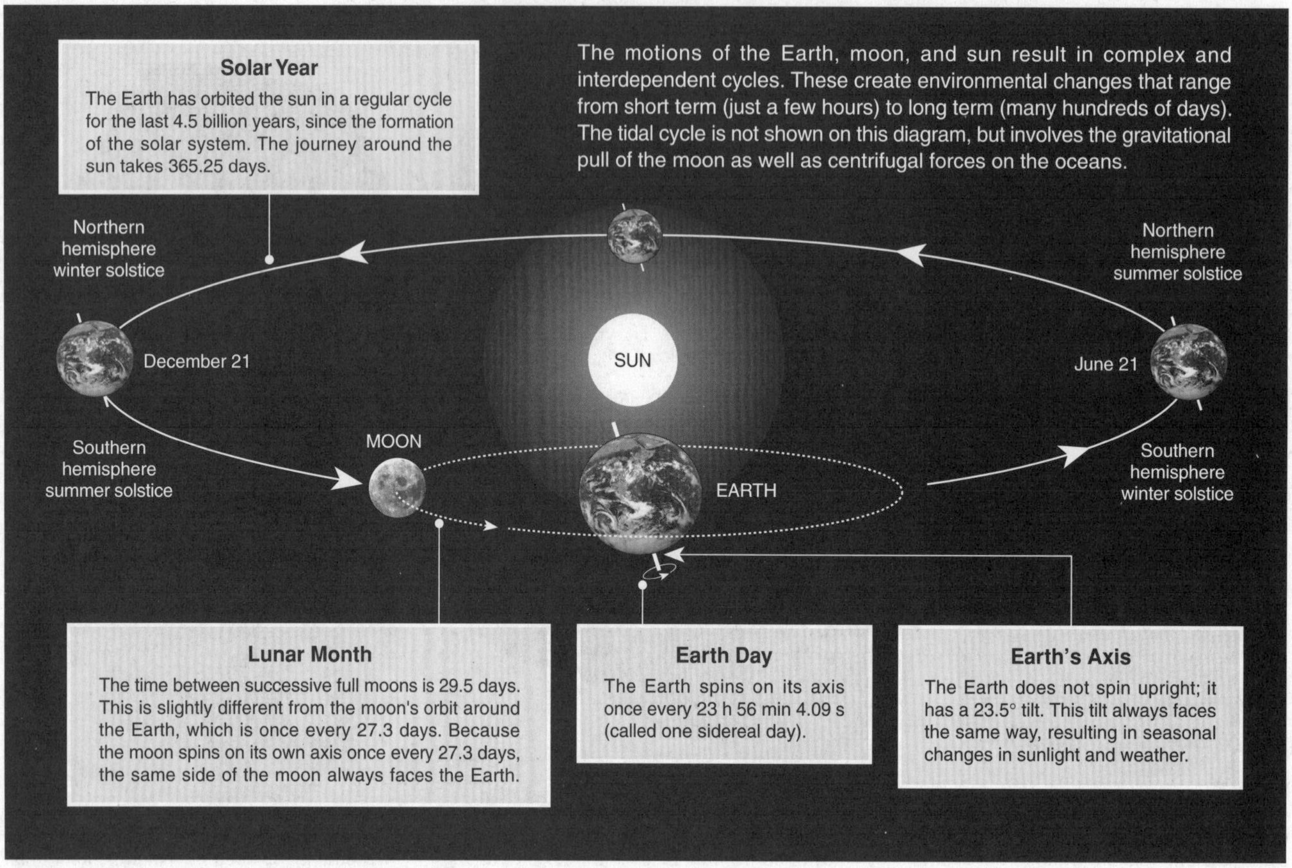
Solar Year
The Earth has orbited the sun in a regular cycle for the last 4.5 billion years, since the formation of the solar system. The journey around the sun takes 365.25 days.
The motions of the Earth, moon, and sun result in complex and interdependent cycles. These create environmental changes that range from short term (just a few hours) to long term (many hundreds of days). The tidal cycle is not shown on this diagram, but involves the gravitational pull of the moon as well as centrifugal forces on the oceans.
Northern hemisphere winter solstice
Northern hemisphere summer solstice
December 21
SUN
June 21
Southern hemisphere summer solstice
MOON
EARTH
Southern hemisphere winter solstice
Lunar Month
The time between successive full moons is 29.5 days. This is slightly different from the moon's orbit around the Earth, which is once every 27.3 days. Because the moon spins on its own axis once every 27.3 days, the same side of the moon always faces the Earth.
Earth Day
The Earth spins on its axis once every 23 h 56 min 4.09 s (called one sidereal day).
Earth's Axis
The Earth does not spin upright; it has a 23.5° tilt. This tilt always faces the same way, resulting in seasonal changes in sunlight and weather.

1. Explain the basis of each of the following:

(a) The Earth's day-night cycle: ______

(b) The Earth's seasons: ______

(c) The summer solstice: ______

(d) The winter solstice: ______

2. (a) Explain why tropical regions receive a greater input of solar radiation than the poles:

______

(b) Describe the consequences of this to the Earth's climate: ______

The Earth's Crust

The Earth has a layered structure comprising a solid inner core, a liquid outer core, a highly viscous mantle, and an outer silicate solid crust. The Earth's crust is thin compared to the bulk of the Earth, averaging just 25-70 km thick below the continents and about 10 km thick below the oceans. Overall, the crust is less dense than the mantle, being relatively rich in lighter minerals such as silicon, calcium, and aluminium. The crust is a dynamic structure, subject to constant change as a result of ocean formation, mountain building, and volcanism. It supports the **biosphere**, the hydrosphere, and the atmosphere.

The **continental crust** is made up of igneous, metamorphic, and sedimentary rocks. It is not recycled within the Earth to the same extent as oceanic crust, so some continental rocks are up to 4 billion years old. Parts of the continental crust go through repeated cycles of erosion, formation into sedimentary rock, and metamorphosis. The crust "floats" in the mantle at a level determined by its thickness and density. The more mass there is above sea-level, the deeper the crust must extend down in support.

Water precipitated from the atmosphere forms rivers and lakes, which flow back to the ocean eroding the landscape in the process.

The Earth's persistent oceans of liquid water cycle moisture through the atmosphere to the land and back again.

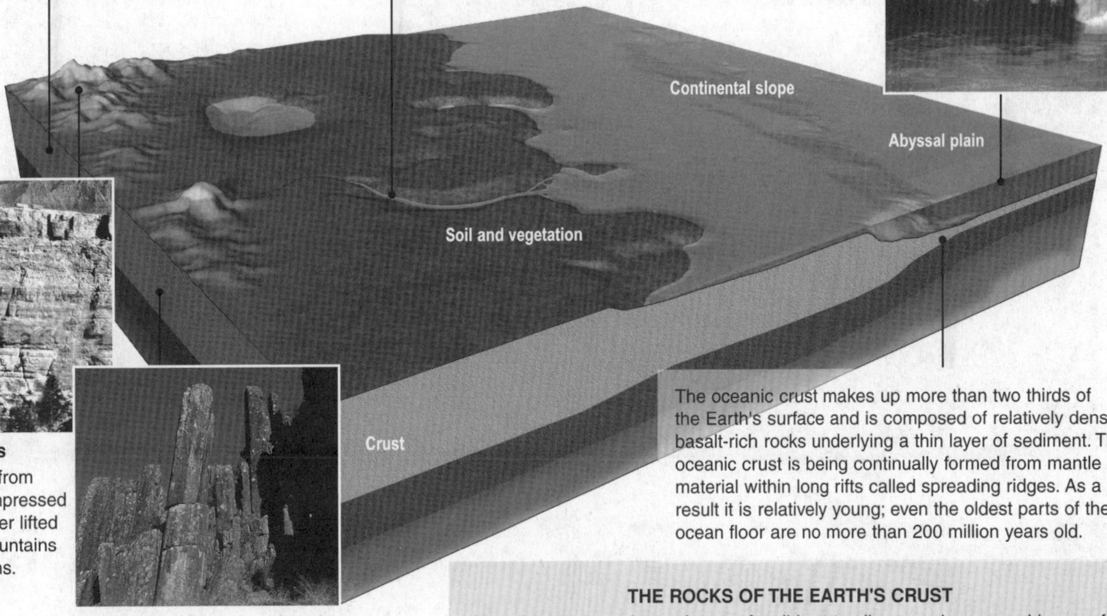

Sedimentary rocks
Sediments eroded from continents and compressed into rock can be later lifted and exposed in mountains and other formations.

Igneous rocks, such as basalt (above), form a major component of the crust and are essentially unchanged since their formation.

The oceanic crust makes up more than two thirds of the Earth's surface and is composed of relatively dense basalt-rich rocks underlying a thin layer of sediment. The oceanic crust is being continually formed from mantle material within long rifts called spreading ridges. As a result it is relatively young; even the oldest parts of the ocean floor are no more than 200 million years old.

THE ROCKS OF THE EARTH'S CRUST

The Earth's crust is made up of solid, naturally occurring assemblages of minerals called rocks. The huge diversity of the Earth's rocks has developed over thousands of millions of years through igneous activity, changes in form (**metamorphism**), and the formation of sediments and sedimentary rocks.

Granite

Obsidian

Igneous rocks solidify from volcanic magma and they vary in composition from basalt to granite and in texture from rapidly cooled glasses, such as obsidian, to slowly cooled coarse grains, such as granite.

Slate

Schist

Metamorphic rocks result when pre-existing rock is transformed by heat and pressure. Metamorphic rocks are classified by texture and composition. Examples include gneiss, slate, marble and schist.

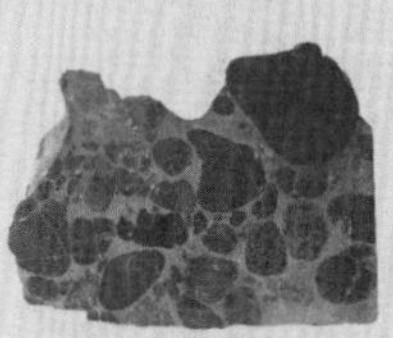

Conglomerate

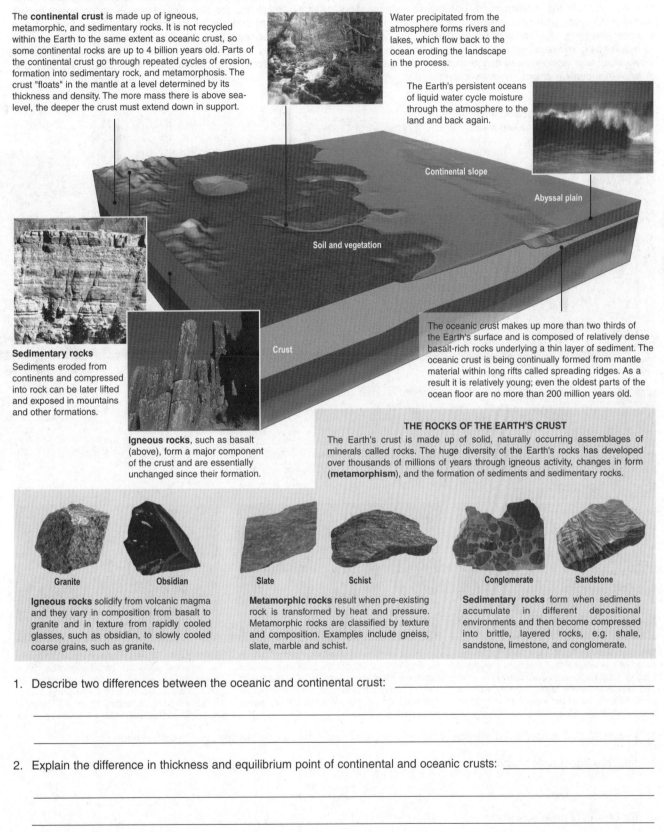

Sandstone

Sedimentary rocks form when sediments accumulate in different depositional environments and then become compressed into brittle, layered rocks, e.g. shale, sandstone, limestone, and conglomerate.

1. Describe two differences between the oceanic and continental crust: ______________________

____

____

2. Explain the difference in thickness and equilibrium point of continental and oceanic crusts: ______________________

____

____

____

3. Explain why the Earth's crust is described as a dynamic structure: ______________________

____

Related activities: The Rock Cycle

Plate Boundaries

The outer rock layer of the Earth, comprising the crust and upper mantle, is called the **lithosphere** and it is broken up into seven large, continent-sized **tectonic plates** and about a dozen smaller plates. Throughout geological time, these plates have moved about the Earth's surface, shuffling continents, opening and closing oceans, and building mountains. The evidence for past plate movements has come from several sources: mapping of plate boundaries, the discovery of sea floor spreading, measurement of the direction and rate of plate movement, and geological evidence such as the distribution of ancient mountain chains, unusual deposits, and fossils. The size of the lithospheric plates is constantly changing, with some expanding and some getting smaller. These changes occur along **plate boundaries**, which are marked by well-defined zones of seismic and volcanic activity. Plate growth occurs at **divergent boundaries** along **sea floor spreading ridges** (e.g. the Mid-Atlantic Ridge and the Red Sea) whereas plate attrition occurs at **convergent boundaries** marked by deep ocean trenches and subduction zones. Divergent and convergent zones make up approximately 80% of plate boundaries. The remaining 20% are called **transform boundaries**, where two plates slide past one another with no significant change in the size of either plate.

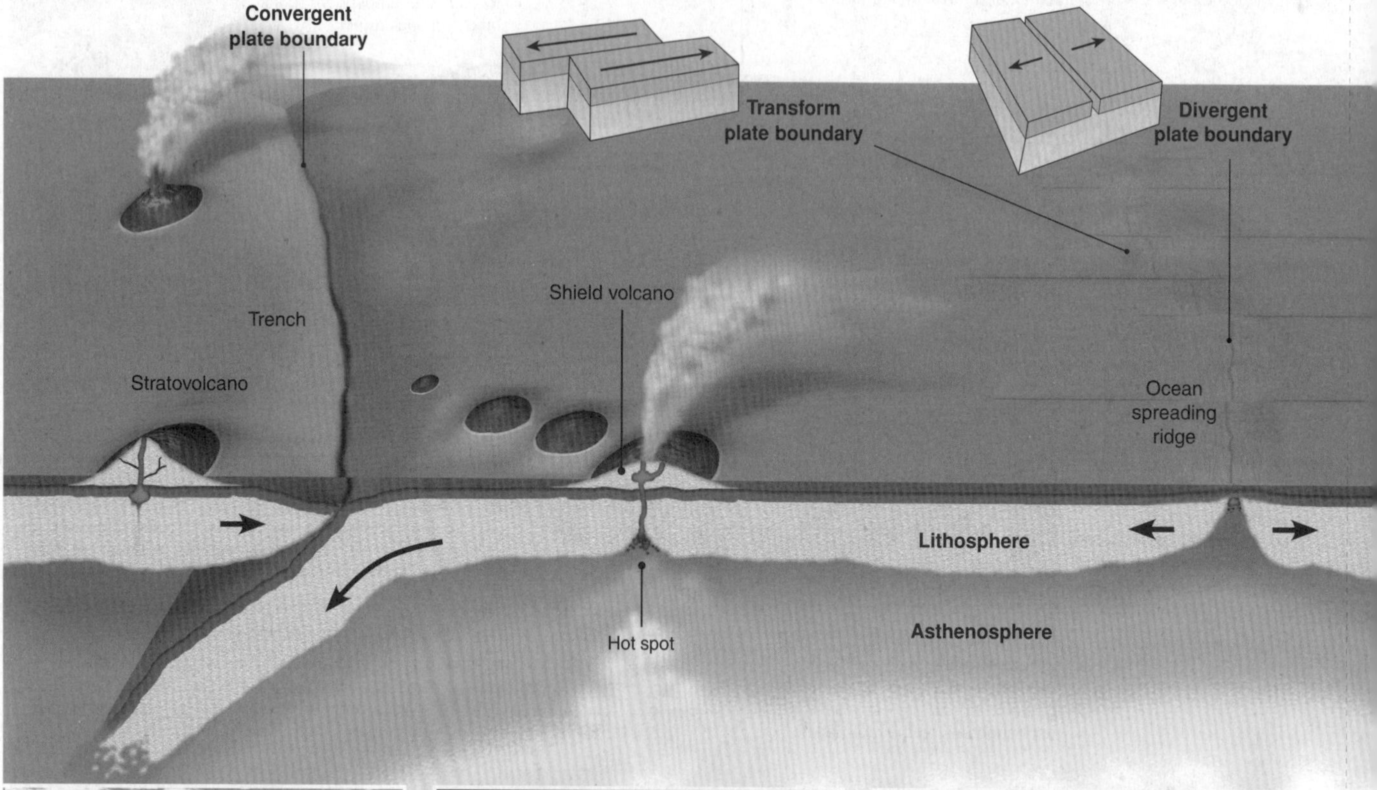

The San Andreas Fault, seen here in an aerial photo, is a geological fault that runs a length of roughly 1300 km through California in the USA. The fault, a strike-slip fault, marks a transform (or sliding) boundary between the Pacific Plate and the North American Plate.

Mount St. Helens is an active stratovolcano in the Pacific Northwest of the USA. It is part the Cascade volcanic arc, a segment of the Pacific Ring of Fire that has formed due to subduction. This volcano is well known for its ash explosions and pyroclastic flows.

Earthquakes cause shaking and ground rupture, as well as landslides and avalanches, fires, tsunamis, and soil liquefaction. Soil liquefaction occurs when, because of the shaking, water-saturated granular material temporarily loses its strength and transforms from a solid to a liquid.

1. Describe what is happening at each of the following plate boundaries and identify an example in each case:

(a) Convergent plate boundary: ______________________

____

(b) Divergent plate boundary: ______________________

____

(c) Transform plate boundary: ______________________

____

The Mechanism of Plate Movement

The relatively cool **lithosphere** overlies the hotter, plastic and more fluid lower part of the mantle, the **asthenosphere**. Heat from the mantle drives two kinds of asthenospheric movement: convection and mantle plumes. However the plates do not ride passively on convection currents as early models proposed. Instead, plate motion is partly driven by the weight of cold, dense plates sinking into the mantle at trenches. This heavier, cooler material sinking under the influence of gravity displaces heated material that rises as mantle plumes.

The movements of the tectonic plates puts the brittle rock of the crust under strain creating **faults** where rocks fracture and slip past each other. Earthquakes are caused by energy release during rapid slippage along faults. Consequently, the Earth's major earthquake (and volcanic) zones occur along plate boundaries.

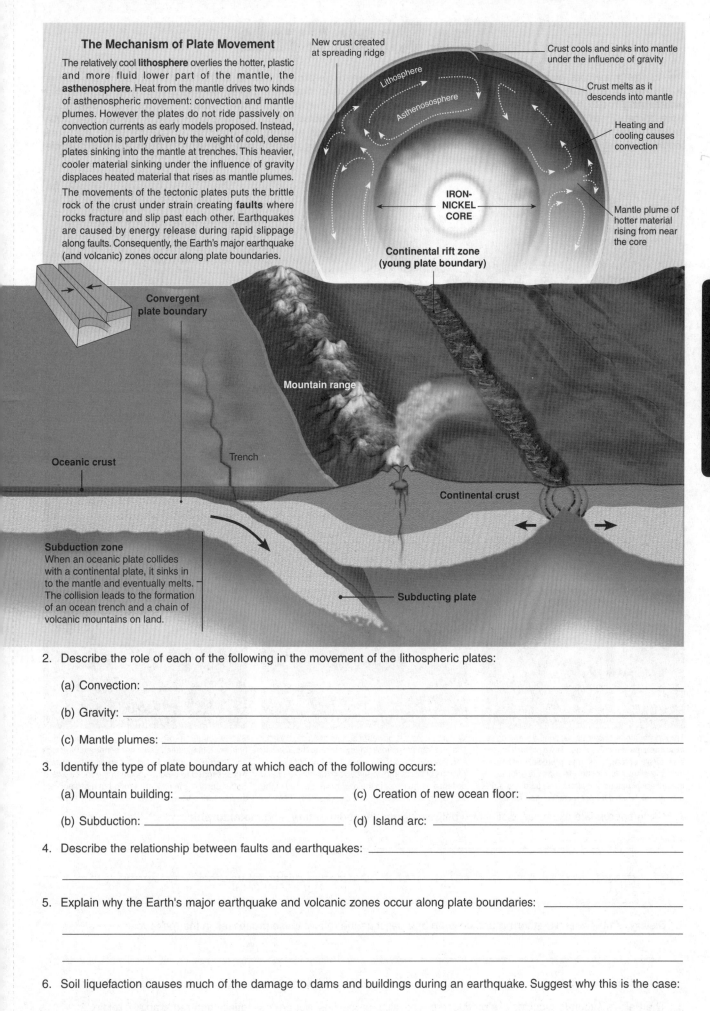

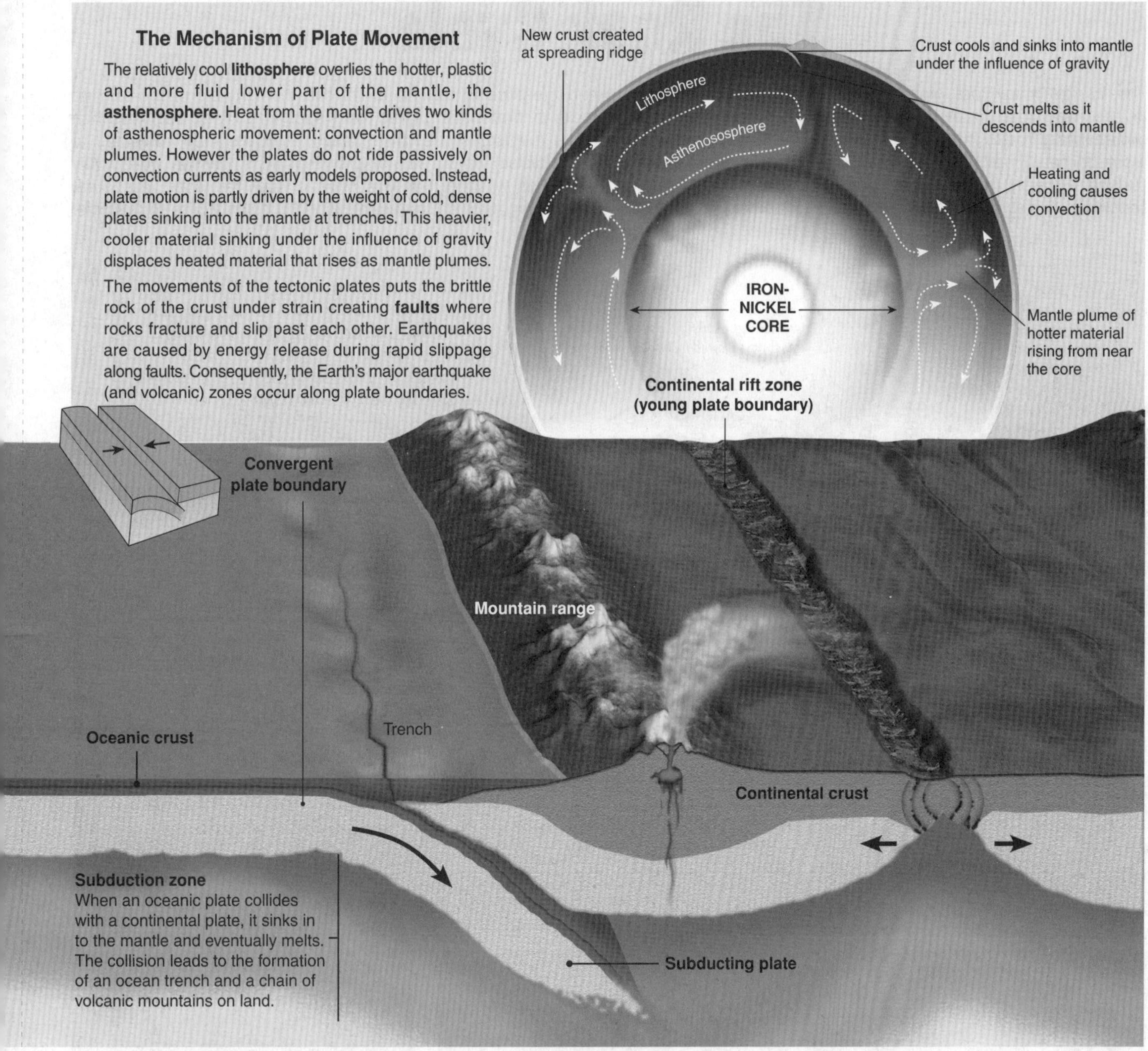

2. Describe the role of each of the following in the movement of the lithospheric plates:

   (a) Convection: ______________________________

   (b) Gravity: ______________________________

   (c) Mantle plumes: ______________________________

3. Identify the type of plate boundary at which each of the following occurs:

   (a) Mountain building: ____________________ (c) Creation of new ocean floor: ____________________

   (b) Subduction: ____________________ (d) Island arc: ____________________

4. Describe the relationship between faults and earthquakes: ______________________________

   ____

5. Explain why the Earth's major earthquake and volcanic zones occur along plate boundaries: ______________________________

   ____

   ____

6. Soil liquefaction causes much of the damage to dams and buildings during an earthquake. Suggest why this is the case:

   ____

   ____

The Rock Cycle

The Earth's many rock types are grouped together according to the way they formed as **igneous**, **metamorphic**, and **sedimentary rocks** (as well as meteorites). These rocks form in a self perpetuating cycle. Volcanism creates rocks at the Earth's surface. Erosion of these and other surface rocks produces sediments, which burial transforms into sedimentary rocks. Heat and pressure within the Earth can then transform pre-existing rocks to form metamorphic rocks such as slate and schist. When rocks are exposed at the surface, they are then subjected to the physical, chemical, and biological processes collectively known as **weathering**. This cycle of rock formation, exposure, weathering, erosion, and deposition is known as the **rock cycle**.

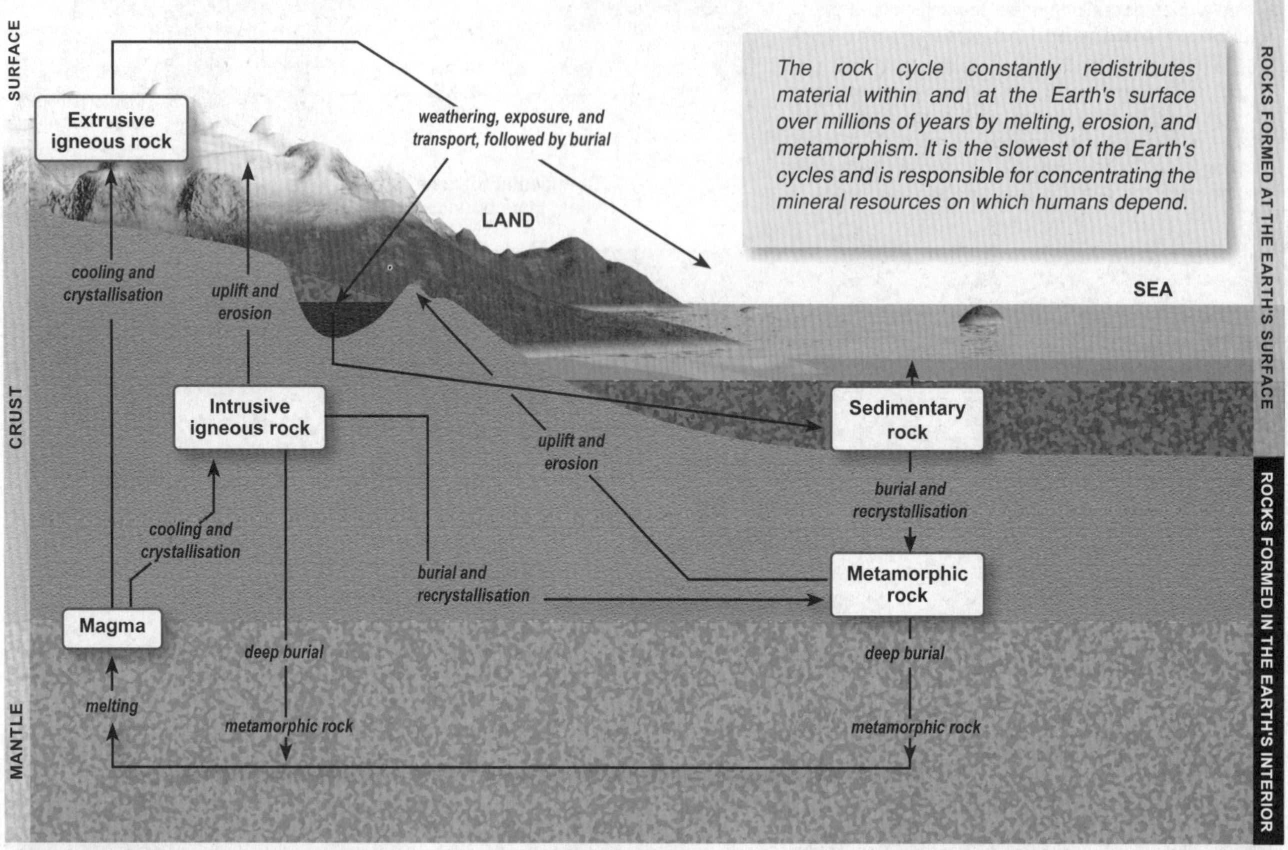

The rock cycle constantly redistributes material within and at the Earth's surface over millions of years by melting, erosion, and metamorphism. It is the slowest of the Earth's cycles and is responsible for concentrating the mineral resources on which humans depend.

These formations, known as hoodoos, are composed of soft sedimentary rock topped by a piece of harder, less easily-eroded rock that protects the column from the elements. Hoodoo shapes are affected by the erosional patterns of alternating hard and softer rock layers, while different minerals produce colour.

Water and ice are powerful agents of erosion. Water lifts and transports rock fragments, and the freezing and thawing splits rocks apart. The flow of seawater millions of years ago, together with ice wedging and collapse along joints in the rock have resulted in the formation of this spectacular arch in Utah, USA.

Salt weathering of rock produces a distinctive honeycombing effect. Seawater penetrates the rock and then evaporates to leave behind salt crystals, which expand to produce holes in the rock surface. Softer parts of the rock are eroded at a greater rate than harder parts.

1. Using appropriate examples, distinguish between **igneous**, **sedimentary**, and **metamorphic** rocks:

2. Distinguish between **weathering** and **erosion** and describe the role of these processes in the rock cycle:

3. The Earth's mineral resources are produced by recycling processes, but are essentially non-renewable. Explain:

Soil and Soil Dynamics

Soils are a complex mixture of unconsolidated weathered rock and organic material. Soils are essential to terrestrial life. Plants require soil and the microbial populations, responsible for recycling organic wastes, live in the soil and contribute to its fertility. Soils are named and classified on the basis of physical and chemical properties in their **horizons** (layers). Soils have three basic horizons (A,B,C). The A horizon is the **topsoil**, which is rich in organic matter. The B horizon is a **subsoil** containing clay and soluble minerals. The C horizon is made up of weathered **parent material** and rock fragments. Soils and their horizons differ widely, and are grouped according to their characteristics, which are determined by the underlying parent rock, the age of the soil, and the conditions under which the soil developed. A few soils weather directly from the underlying rocks and these **residual soils** have the same general chemistry as the original rocks. More commonly, soils form in materials that have moved in from elsewhere.

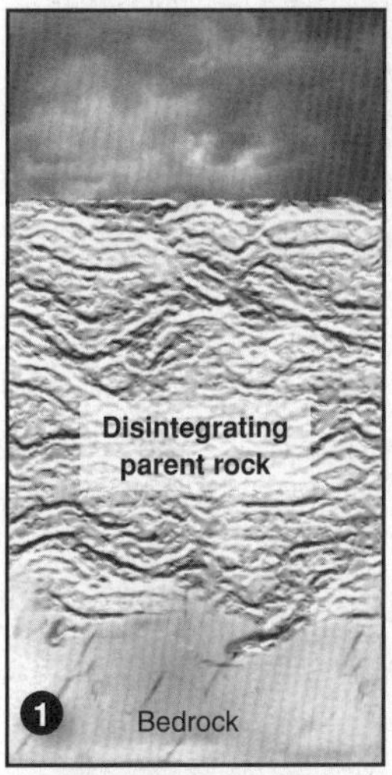

The parent rock is broken down by weathering to form a **regolith** which overlies the solid bedrock. The soil that forms is part of the regolith.

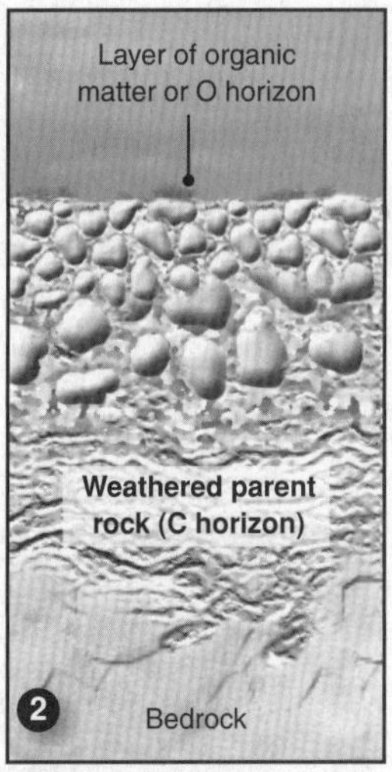

Plants establish and organic material builds up on the surface. The organic material aids the disintegration of the parent material.

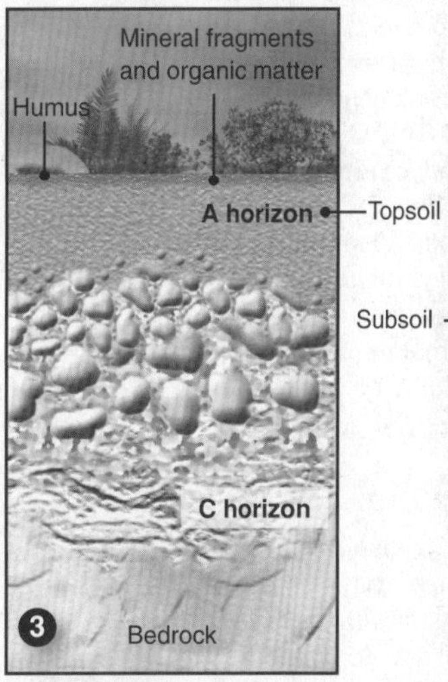

As the mineral and organic content mix, horizons begin to form, with humus-rich layers at the surface and mineral-rich layers at the base.

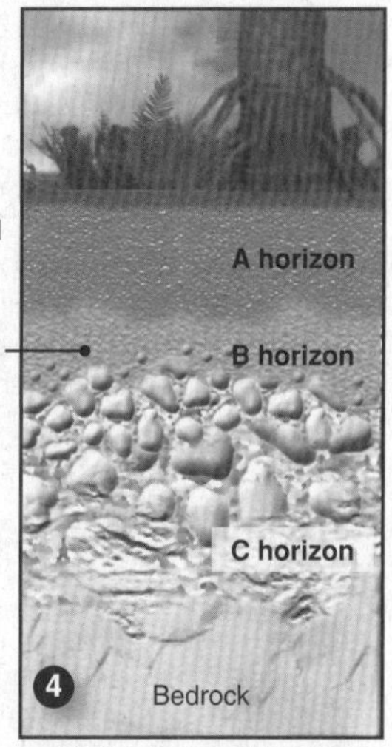

Horizons are well developed in mature soils. The final characteristics of the soil are determined by the regional conditions and the rock type.

Influences on Soil Development

The character and composition of the parent material is important in determining the properties of a soil. Parent materials include volcanic deposits, and sediments deposited by wind, water or glaciers.

The occurrence of freeze-thaw and wet-dry cycles, as well as average temperature and moisture levels are important in soil development. Climate also affects vegetation, which in turn influences soil development.

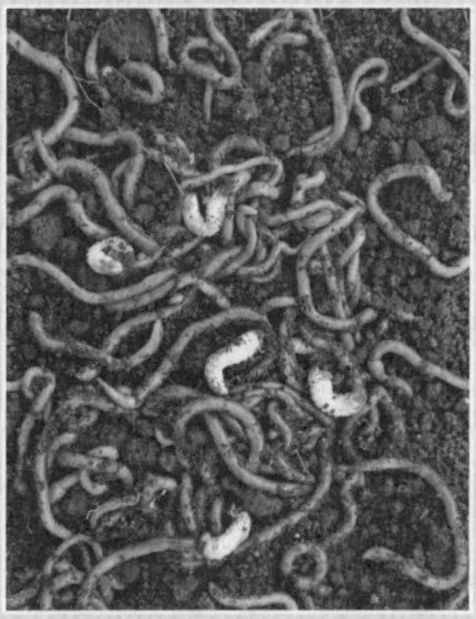

Plants, animals, fungi, and bacteria help to create a soil both through their activities and by adding to the soil's organic matter when they die. Moist soils with a high organic content tend to be higher in biological activity.

The topography (hilliness) of the land influences soil development by affecting soil moisture and tendency towards erosion. Soils in steep regions are more prone to loss of the topsoil and erosion of the subsoil.

1. Explain the role of weathering in soil formation: ______________________

____

____

2. Discuss the influence of climate, rock type, and topography on the characteristics of a mature soil:

____

____

____

Related activities: The Earth's Crust, The Rock Cycle, Soil Degradation
Web links: Soil Classification and Formation

Soils In Different Climates

Soils are formed by the break down of rock and the mixing of inorganic and organic material. The soil profile is a series of horizontal layers that differ in composition and physical properties. Each recognisable layer is called a horizon. Soils have three basic horizons (A,B,C). The A horizon is the **topsoil**, which is rich in organic matter. If there is also a layer of litter (undecomposed or partly decomposed organic matter), this is called the O horizon, but it is often absent. The B horizon is a **subsoil** containing clay and soluble minerals. The C horizon is made up of weathered **parent material** and rock fragments. These horizons may be variously developed depending on whether or not the soil is mature. Mature soils have had enough time to develop distinct horizons. Immature soils have horizons that are lacking.

ARID REGIONS
Desert soils are alkaline mineral soils with variable amounts of clay, low levels of organic matter, and poorly developed vertical profiles.

HUMID TROPICS
Tropical soils: Leaching and chemical weathering make these soils acidic. Aluminium and iron oxides accumulate in the deep B horizon.

MID-LATITUDES
Grassland soils: Mature, alkaline, deep, well drained soils. They are typically nutrient-rich and productive with a high organic content.

Soil Texture

Soil texture depends on the amount of each size of mineral particle in the soil (sand, silt, and clay sized particles). Coarse textured soils are dominated by sand, medium textures by silt, and fine textured soils by clay.

SAND ...feels gritty | SILT ...feels silky | CLAY ...feels sticky

decreasing particle size →

POLAR REGIONS
Very low temperatures slow the decomposition of organic matter and maintain the permafrost layer in these frozen soils.

TEMPERATE
Weathered forest soils: Well developed soils with a deep organic layer and accumulated clay at lower levels.

SEASONALLY WET
Swelling soils: Marked seasonal rainfall results in deep cracks as the soil alternately swells and shrinks.

3. Describe the role of soil organisms in soil structure and development: ______________________

____

____

4. Identify which feature of a soil would most influence its:

   (a) Fertility: ______________ (b) Water-holding capacity: ______________

5. Explain how the characteristics described below arise in each of the following soil types:

   (a) Accumulation of organic matter in the frozen soils of the Arctic: ______________________

   ____

   (b) Shallow A horizon and poorly developed vertical profile of a desert soil: ______________________

   ____

6. Suggest why soils with roughly equal proportions of sand, silt, and clay are considered to be the best soils for cultivation:

____

____

The Atmosphere and Climate

The Earth's atmosphere is a layer of gases surrounding the globe and retained by gravity. It contains roughly 78% nitrogen, 20.95% oxygen, 0.93% argon, 0.038% carbon dioxide, trace amounts of other gases, and a variable amount (average around 1%) of water vapour. This mixture of gases, known as **air**, protects life on Earth by absorbing ultraviolet radiation and reducing temperature extremes between day and night. The atmosphere consists of layers around the Earth, each one defined by the way temperature changes within its limits. The outermost troposphere thins slowly, fading into space with no boundary. The air of the atmosphere moves in response to heating from the sun and, globally, the **atmospheric circulation** transports warmth from equatorial areas to high latitudes and returning cooler air to the tropics. It is the interaction of the atmosphere and the oceans that creates the Earth's the longer term pattern of atmospheric conditions we call **climate** (as opposed to shorter-term weather). The world's climates are not static; they have been both warmer and cooler in the past. At present, the average global temperature is increasing, but this rise is not evenly spread around the globe. The present climate warming is most likely to be due, at least in part, to an enhanced **greenhouse effect**, caused by higher concentrations of greenhouse gases in the atmosphere.

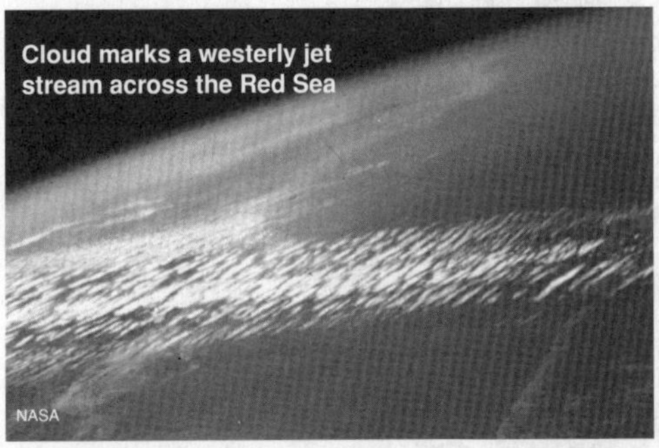

Jet streams are narrow, winding ribbons of strong wind in the upper troposphere. They mark the boundary between air masses at different temperatures. Cloud forms in the air that is lifted as it is driven into the core of the jet stream.

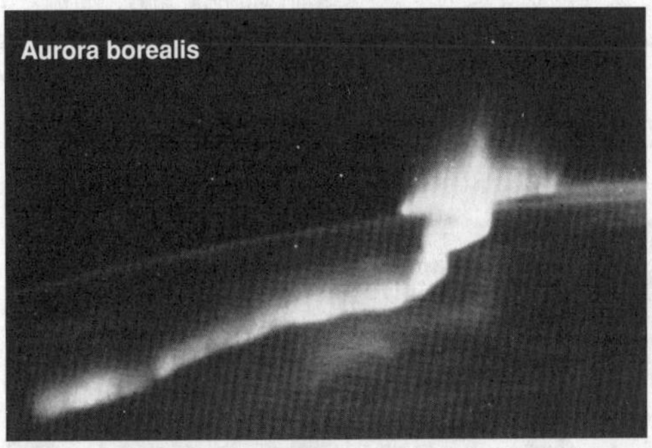

The Northern and Southern Lights (the Aurora Borealis and the Aurora Australis respectively), appear in the thermosphere. Typically, auroras appear either as a diffuse glow or as "curtains" that extend in an approximately east to west direction.

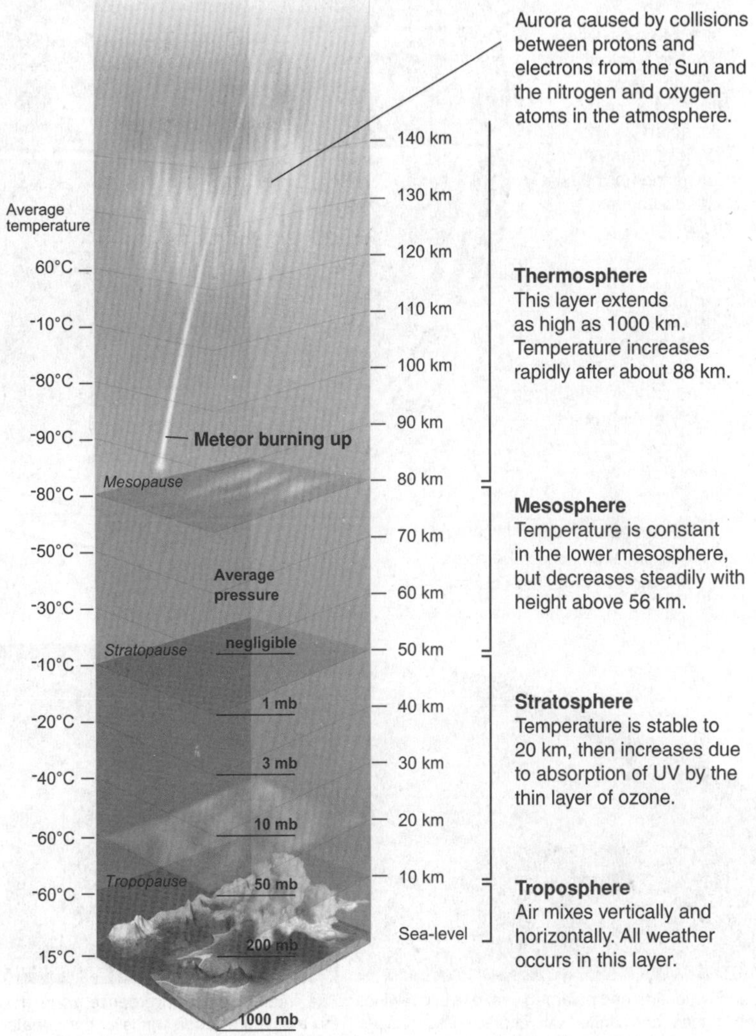

1. Describe two important roles of the atmosphere: ______________________________

____

____

2. Explain what drives the atmospheric circulation: ______________________________

____

____

3. Describe one characteristic feature of each of the following layers of the atmosphere:

(a) Troposphere: ______________________________

(b) Stratosphere: ______________________________

(c) Mesosphere: ______________________________

(d) Thermosphere: ______________________________

Related activities: Ocean Circulation and Currents, Global Warming
Web links: Consequences of Rotation for Weather, Jetstream

The Tricellular Model of Atmospheric Circulation

High temperatures over the equator and low temperatures over the poles, combined with the rotation of the Earth, produce a series of cells in the atmosphere. This model of atmospheric circulation, with three cells in each hemisphere, is known as the **tricellular model**.

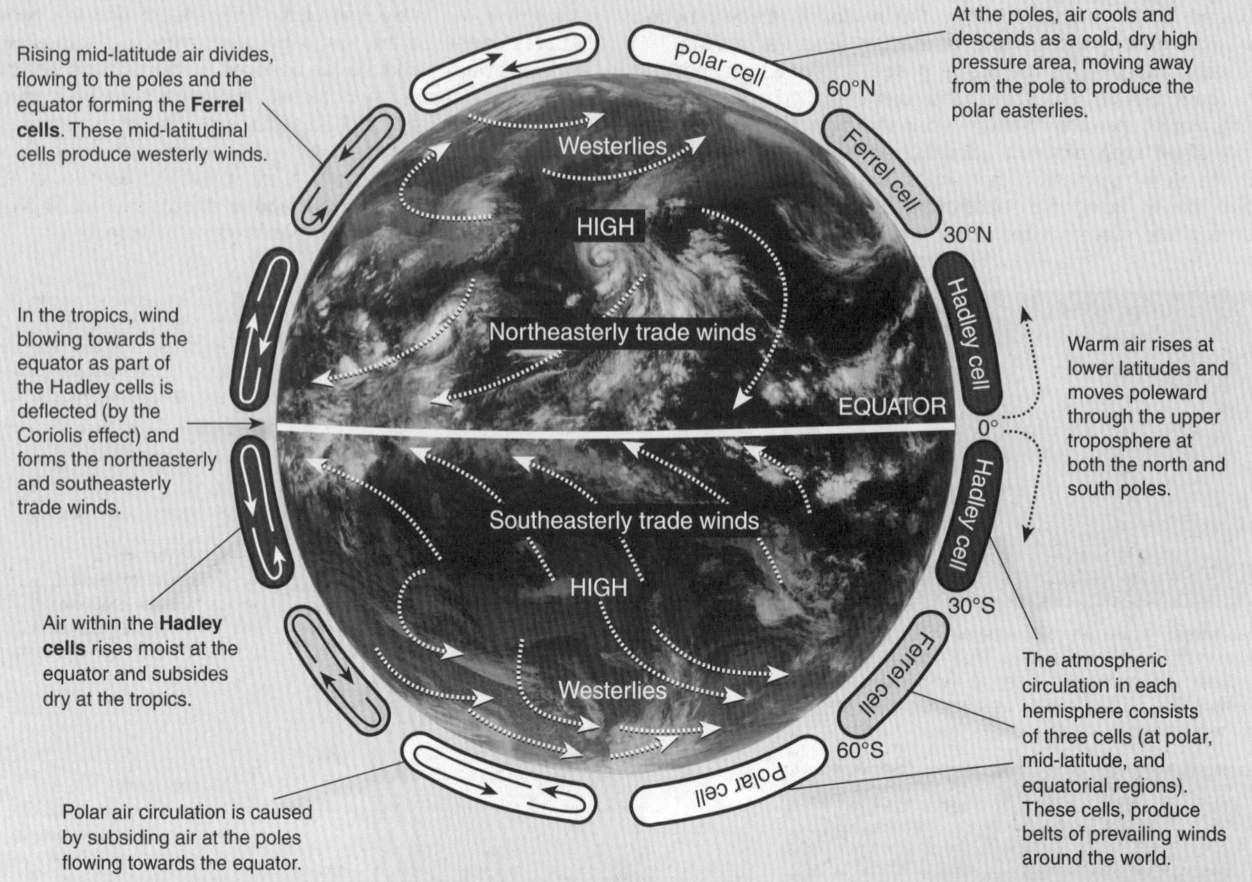
Rising mid-latitude air divides, flowing to the poles and the equator forming the **Ferrel cells**. These mid-latitudinal cells produce westerly winds.
At the poles, air cools and descends as a cold, dry high pressure area, moving away from the pole to produce the polar easterlies.
Polar cell
60°N
Ferrel cell
30°N
Westerlies
HIGH
Northeasterly trade winds
Hadley cell
EQUATOR
0°
Hadley cell
Southeasterly trade winds
HIGH
Westerlies
30°S
Ferrel cell
60°S
Polar cell
In the tropics, wind blowing towards the equator as part of the Hadley cells is deflected (by the Coriolis effect) and forms the northeasterly and southeasterly trade winds.
Warm air rises at lower latitudes and moves poleward through the upper troposphere at both the north and south poles.
Air within the **Hadley cells** rises moist at the equator and subsides dry at the tropics.
The atmospheric circulation in each hemisphere consists of three cells (at polar, mid-latitude, and equatorial regions). These cells, produce belts of prevailing winds around the world.
Polar air circulation is caused by subsiding air at the poles flowing towards the equator.

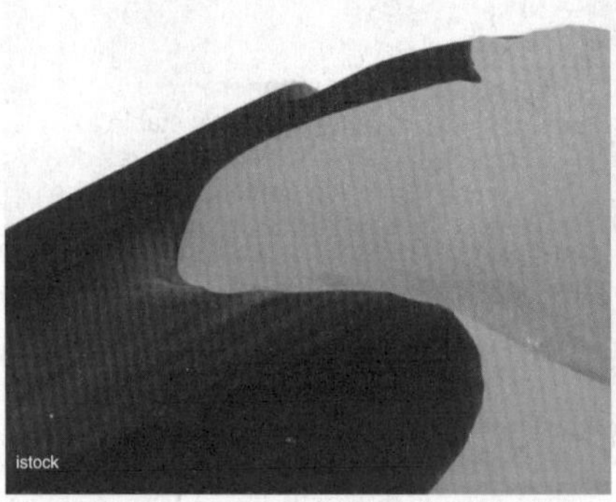
istock

In sandy deserts and polar regions, prevailing winds form dunes and drifts with characteristic shapes. Sand or snow grains are blown up the slope and fall down the far side to create sinuous crests extending for great distances.

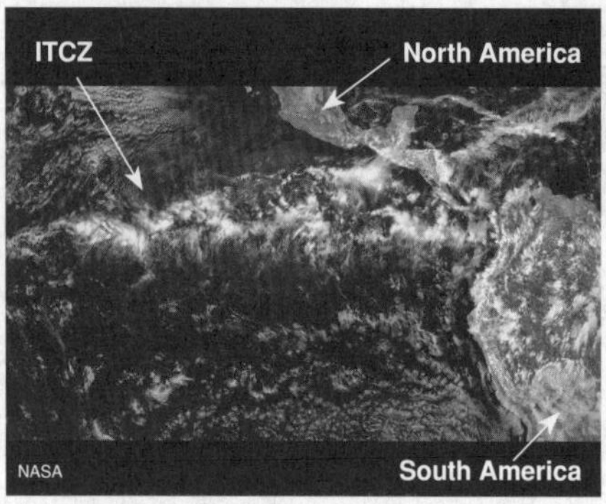
ITCZ
North America
NASA
South America

The Intertropical Convergence zone (ITCZ) marks the meeting of trade winds at the equator. Variation in the location of the ITCZ drastically affects rainfall in many equatorial nations, causing the wet and dry seasons of the tropics.

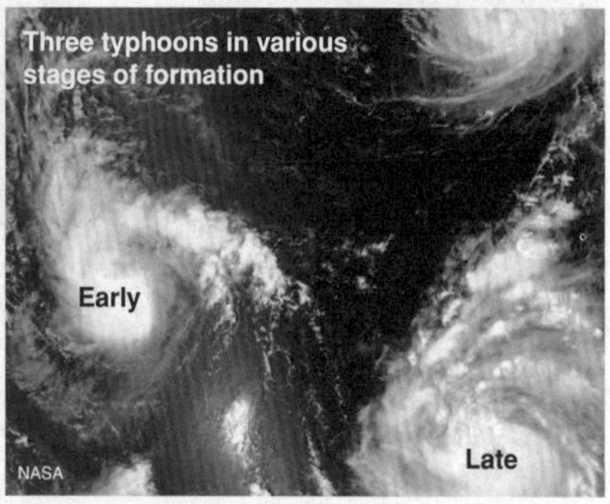
Three typhoons in various stages of formation
Early
NASA
Late

Tropical cyclones (also called typhoons or hurricanes), are low pressure systems that develop mainly over warm seas where winds start the air spiralling, producing low surface pressure into which air accelerates.

4. (a) Explain what is meant by a **prevailing wind**: ______

(b) Describe some of the physical and biological effects of prevailing winds: ______

5. The ITCZ was also called *the doldrums* by early sailors. Suggest why it was given this name: ______

Variation and Oscillation

Energy from the sun is distributed through a global system of atmospheric and ocean circulation that creates the Earth's **climate**. Heated air moving towards the poles from the equator does not flow in a single uniform convection current. Friction, drag, and momentum cause air close to the Earth's surface to be pulled in the direction of the Earth's rotation. This deflection is called the **Coriolis effect** and it is responsible for the direction of movement of large-scale weather systems in both hemispheres. The interactions of atmospheric systems are so complex that climatic conditions are never exactly the same from one year to the next. However, it is possible to find periodic patterns or **oscillations** in climate. The **El-Niño-Southern Oscillation**, which has a periodicity of three to seven years, is one such climate cycle. El-Niño years cause a reversal of the ordinary climate regime and are connected to such economically disastrous events as the collapse of fisheries stocks (e.g the Peruvian anchovy stock), severe flooding in the Mississippi Valley, drought-induced crop failures and forest fires in Australia and Indonesia.

The Coriolis Effect

Air flowing towards, or away from, the equator follows a curved path that swings it to the right in the northern hemisphere and to the left in the southern hemisphere (right). This phenomenon, known as the **Coriolis effect**, is caused by the anticlockwise rotation of the Earth about its axis, so as air moves across the Earth's surface, the surface itself is moving but at a different speed. The magnitude of the Coriolis effect depends on the latitude and the speed of the moving air. It is greatest at the poles and is responsible for the direction of the rotation of large hurricanes.

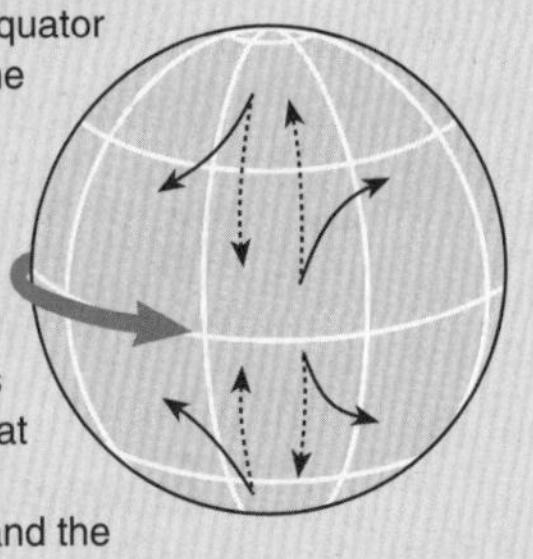

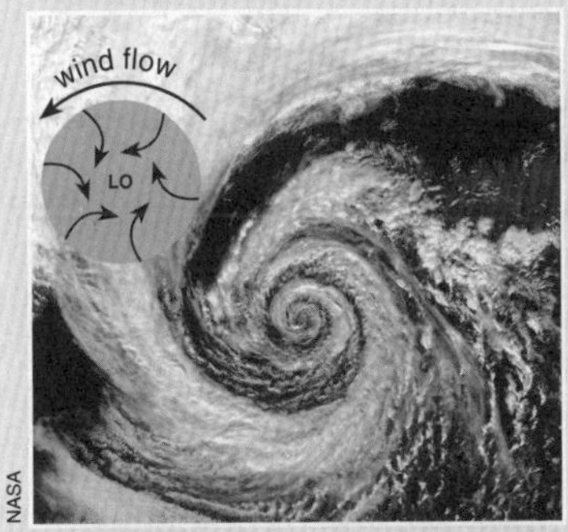

Air flows from high pressure to low pressure (see grey inset). In the northern hemisphere, the Coriolis effect deflects this moving air to the right, causing cyclonic (low pressure) systems to rotate counter-clockwise as seen here in a low pressure system over Iceland. Cyclonic weather is usually dull, with grey cloud and persistent rain.

In the southern hemisphere, cyclonic systems spiral in a clockwise direction, seen in this photograph of cyclone Catarina, a rare South Atlantic tropical cyclone which hit Brazil in March 2004. As air rushes into the low pressure area, it is defected to the left, causing a clockwise spiral.

Frontal Weather

A **weather front** marks the boundary between two air-masses at different densities. A front is about 100-200 km wide and slopes where warm and cool air masses collide. A front appears on a weather map as a line with triangles (cold front) or semicircles (warm front) attached.

In a **cold front**, cold air undercuts warm air, forcing it steeply upwards along the line of the front and triggering the formation of towering cumulus clouds. Cold fronts are often associated with low-pressure systems and unsettled weather conditions.

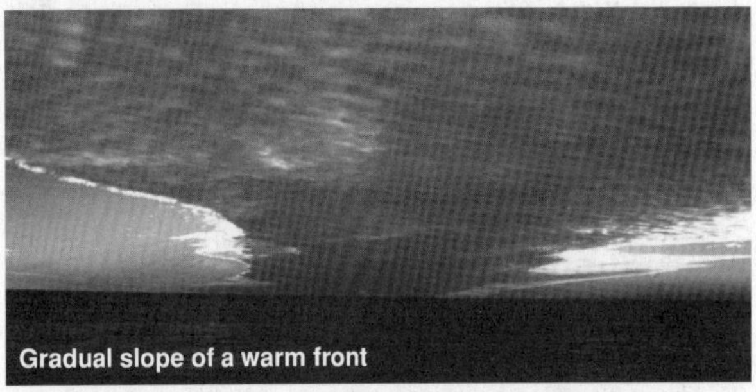

Gradual slope of a warm front

In a **warm front**, warm air rises over cold air more gradually, producing flattened, stratus-type clouds. Warm fronts produce low intensity rainfall that may last for some time and preceed warm weather. Because it moves more quickly, a cold front will eventually overtake a warm front, creating an occlusion.

1. Explain the role of the Coriolis effect in creating the prevailing winds in different regions of the globe:

2. In the spaces provided below, draw schematic diagrams, similar to that in the top left of the photograph above, to show:

(a) Movement of air from a high pressure to a low pressure system if there was no Coriolis effect:

(b) Movement of air from a high pressure to a low pressure system in the southern hemisphere:

(a)	(b)

Variation and Oscillation

Interactions between the atmosphere and the oceans are at the core of most global climate patterns. These climate patterns are referred to as **oscillations** because they fluctuate on time scales ranging from days to decades. The **El-Niño-Southern Oscillation cycle** (ENSO) is the most prominent of these global oscillations, causing weather patterns involving increased rain in specific places but not in others. It is one of the many causes of drought.

In non-El-Niño conditions, a low pressure system over Australia draws the southeast trade winds across the eastern Pacific from a high pressure system over South America. These winds drive the warm South Equatorial Current towards Australia's coast. Off the coast of South America, upwelling of cold water brings nutrients to the surface.

Normal climatic conditions

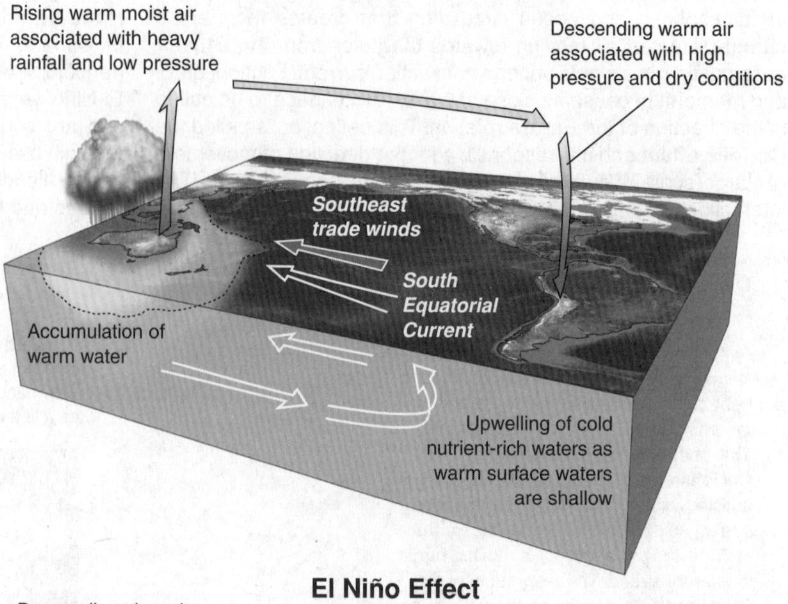

In an El Niño event, the pressure systems that normally develop over Australia and South America are weakened or reversed, beginning with a rise in air pressure over the Indian Ocean, Indonesia, and Australia. Warm water extends deeper and flows eastwards, blocking the nutrient upwelling along the west coast of the Americas. This has a devastating effect on fish stocks. El Niño brings drought to Indonesia and northeastern South America, while heavy rain over Peru and Chile causes the deserts to bloom.

El Niño Effect

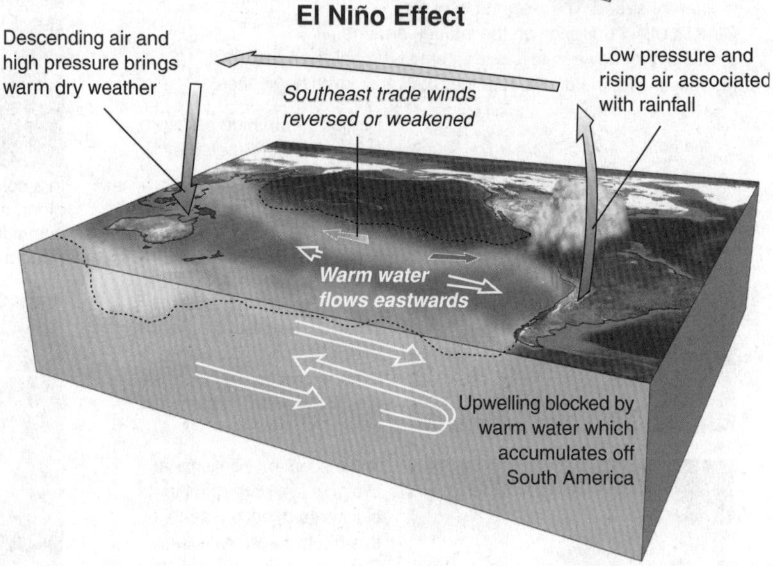

3. Using the diagrams you have just drawn to help you, explain why cyclonic weather systems spiral counter-clockwise in the northern hemisphere but clockwise in the southern hemisphere:

4. Explain the effect of an El Niño year on:

   (a) The climate of the western coast of South America:

   (b) The climate of Indonesia and Australia:

Ocean Circulation and Currents

Throughout the oceans, there is a constant circulation of water, both across the surface and at depth. Surface circulation, much of which is in the form of circular **gyres**, is driven by winds. In contrast, the deep-water ocean currents (the **thermohaline circulation**) is driven by the cooling and sinking of water masses in polar and subpolar regions. Cold water circulates through the Atlantic, penetrating the Indian and Pacific oceans, before returning as warm upper ocean currents to the South Atlantic.

Deep water currents move slowly and once a body of water sinks it may spend hundreds of years away from the surface. The polar oceans comprise the Arctic Ocean in the northern hemisphere and the Southern Ocean in the south. They differ from other oceans in having vast amounts of ice, in various forms, floating in them. This ice coverage has an important stabilising effect on global climate, insulating large areas of the oceans from solar radiation in summer and preventing heat loss in winter.

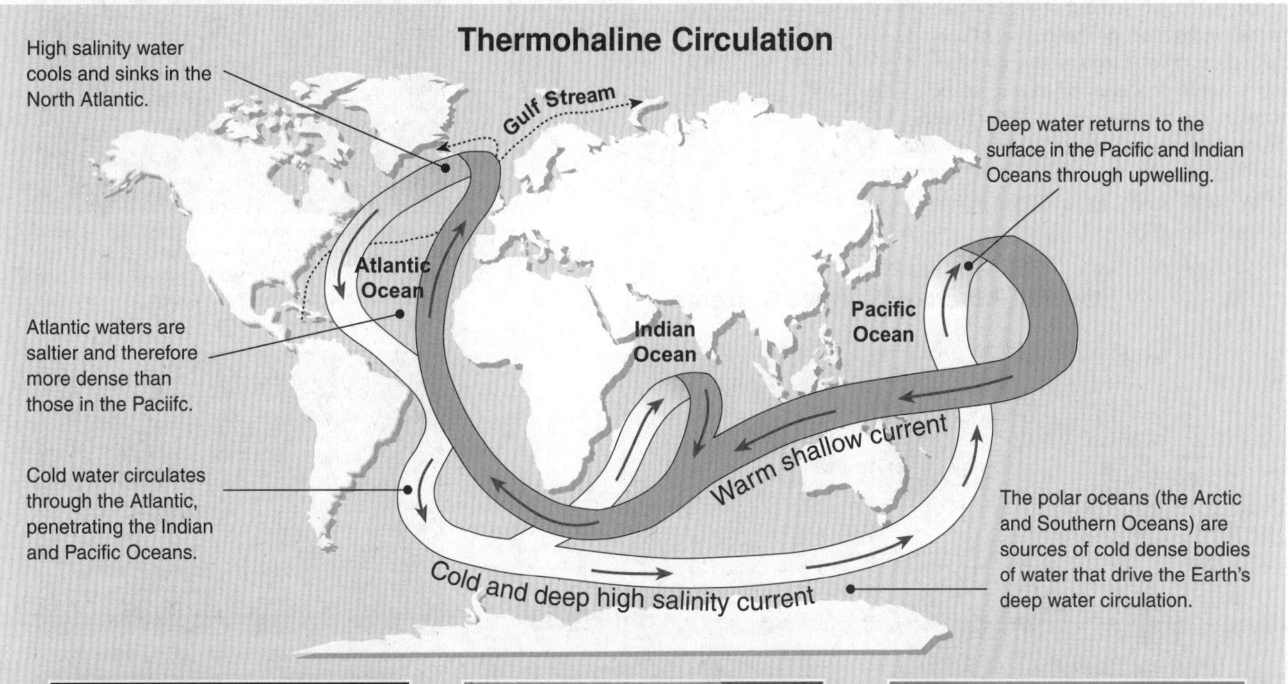
Thermohaline Circulation
High salinity water cools and sinks in the North Atlantic.
Gulf Stream
Deep water returns to the surface in the Pacific and Indian Oceans through upwelling.
Atlantic Ocean
Atlantic waters are saltier and therefore more dense than those in the Paciifc.
Indian Ocean
Pacific Ocean
Warm shallow current
Cold water circulates through the Atlantic, penetrating the Indian and Pacific Oceans.
The polar oceans (the Arctic and Southern Oceans) are sources of cold dense bodies of water that drive the Earth's deep water circulation.
Cold and deep high salinity current

Antarctica
Southern Ocean
Wikimedia Commons NOAA

The Southern Ocean encircles Antarctica and is covered in ice for much of the year. Complex currents in the Southern Ocean produce rich upwelling zones that support abundant plankton and complex food webs.

Arctic Ocean

The vast amounts of ice associated with the polar oceans has an important stabilising effect on the global climate, insulating large areas of oceans from solar radiation in the summer and preventing heat loss in winter.

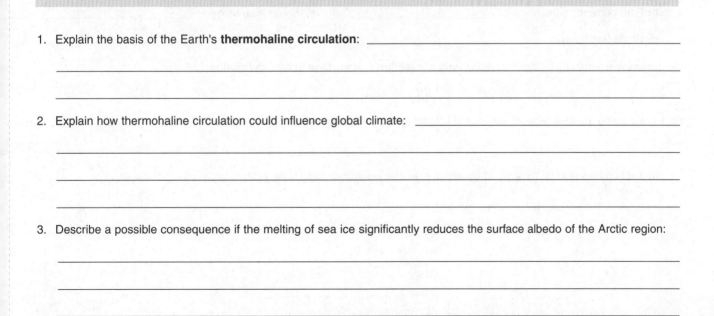

Icebreaker, Arctic sea ice
istock

Satellite observations show that the Arctic sea ice is melting earlier and more rapidly than previously reported. The loss of ice cover will dramatically reduce the surface **albedo** (reflectivity) in the Arctic region.

1. Explain the basis of the Earth's **thermohaline circulation**: ______________________________

2. Explain how thermohaline circulation could influence global climate: ______________________________

3. Describe a possible consequence if the melting of sea ice significantly reduces the surface albedo of the Arctic region: ______________________________

Surface Circulation in the Oceans

The surface circulation of the oceans is driven by winds, but modified by the **Coriolis effect**. In the northern hemisphere, the Coriolis effect deflects the wind-driven water movements slightly to the right and in the southern hemisphere to the left. Drag accentuates the Coriolis effect so that the average water motion in the top few hundred metres of the ocean surface is almost at right angle to the wnd direction. The overall effect is a pattern of large scale circular movements of water, or **gyres**, which rotate clockwise in the northern hemisphere and anticlockwise in the southern (below and right). These currents carry warm water away from the equator and colder water towards it.

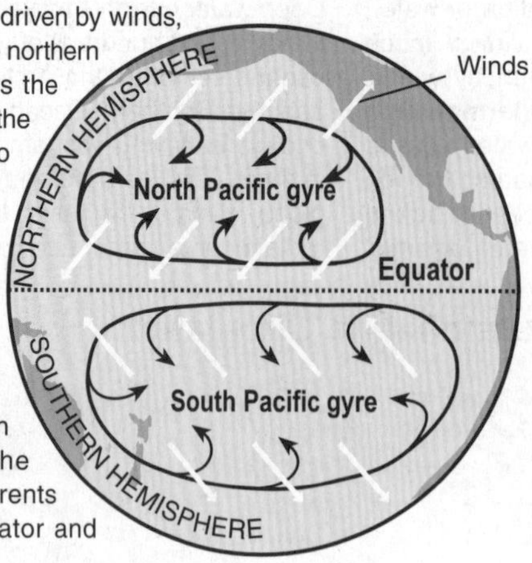

Overall Pattern of Surface Currents

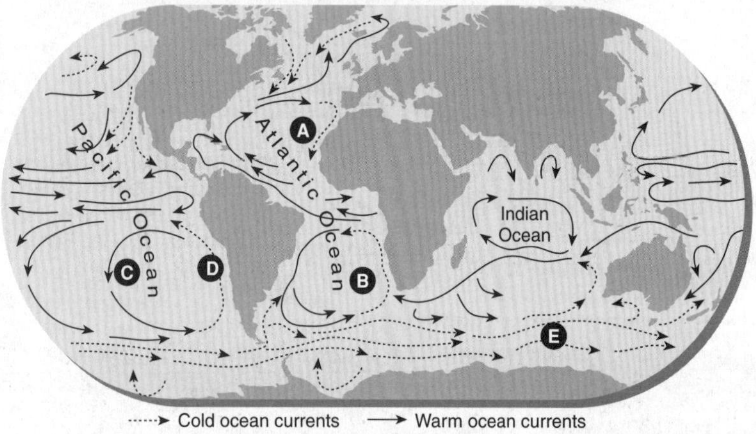

Local Currents and Upwelling

Local currents and vertical transport of water (upwelling and downwelling) are important phenomena around coastal regions. Upwelling in particular has important biological effects because it returns nutrients to surface waters, which promotes the growth of plankton.

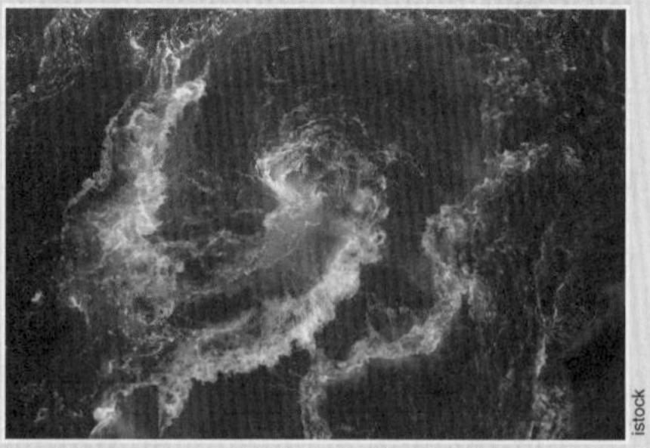

Local currents are the result of interactions between tidal forces and coastlines. The whirlpools (vortices), seen above at Saltstraumen in Norway, are created by exceptionally strong tidal movements as water forces its way through a long narrow strait.

Plankton blooms, seen here as bright spots around the coast of England and Ireland, often occur in **upwelling zones**, as nutrients are brought to the surface. Upwelling occurs to replace the seawater that is moved offshore by surface circulation.

4. Contrast the mechanisms operating to drive deep water and surface water circulation: ____________________

____

____

5. Match each description below with its appropriate letter on the above diagram "*Overall Pattern of Surface Currents*":

(a) Antarctic circumpolar current: ____________ (d) North Atlantic gyre: ____________

(b) Peru current: ____________ (e) South Atlantic gyre: ____________

(c) South Pacific gyre: ____________

6. Describe a similarity between **atmospheric circulation** and **surface ocean circulation** patterns: ____________________

____

7. (a) Describe the biological importance of upwelling in coastal regions: ____________________

____

(b) Explain how normal patterns of upwelling are affected during an El Niño year: ____________________

____

____

____

____

____

Global Water Resources

The Earth is an **aqueous planet**; 71% of its surface is covered by water. The majority of the Earth's water (a little over 97%) is stored within the oceans, and less than 3% of the total water supply is freshwater. A small amount (0.0071%) of the world's water exists as usable freshwater at the Earth's surface (in lakes, rivers, and wetlands). The remaining freshwater is contained within ice caps or glaciers, groundwater, or atmospheric moisture. The total amount of water on Earth is fairly constant at any one time, cycling constantly between liquid, vapour, and ice. This cyclic process is termed the water or **hydrological cycle**.

The **Ogallala aquifer** is a vast water-table aquifer located beneath the Great Plains in the US. It covers portions of eight states and is extensively used for irrigation. At current usage rates it may be depleted by 2020, and is essentially non-renewable as it will take thousands of years to recharge.

The **Volga River** and its many tributaries form an extensive river system, which drains an area of about 1.35 million km^2 in the most heavily populated part of Russia. High levels of chemical pollution currently give cause for environmental concern.

From glacial origins, the **Yangtze River** flows 6300 km eastwards into the East China Sea. The Yangtze is subject to extensive flooding, which is only partly controlled by the massive Three Gorges Dam, and it is heavily polluted.

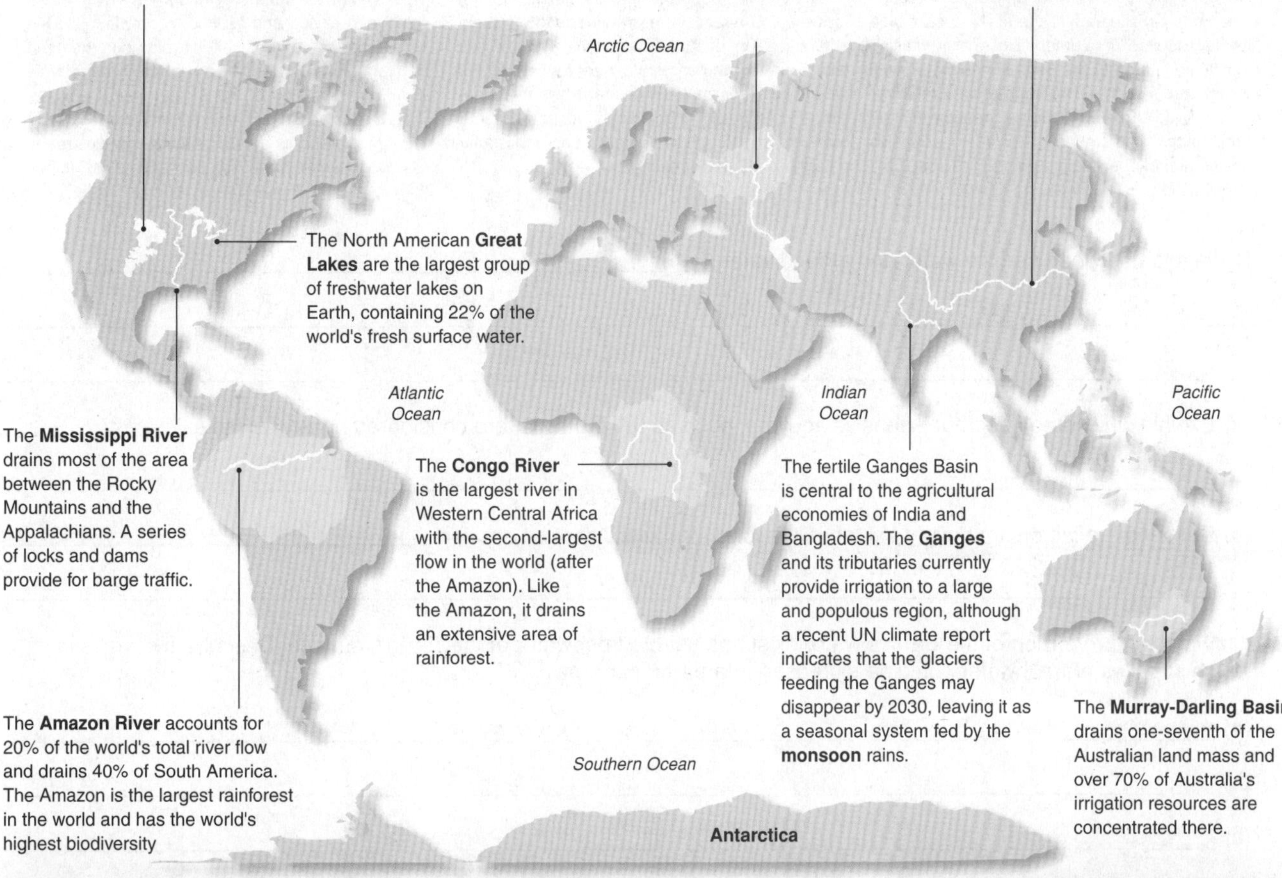

The North American **Great Lakes** are the largest group of freshwater lakes on Earth, containing 22% of the world's fresh surface water.

The **Mississippi River** drains most of the area between the Rocky Mountains and the Appalachians. A series of locks and dams provide for barge traffic.

The **Congo River** is the largest river in Western Central Africa with the second-largest flow in the world (after the Amazon). Like the Amazon, it drains an extensive area of rainforest.

The fertile Ganges Basin is central to the agricultural economies of India and Bangladesh. The **Ganges** and its tributaries currently provide irrigation to a large and populous region, although a recent UN climate report indicates that the glaciers feeding the Ganges may disappear by 2030, leaving it as a seasonal system fed by the **monsoon** rains.

The **Amazon River** accounts for 20% of the world's total river flow and drains 40% of South America. The Amazon is the largest rainforest in the world and has the world's highest biodiversity

The **Murray-Darling Basin** drains one-seventh of the Australian land mass and over 70% of Australia's irrigation resources are concentrated there.

Rivers form when rain and meltwater are channelled downhill along surface irregularities. They typically end in either a lake or at the sea. Rivers are used for transportation, recreation and irrigation, and supply food and hydroelectric power. They shape the landscape through erosion and deposition.

Lakes form naturally in surface depressions or a result of damming and, when large enough, can have a profound effect on regional weather. The Great Lakes by moderate seasonal temperatures somewhat, by absorbing heat and cooling the air in summer, then slowly radiating that heat in autumn.

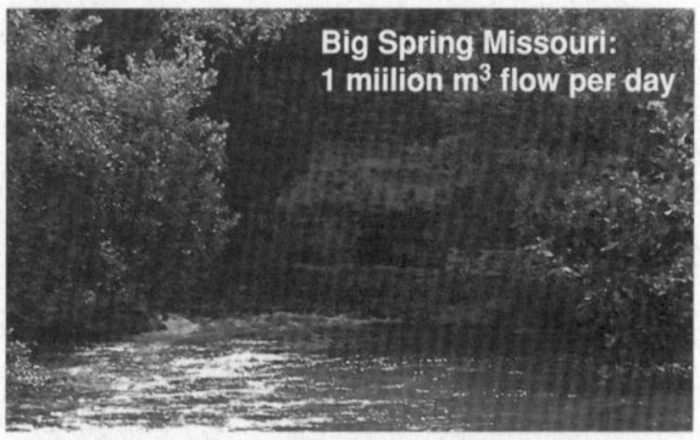

Aquifers are typically saturated regions of the subsurface that produce an economically feasible quantity of water to a well or spring. Aquifers can occur at various depths but those closer to the surface are more likely to be exploited for water supply and irrigation.

Meltwater is the water released by the melting of snow or ice, including glacial ice. Meltwater can destabilise glacial lakes and snowpack causing floods and avalanches. Meltwater also acts as a lubricant in the basal sliding of glaciers.

Polar icecaps and glaciers store the majority of the world's freshwater. While the Antarctic ice-sheet is growing, Arctic ice is thinning. Because snow and ice form a protective, cooling layer over the Arctic, this thinning accelerates temperature rise.

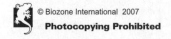

Consecutive years of poor rainfall are behind the **East African** drought. Readily available water reserves have been used up in many areas, leaving insufficient water for irrigation and stock. Fires and desertification are also consequences of drought.

Related activities: The Water Cycle, Water Use, Humans and Resources
Web links: The Earth In Our Hands: Groundwater, World Water Hotspots

The World's Oceans

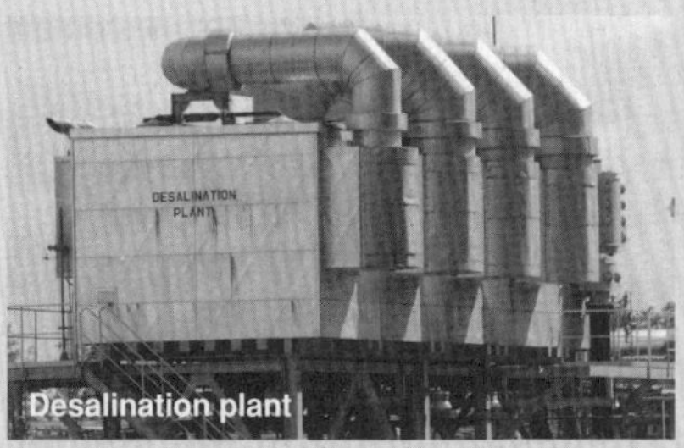

Desalination plant

The major minerals in seawater (sodium and chloride) make it too salty for drinking or irrigation unless it is desalinated. Large-scale desalination typically requires large amounts of energy, making it costly compared to the use of fresh water from rivers or groundwater. The large energy reserves of many Middle Eastern countries, along with their relative scarcity of water, have led to extensive construction of desalination plants in this region.

Fish harvest

The ocean supplies humans with a vast range of resources and opportunities, including fisheries (above) and aquaculture, transportation, tourism and recreation, offshore extraction (of oil and gas, gravel, sand, and minerals), energy production (tidal, wave, thermal, salinity, and wind), marine biotechnology (pharmaceuticals), conservation of biodiversity (through protected areas and marine reserves), waste disposal.

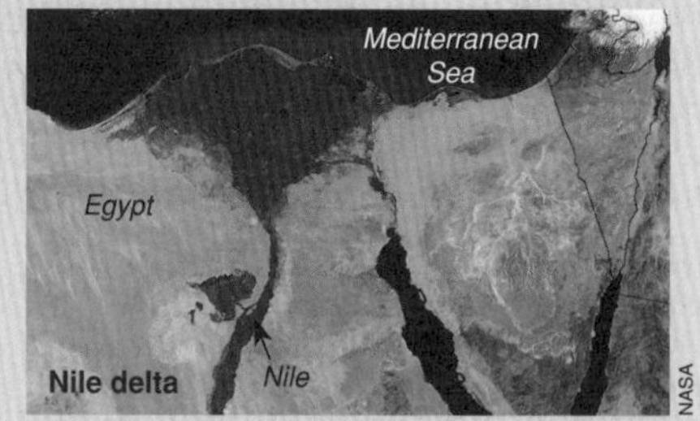

Nile delta

Oceans receive the outflow from the world's major river systems. A delta is formed when a river enters an ocean or sea and sediments are deposited in a fan-shaped pattern. Deltas are among the most fertile regions on Earth and are heavily populated. Since construction of the Aswan High Dam, the Nile Delta (above) no longer receives an annual supply of nutrients and sediments from upstream. Its floodplain soils are now poorer as a result.

1. Distinguish between surface water and groundwater: ___

2. (a) Explain why some deep but extensive aquifers, such as the Ogallala, are considered non-renewable: ___

   (b) Describe the factors that might influence the long term viability of an aquifer: ___

3. Only a small proportion of the Earth's water exists as usable freshwater at the Earth's surface. Describe five ways in which supplies of freshwater could be increased in a particular area:

   (a) ___

   (b) ___

   (c) ___

   (d) ___

   (e) ___

4. Much of the Earth's freshwater is locked up in ice. Describe the critical role of this "unavailable" water: ___

5. (a) Explain why delta regions are among the most fertile in the world: ___

   (b) Describe how flood control schemes can detrimentally affect this fertility: ___

6. Discuss the association between the location of the world's major water resources, areas of high biodiversity, and human population density. You may need to consult other resources to bring together your lines of evidence:

Water Use

Water is an essential commodity for sustaining life, it plays an important role in shaping the Earth's surface, moderating the climate and is used by humans to meet requirements for food, shelter and recreation. Despite the importance and limited availability of water, it is a resource that has been poorly managed. Pollution, overuse, and misuse of water resources mean that water resources are being used up at a faster rate that than the water cycle can replenish it at. Consequently water shortages are affecting many regions, and may ultimately lead to disputes or "water wars" in the future.

Spray irrigation

Agricultural use of water

Intensive agriculture uses large volumes of water. Crop irrigation accounts for 65% of the world's water use yet it is largely inefficient, with only 40% of the water reaching the crops and the rest being lost through evaporation, seepage, and runoff. Improved irrigation practices, such as drip irrigation, could double the amount of water delivered to crops and free large amounts of water for other uses.

Fish processing

Industrial use of water

Industrial related water usage increases as the human population increases. The processing of food, and the manufacture of metal, wood and paper products, gasoline and oils all consume large amounts of water. High consumer demand and low levels of recycling exacerbate the problem, but there are large savings to be made by using recycled water and improving the efficiency of water use.

Waste water (sewage) treatment

Domestic water use

Nearly half of the water supplied by municipal water systems in the US is used to flush toilets and water lawns, and another 20-35% is lost through leakages. Treatment of waste water places major demands on cities yet there are few incentives to reduce water use and recycle. Providing a reliable supply of clean (**potable**) water remains a major public health issue in many poorer regions of the world.

Using the World's Freshwater

With increased population growth and economic development, global rates of water withdrawal from surface and groundwater sources are projected to more than double in the next two decades. This will exceed available surface runoff in a number of regions.

Manufacturing and production processes are usually water intensive. Some everyday items (right) use surprisingly large amounts of water in their production. As urbanisation and industrialisation grow, increasing efficiency of water use will be paramount.

How much water to manufacture...?

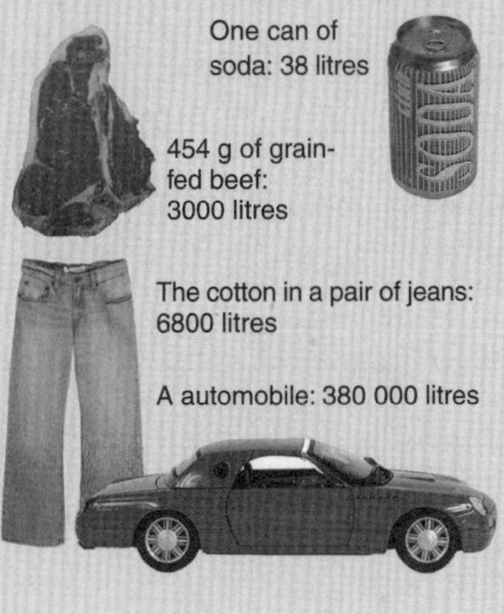

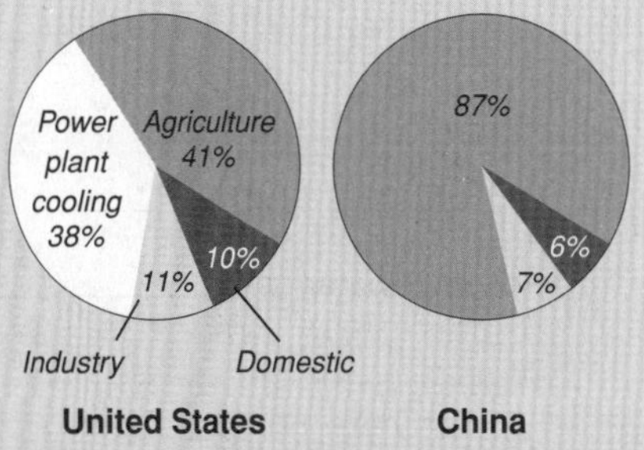

Uses of withdrawn water vary both regionally and between countries. The USA, for example, uses more water for power generation and industry than China, where agricultural uses account for most of the withdrawals.

1. Account for the differences between China and the USA in terms of how water use is apportioned: ______

2. Explain why water availability is considered to be one of the critical global resource issues in the 21st Century:

3. Describe ways in which the efficiency of water use could be improved in the following areas:

   (a) Crop irrigation: ______

Damming the Nile

Two dams straddle the Nile River at Aswan, Egypt. The newer (and by far larger) of the two, the **Aswan High Dam**, was completed in 1970 and formed Lake Nasser, a 550 km long reservoir capable of holding two years of the Nile's annual water flow.

The main objectives of the project were energy generation, flood control, and the provision of water for agriculture. These goals have have been achieved, but construction of the Aswan Dams has also had a number of detrimental effects. Before impoundment, the Nile flooded annually, bringing minerals and nutrient-rich sediments to the floodplain and flushing out accumulated salts. Without flooding, fertilisers must be applied to the land and salts build up in the soils, causing crops to fail. Moreover, without annual deposition of river sediments, the land is eroding, allowing the sea to encroach up the river delta. If global warming causes a rise in sea levels, 60% of Egypt's habitable land may be flooded. Damming has also caused 64% of commercially fished species in the Nile to disappear. Time will tell if better management will help to reverse the problems currently being experienced in the Nile Delta region.

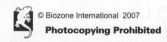

Satellite image (above) and panoramic view (below) of the Aswan High Dam

(b) Domestic water use:

(c) Industrial water use:

4. Using a local or international example, discuss the environmental and economic problems generally associated with damming a large river:

5. Investigate water use in your community, identifying major surface and groundwater sources, how water usage is apportioned, and projected water problems. Summarise your findings in a brief report and attach it to this activity page:

Ecosystems

Understanding ecosystems, energy flow, and biogeochemical cycles

Ecosystems and the influence of abiotic factors. Habitat and niche. Energy flow and nutrient cycling. Ecosystem stability and succession.

Learning Objectives

- ☐ 1. Compile your own glossary from the **KEY WORDS** displayed in **bold type** in the learning objectives below.

Ecosystems *(pages 33-35)*

- ☐ 2. Define the terms: **ecology**, **ecosystem**, **community**, **population**, **species**, and **environment**, and provide examples of each. Distinguish between **biotic factors** and **abiotic factors** in terms of your definition of an ecosystem. Recognise that the **biosphere** consists of interdependent ecosystems and that ecosystems are dynamic entities, subject to change.
- ☐ 3. Recognise major **biomes** on Earth and explain how they are classified according to major vegetation type. Appreciate the influence of latitude and local climate in determining the distribution of world biomes.
- ☐ 4. Identify the **biotic** and **abiotic** components of a named ecosystem. Include reference to the predominant **vegetation** type and the physical factors that determine the ecosystem's characteristics. Explain how the different parts of an ecosystem influence each other.

Habitats and Microclimates *(pages 36-40)*

- ☐ 5. Explain how **environmental gradients** can occur over relatively short distances, e.g. on rocky shores, forests, lakes, or deserts. Explain how gradients in the abiotic environment contribute to community patterns.
- ☐ 6. Describe aspects of the **abiotic environment**, and explain how they influence the distribution of species in a local environment. Explain how **limiting factors** determine species distribution. Explain why limiting factors are often different for plants and animals.
- ☐ 7. List the factors commonly used to describe a **habitat**. Distinguish between habitat and **microhabitat** and explain how microhabitats arise as a result of differences in **microclimates**. Recognise the habitat as part of the **niche** of a **species**. Distinguish between the **tolerance range** and **optimum range** for a species.

Niche and Adaptation *(page 41)*

- ☐ 8. Describe the components of an organism's **ecological niche**. Recognise the constraints that are placed on the niche occupied by an organism. Distinguish between the **fundamental** and the **realised niche**.
- ☐ 9. Understand the significance of Gause's **competitive exclusion** principle with respect to niche overlap between species. Explain the effect of **interspecific competition** on **niche breadth**.
- ☐ 10. Describe examples of **adaptive features** (**adaptations**) in named organisms. Recognise **physiological**, **structural**, and **behavioural adaptations** for survival in a given niche and appreciate that these are the result of changes that occur to the species as a whole, but not to individuals within their own lifetimes.

Energy in Ecosystems *(page 42)*

- ☐ 11. Recognise that the first and second **laws of thermodynamics** govern energy flow in ecosystems. In practical terms, explain what the energy laws mean with respect to energy conversions in ecosystems.
- ☐ 12. Appreciate that light is the initial energy source for almost all ecosystems and that photosynthesis is the main route by which energy enters most food chains. Recognise that energy is dissipated as it is transferred through trophic levels.
- ☐ 13. Understand the interrelationship between nutrient cycling and energy flow in ecosystems:
  - Energy flows through ecosystems in the high energy chemical bonds within **organic matter**.
  - Nutrients move within and between ecosystems in **biogeochemical cycles** involving exchanges between the atmosphere, the Earth's crust, water, and living organisms.

Trophic relationships *(pages 42-44)*

- ☐ 14. Describe how the energy flow in ecosystems is described using **trophic levels**. Distinguish between **producers**, **primary and secondary consumers**, **detritivores**, and **saprotrophs** (**decomposers**), identifying the relationship between them. Describe the role of each of these in energy transfer.
- ☐ 15. Describe how energy is transferred between different trophic levels in **food chains** and **food webs**. Compare the amount of energy available to each trophic level and recognise that the **efficiency** of this energy transfer is only 10-20%. Explain what happens to the remaining energy and why energy cannot be recycled.
- ☐ 16. With respect to the **efficiency** of energy transfer in food chains, explain the small biomass and low numbers of organisms at higher trophic levels.
- ☐ 17. Provide three examples of food chains, each with at least three linkages. Show how the (named) organisms are interconnected through their feeding relationships. Assign trophic levels to the organisms in a food chain.
- ☐ 18. Construct a food web for a named community, showing how the (named) organisms are interconnected through their feeding relationships.

Measuring energy flow *(pages 45-50)*

- ☐ 19. Explain the terms: **productivity**, **gross primary production**, **net primary production**, and **biomass**. Recognise how these relate to the transfer of energy to the next trophic level.
- ☐ 20. Explain how the energy flow in an ecosystem can be described quantitatively using an **energy flow diagram**. Include reference to the following:

- Trophic levels (scaled boxes to illustrate relative amounts of energy at each level)
- Direction of energy flow
- Processes involved in energy transfer
- Energy sources and energy sinks.

☐ 21. Describe food chains quantitatively using **ecological pyramids**. Explain how these may be based on numbers of organisms, biomass of organisms, or energy content of organisms at each trophic level. Identify problems with the use of number pyramids, and understand why pyramids of biomass or energy are usually preferable.

☐ 22. Given appropriate information, construct and interpret pyramids of numbers, energy, and biomass for different communities. Express the energy available at each trophic level in appropriate units. Identify the relationship between different types of pyramids and their corresponding food chains and webs. Explain why the shape of each graph is a pyramid (or sometimes an inverse pyramid).

Biogeochemical Cycles *(pages 51-57)*

☐ 23. Explain the terms **nutrient cycle** and **environmental reservoir**. Draw and interpret a generalised model of a nutrient cycle, identifying the roles of **primary productivity**, **decomposition**, and **saprotrophs** and **detritivores** in nutrient cycling.

☐ 24. Explain how humans may intervene in nutrient cycles and describe the effects of these interventions.

☐ 25. Using a diagram, describe the stages in the **carbon cycle**, identifying the form of carbon at the different stages, and using arrows to show the direction of nutrient flow and labels to identify the processes involved. Identify the role of microorganisms, carbon sinks, and carbonates in the cycle.

☐ 26. Identify factors influencing the rate of carbon cycling. Recognise the role of respiration and photosynthesis in the short-term fluctuations and in the long-term global balance of oxygen and carbon dioxide.

☐ 27. Using a diagram, describe the stages in the **nitrogen cycle**, identifying the form of nitrogen at the different stages, and using arrows to show the direction of nutrient flow and labels to identify the processes involved. Identify and explain the role of microorganisms in the cycle, as illustrated by:

(a) **Nitrifying bacteria** (*Nitrosomonas*, *Nitrobacter*)

(b) **Nitrogen-fixing bacteria** (*Rhizobium*, *Azotobacter)*

(c) **Denitrifying bacteria** (*Pseudomonas*, *Thiobacillus*)

☐ 28. Describe the features of the **water** (hydrologic) **cycle**. Understand the ways in which water is cycled between various reservoirs and describe the major processes involved, including **evaporation**, **condensation**, **precipitation**, and **runoff**.

☐ 29. Using diagrams, describe the **phosphorus cycle**, using arrows to show the direction of nutrient flow and labels to identify the processes involved. Identify the role of microorganisms in the cycle and contrast the phosphorus cycle with other nutrient cycles.

☐ 30. Using a diagram, describe stages in the **sulfur cycle**, identifying the form of sulfur at the different stages, and using arrows to show the direction of nutrient flow and labels to identify the processes involved.

Ecosystem Stability *(pages 58-62)*

☐ 31. Explain what is meant by the **stability** of an ecosystem and identify its components. Explain the relationship between **ecosystem stability** and **diversity**, including the role of **keystone species**.

☐ 32. Explain what is meant by **ecological succession** and identify factors contributing to successional change.

☐ 33. Describe **primary succession** from **pioneer species** to a **climax community**. Describe the characteristics of species typical of each successional stage. Describe how community diversity changes during the course of a succession. Comment on the stability of the pioneer and climax communities and relate this to the relative importance of abiotic and biotic factors at each stage.

☐ 34. Using examples, distinguish between **primary** and **secondary succession**, outlining the features of each type. Appreciate how the **climax** vegetation varies according to local climate, latitude, and altitude.

Supplementary Texts

See page 7 for additional details of these texts:

- Christopherson, R.W, 2007. **Elemental Geosystems**, chpt. 16.
- Miller, G.T, 2007. **Living in the Environment: Principles, Connections & Solutions**, chpt. 2-3, 7.
- Raven *et. al.*, 2002. **Environment**, chapt. 3-6.
- Reiss, M. & J. Chapman, 2000. **Environmental Biology** (Cambridge University Press), pp. 1, 3, 76.
- Smith, R. L. & T.M. Smith, 2001. **Ecology and Field Biology**, reading as required.

Periodicals

See page 7 for details of publishers of periodicals:

STUDENT'S REFERENCE

- **Ecosystems** Biol. Sci. Rev., 9(4) March 1997, pp. 9-14. *Ecosystems: food chains & webs, energy flows, nutrient cycles, and ecological pyramids.*
- **Big Weather** New Scientist, 22 May 1999 (Inside Science). *Global weather patterns and their role in shaping the ecology of the planet.*
- **The Ecological Niche** Biol. Sci. Rev., 12(4), March 2000, pp. 31-35. *An excellent account of the niche - an often misunderstood concept that is never-the-less central to ecological theory.*
- **The Other Side of Eden** Biol. Sci. Rev., 15(3) February 2003, pp. 2-7. *An account of the Eden Project: the collection of artificial ecosystems in Cornwall. Its aims, future directions, and its role in the study of natural ecosystems are discussed.*
- **The Nitrogen Cycle** Biol. Sci. Rev., 13(2) November 2000, pp. 25-27. *An excellent account of the the nitrogen cycle: conversions, role in ecosystems, and the influence of human activity.*
- **The Case of the Missing Carbon** National Geographic, 205(2), Feb. 2004, pp. 88-117. *8 billion tonnes of CO_2 is dumped into the atmosphere annually, but only 3.5 billion tonnes remains there. The rest is absorbed into the Earth's carbon sinks.*
- **Microbes and Nutrient Cycling** Biol. Sci. Rev., 19(1) Sept. 2006, pp. 16-20. *The roles of microorganisms in nutrient cycling.*
- **The Carbon Cycle** New Scientist, 2 Nov. 1991 (Inside Science). *The role of carbon in ecosystems.*
- **Ultimate Interface** New Scientist, 14 Nov. 1998 (Inside Science). *Biogeochemical cycles and the role of the soil in cycling processes.*
- **Biodiversity and Ecosystems** Biol. Sci. Rev., 11(4), March 1999, pp. 18-23. *Species richness and the breadth and overlap of niches. An account of how biodiversity influences ecosystem dynamics.*
- **Climate Change and Biodiversity** Biol. Sci. Rev., 16(1) Sept. 2003, pp. 10-14. *While the focus of this account is on climate change, it provides useful coverage of ecosystem structure and processes and how these are studied.*
- **Plant Succession** Biol. Sci. Rev., 14 (2) November 2001, pp. 2-6. *Thorough coverage of primary and secondary succession, including the causes of different types of succession.*

Internet

See pages 4-5 for details of how to access **Bio Links** from our web site: **www.thebiozone.com** From Bio Links, access sites under the topics:

Glossaries: • Ecology glossary > **General Online Biology Resources**: • Access excellence • Ken's bioweb resources • Virtual library: bioscience **ECOLOGY**: • Introduction to biogeography and ecology > **Ecosystems**: • The ecology help site • Freshwater ecosystems • Marine science • The rocky intertidal zone • What are ecosystems?... *and others* > **Energy Flows and Nutrient Cycles**: • A marine food web • Bioaccumulation • Human alteration of the global nitrogen cycle • Nitrogen: the essential element • The carbon cycle • The nitrogen cycle • Trophic pyramids and food webs

Presentation MEDIA to support this topic:

ECOLOGY:
- Ecosystems • Niche
- Communities

Components of an Ecosystem

The concept of the ecosystem was developed to describe the way groups of organisms are predictably found together in their physical environment. A community comprises all the organisms within an ecosystem. Both physical (abiotic) and biotic factors affect the organisms in a community, influencing their distribution and their survival, growth, and reproduction.

The Biosphere

The **biosphere**, which contains all the Earth's living organisms, amounts to a narrow belt around the Earth extending from the bottom of the oceans to the upper atmosphere. Broad scale life-zones or **biomes** are evident within the biosphere, characterised according to the predominant vegetation. Within these biomes, **ecosystems** form natural units comprising the non-living, physical environment (the soil, atmosphere, and water) and the **community** (all the organsims living in a particular area).

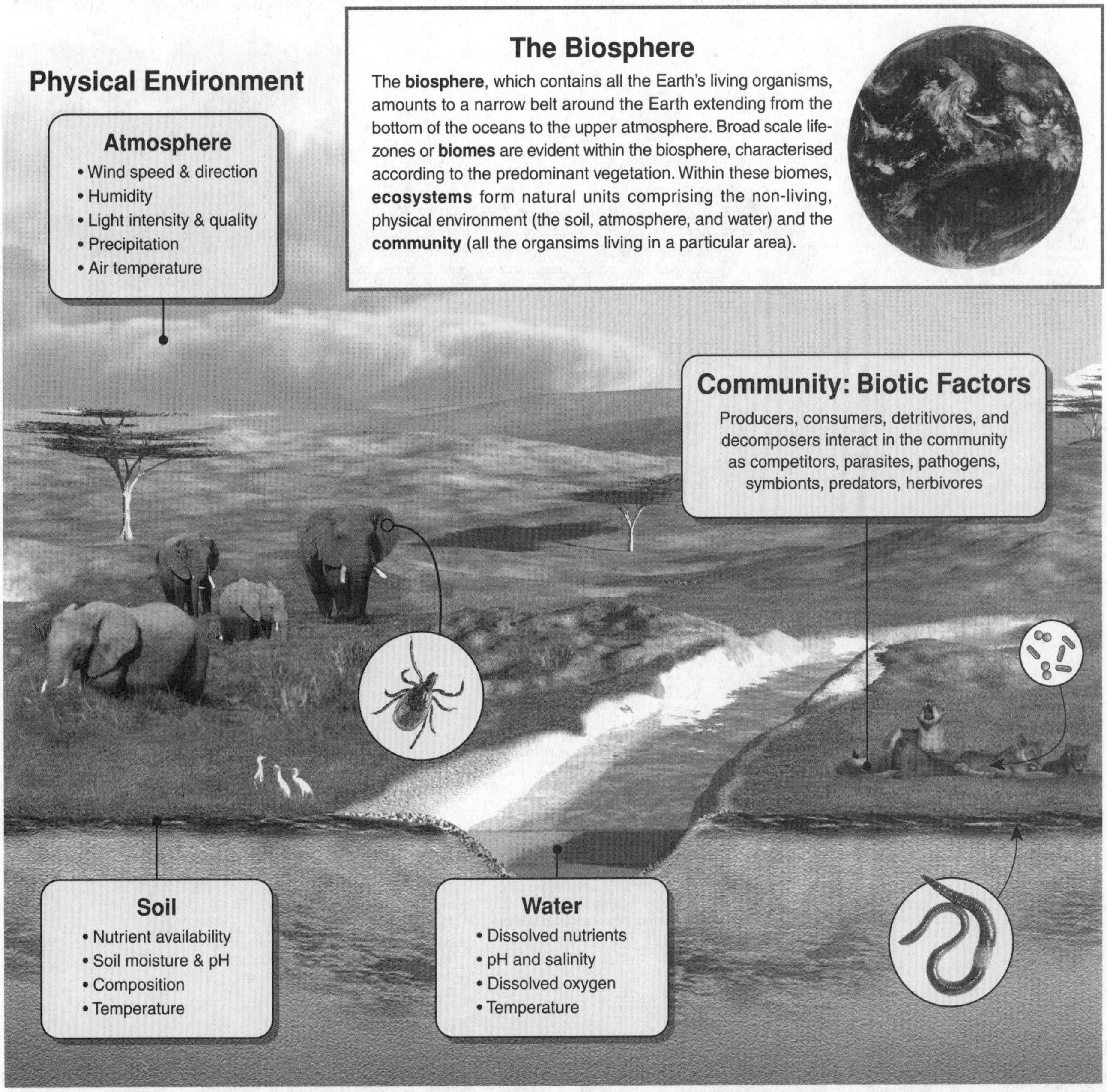

1. Distinguish clearly between a community and an ecosystem: ______

2. Distinguish between biotic and abiotic factors: ______

3. Use one or more of the following terms to describe each of the features of a rainforest listed below:
Terms: *population, community, ecosystem, physical factor.*

(a) All the green tree frogs present: ______

(b) The entire forest: ______

(c) All the organisms present: ______

(d) The humidity: ______

Related activities: Biomes

Biomes

Global patterns of vegetation distribution are closely related to climate. Although complex, major vegetation **biomes** can be recognised. These are large areas where the vegetation type shares a particular suite of physical requirements. Biomes have characteristic features, but the boundaries between them are not distinct. The same biome may occur in widely separated regions of the world wherever the climatic and soil conditions are similar. See the continuation of the world map below over the next page.

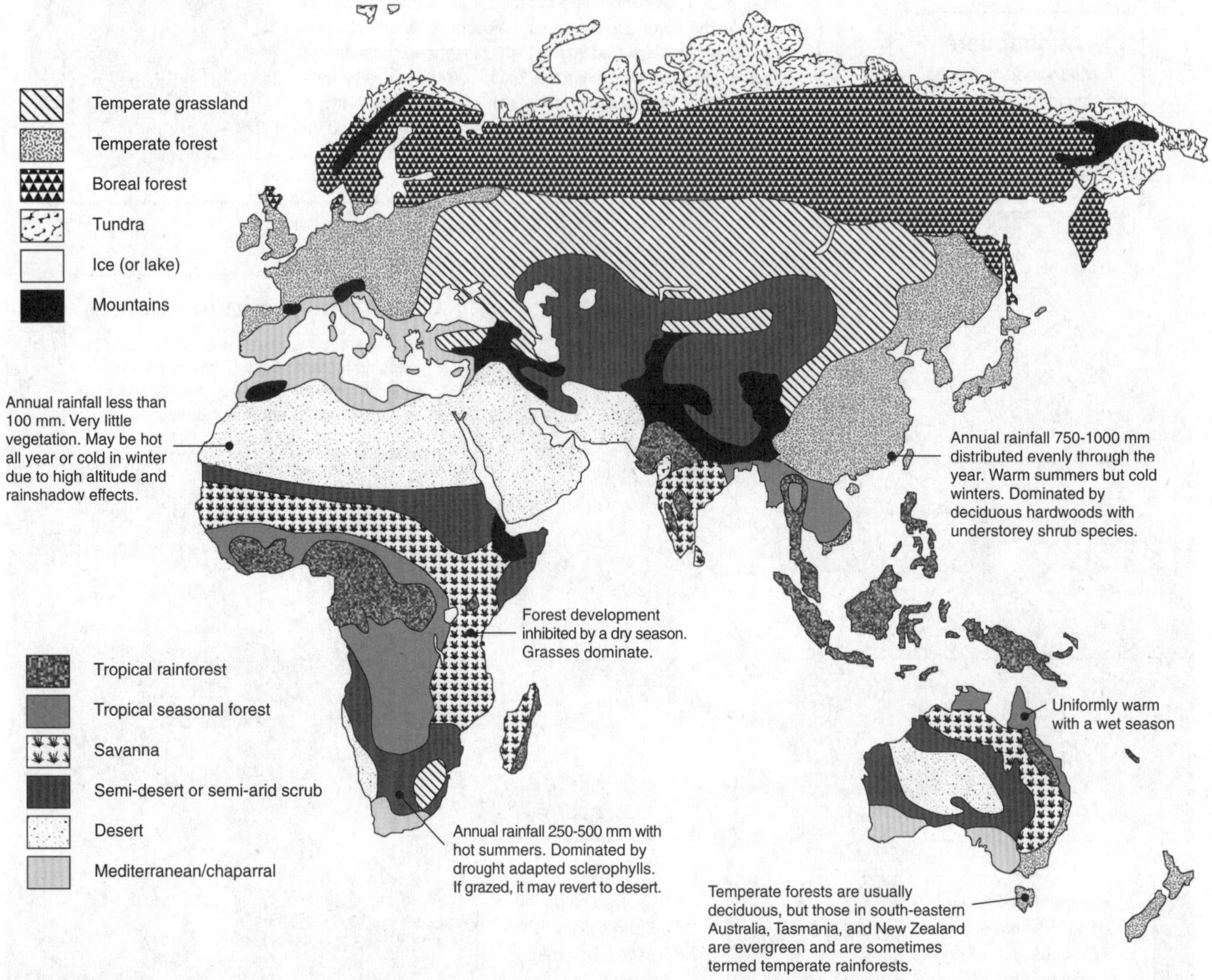
Temperate grassland
Temperate forest
Boreal forest
Tundra
Ice (or lake)
Mountains
Annual rainfall less than 100 mm. Very little vegetation. May be hot all year or cold in winter due to high altitude and rainshadow effects.
Annual rainfall 750-1000 mm distributed evenly through the year. Warm summers but cold winters. Dominated by deciduous hardwoods with understorey shrub species.
Forest development inhibited by a dry season. Grasses dominate.
Tropical rainforest
Tropical seasonal forest
Savanna
Semi-desert or semi-arid scrub
Desert
Mediterranean/chaparral
Uniformly warm with a wet season
Annual rainfall 250-500 mm with hot summers. Dominated by drought adapted sclerophylls. If grazed, it may revert to desert.
Temperate forests are usually deciduous, but those in south-eastern Australia, Tasmania, and New Zealand are evergreen and are sometimes termed temperate rainforests.

Tundra

Semi-arid scrub

Mediterranean/chaparral

Desert

1. Suggest what abiotic factor(s) limit the northern extent of boreal forest: ______________________

2. Grasslands have about half the productivity of tropical rainforests, yet this is achieved with less than a tenth of the biomass; grasslands are more productive per unit of biomass. Suggest how this greater efficiency is achieved:

Related activities: Atmosphere and Climate, Physical Factors and Gradients
Web links: The World's Biomes

Vegetation patterns are determined largely by climate but can be modified markedly by human activity. Semi-arid areas that are overgrazed will revert to desert and have little ability to recover. Similarly, many chaparral regions no longer support their original vegetation, but are managed for vineyards and olive groves. Wherever they occur, mountainous regions are associated with their own altitude adapted vegetation. The rainshadow effect of mountains governs the distribution of deserts in some areas too, as in Chile and the Gobi desert in Asia. The classification of biomes may vary slightly; some sources distinguish hot deserts (such as the Sahara) from cold deserts and semi-deserts (such as the Gobi). However, most classifications recognise desert, tundra, grassland and forest types and differentiate them on the basis of latitude.

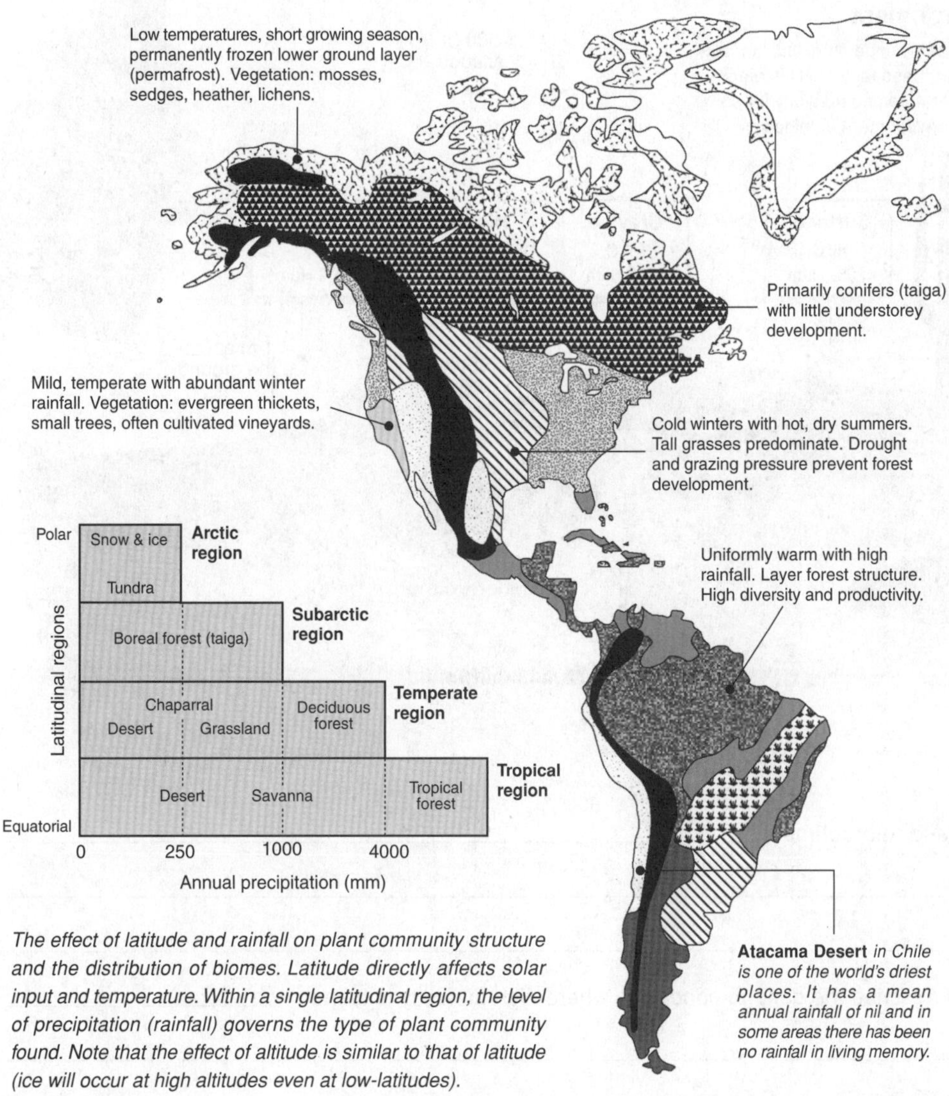

Low temperatures, short growing season, permanently frozen lower ground layer (permafrost). Vegetation: mosses, sedges, heather, lichens.
Primarily conifers (taiga) with little understorey development.
Mild, temperate with abundant winter rainfall. Vegetation: evergreen thickets, small trees, often cultivated vineyards.
Cold winters with hot, dry summers. Tall grasses predominate. Drought and grazing pressure prevent forest development.
Uniformly warm with high rainfall. Layer forest structure. High diversity and productivity.
Polar
Snow & ice
Tundra
Arctic region
Boreal forest (taiga)
Subarctic region
Chaparral
Desert
Grassland
Deciduous forest
Temperate region
Desert
Savanna
Tropical forest
Tropical region
Equatorial
Latitudinal regions
0
250
1000
4000
Annual precipitation (mm)
Atacama Desert *in Chile is one of the world's driest places. It has a mean annual rainfall of nil and in some areas there has been no rainfall in living memory.*

The effect of latitude and rainfall on plant community structure and the distribution of biomes. Latitude directly affects solar input and temperature. Within a single latitudinal region, the level of precipitation (rainfall) governs the type of plant community found. Note that the effect of altitude is similar to that of latitude (ice will occur at high altitudes even at low-latitudes).

Tropical rainforest

Tropical seasonal forest

Savanna

Temperate forest

Temperate grassland

Ice

Mountains

Boreal forest

3. Suggest a reason for the distribution of deserts and semi-desert areas in northern parts of Asia and in the west of North and South America (away from equatorial regions):

4. Compared with its natural extent (on the map), little unaltered temperate forest now exists. Explain why this is the case:

Physical Factors and Gradients

Gradients in abiotic factors are found in almost every environment; they influence habitats and microclimates, and determine patterns of species distribution. This activity, covering the next four pages, examines the physical gradients and microclimates that might typically be found in four, very different environments. Note that **dataloggers** (pictured right), are being increasingly used to gather such data. The principles of their use are covered in the topic *Practical Ecology*.

A Desert Environment

Desert environments experience extremes in temperature and humidity, but they are not uniform with respect to these factors. This diagram illustrates hypothetical values for temperature and humidity for some of the microclimates found in a desert environment at midday.

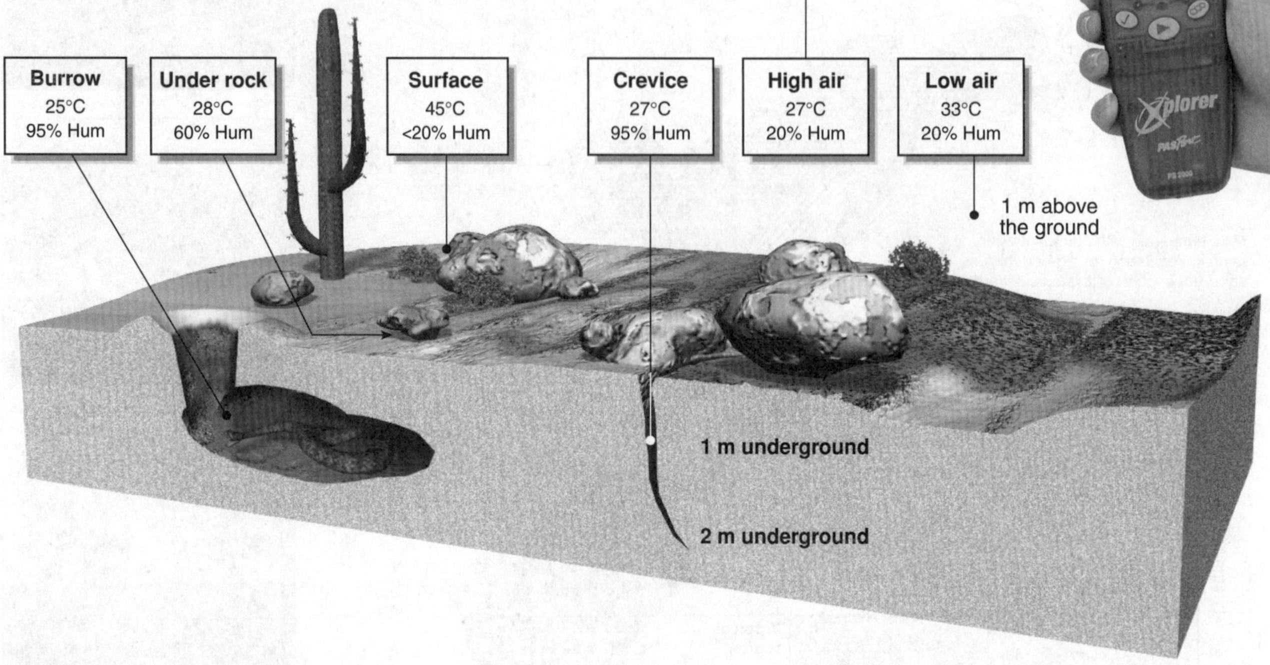

1. Distinguish between **climate** and **microclimate**: ______________________

2. Study the diagram above and describe the general conditions where high humidity is found: ______________________

3. Identify the three microclimates that a land animal might exploit to avoid the extreme high temperatures of midday: ______________________

4. Describe the likely consequences for an animal that was unable to find a suitable microclimate to escape midday sun: ______________________

5. Describe the advantage of high humidity to the survival of most land animals: ______________________

6. Describe the likely changes to the temperature and relative humidity that occur during the night: ______________________

Related activities: Monitoring Physical Factors
Web links: Tide Pool Ecology

Physical Factors at Low Tide on a Rock Platform

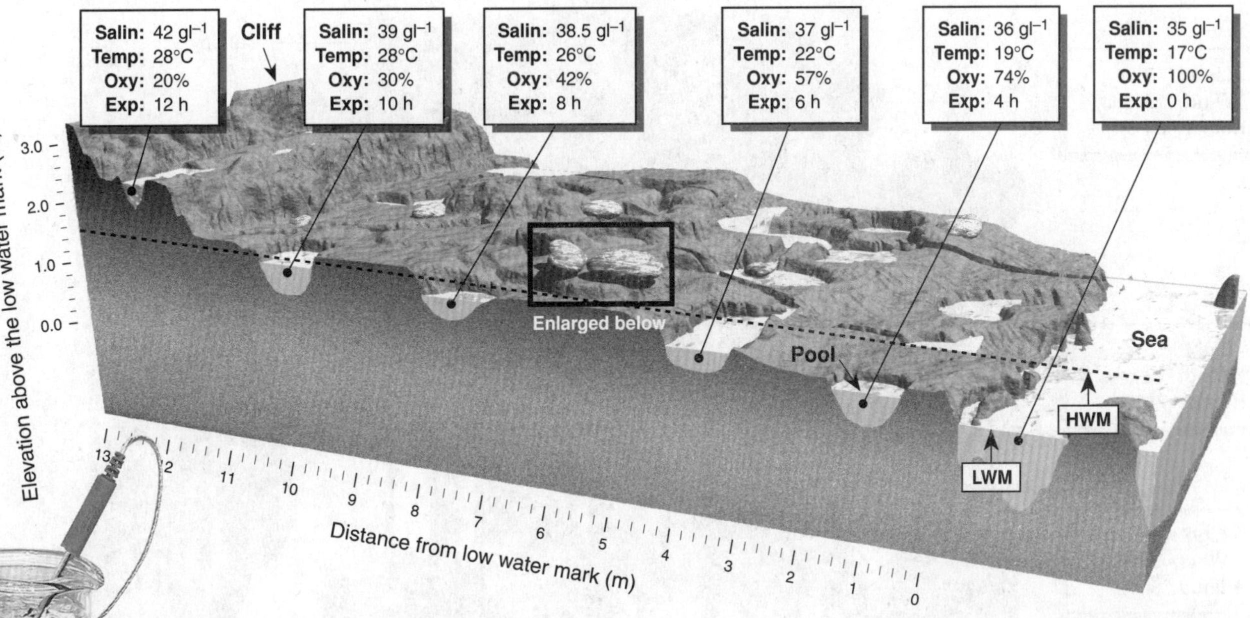

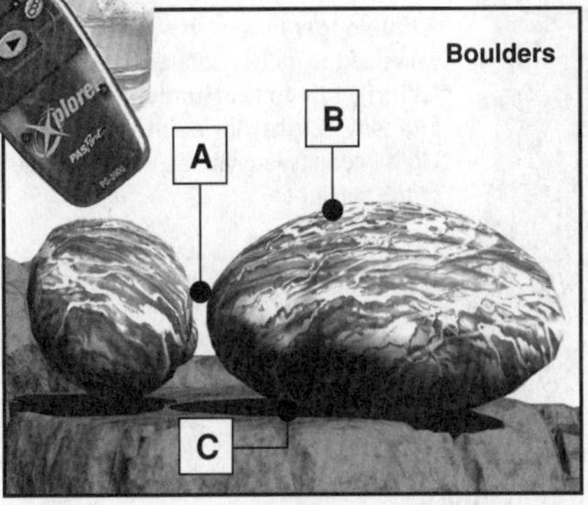

The diagram above shows a profile of a rock platform at low tide. The **high water mark** (HWM) shown here is the average height the spring tide rises to. In reality, the high tide level will vary with the phases of the moon (i.e. spring tides and neap tides). The **low water mark** (LWM) is an average level subject to the same variations due to the lunar cycle. The rock pools vary in size, depth, and position on the platform. They are isolated at different elevations, trapping water from the ocean for time periods that may be brief or up to 10 – 12 hours duration. Pools near the HWM are exposed for longer periods of time than those near the LWM. The difference in exposure times results in some of the physical factors exhibiting a **gradient**; the factor's value gradually changes over distance. Physical factors sampled in the pools include salinity, or the amount of dissolved salts (g) per liter (**Salin**), temperature (**Temp**), dissolved oxygen compared to that of open ocean water (**Oxy**), and exposure, or the amount of time isolated from the ocean water (**Exp**).

7. Describe the environmental gradient (general trend) from the low water mark (LWM) to the high water mark (HWM) for:

(a) Salinity: ______________________________

(b) Temperature: ______________________________

(c) Dissolved oxygen: ______________________________

(d) Exposure: ______________________________

8. Rock pools above the normal high water mark (HWM), such as the uppermost pool in the diagram above, can have wide extremes of salinity. Explain the conditions under which these pools might have either:

(a) Very low salinity: ______________________________

____

(b) Very high salinity: ______________________________

____

9. (a) The inset diagram (above, left) is an enlarged view of two boulders on the rock platform. Describe how the physical factors listed below might differ at each of the labelled points **A**, **B**, and **C**:

Mechanical force of wave action: ______________________________

____

Surface temperature when exposed: ______________________________

____

(b) State the term given to these localised variations in physical conditions: ______________________________

Physical Factors in a Tropical Rainforest

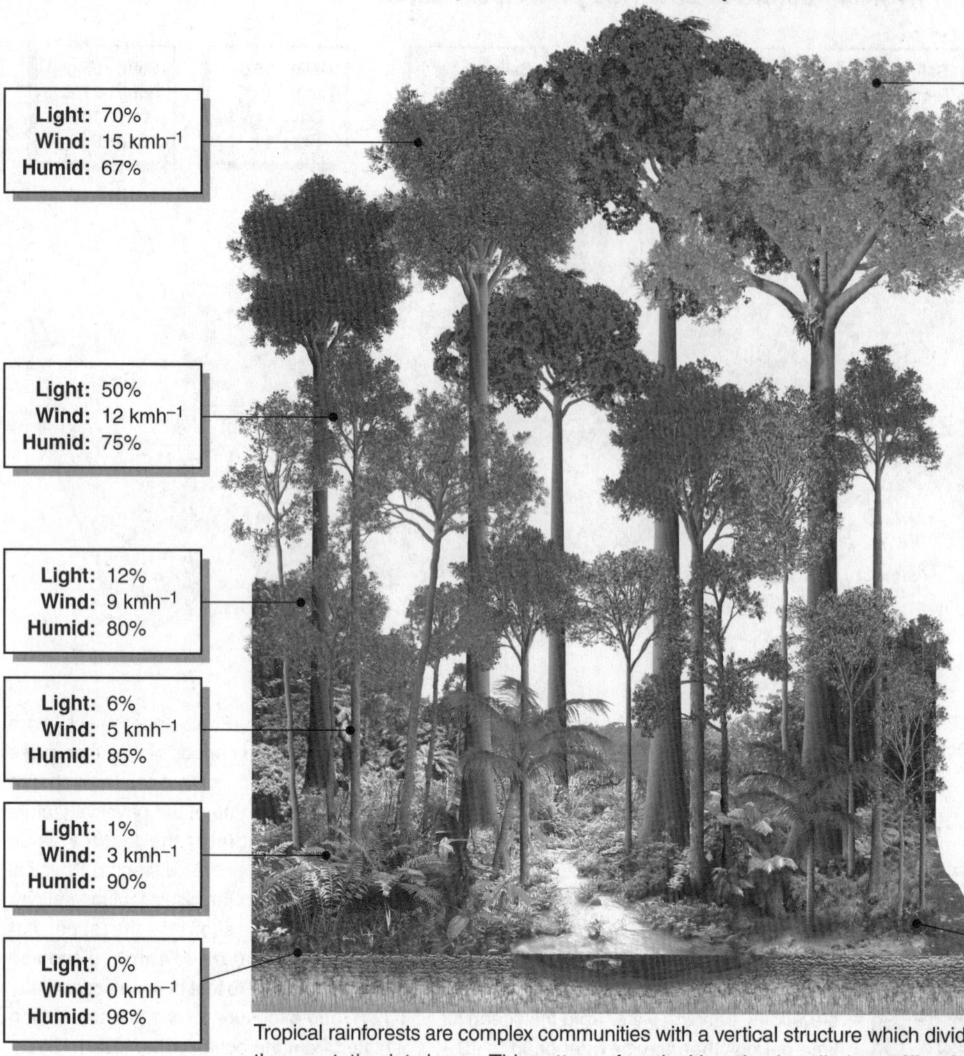

Tropical rainforests are complex communities with a vertical structure which divides the vegetation into layers. This pattern of vertical layering is called **stratification**.

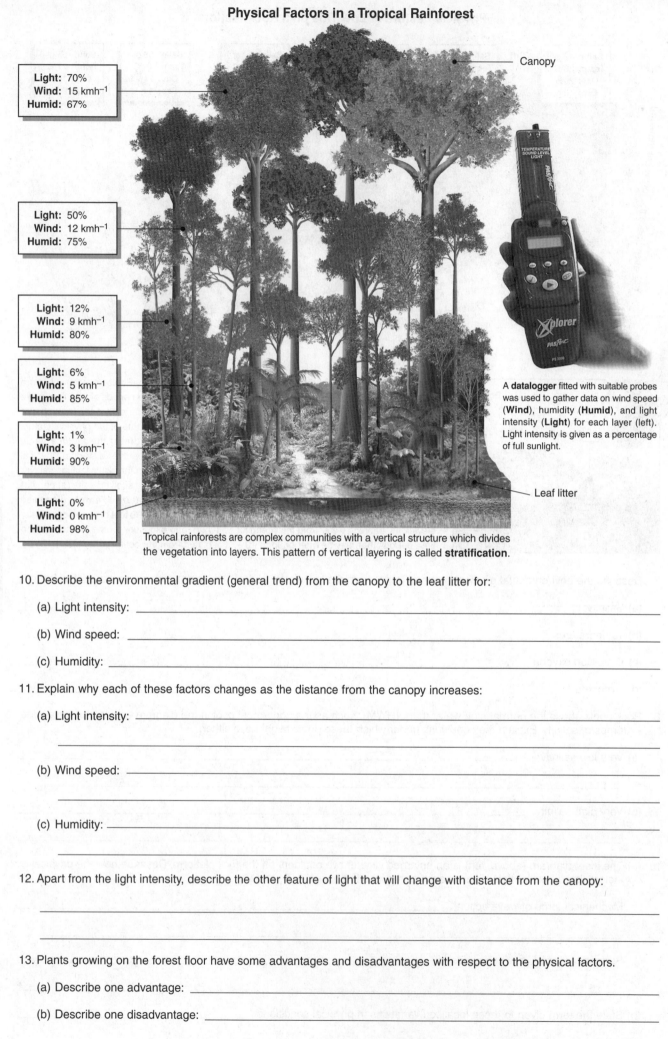

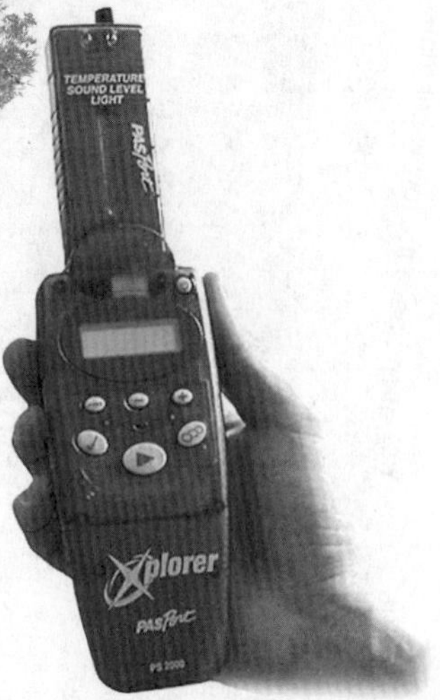

A **datalogger** fitted with suitable probes was used to gather data on wind speed (**Wind**), humidity (**Humid**), and light intensity (**Light**) for each layer (left). Light intensity is given as a percentage of full sunlight.

10. Describe the environmental gradient (general trend) from the canopy to the leaf litter for:

(a) Light intensity: ____________________

(b) Wind speed: ____________________

(c) Humidity: ____________________

11. Explain why each of these factors changes as the distance from the canopy increases:

(a) Light intensity: ____________________

____

(b) Wind speed: ____________________

____

(c) Humidity: ____________________

____

12. Apart from the light intensity, describe the other feature of light that will change with distance from the canopy:

____

____

13. Plants growing on the forest floor have some advantages and disadvantages with respect to the physical factors.

(a) Describe one advantage: ____________________

(b) Describe one disadvantage: ____________________

Physical Factors in an Oxbow Lake in Summer

Oxbow lakes are formed from old river meanders which have been cut off and isolated from the main channel following the change of the river's course. Commonly, they are very shallow (about 2-4 metres deep) but occasionally they may be deep enough to develop temporary, but relatively stable, temperature gradients from top to bottom (below). Small lakes are relatively closed systems and events in them are independent of those in other nearby lakes, where quite different water quality may be found. The physical factors are not constant throughout the water in the lake. Surface water and water near the margins can have quite different values for such factors as water temperature (**Temp**), dissolved oxygen (**Oxygen**) measured in milligrams per litre (mgl^{-1}), and light penetration (**Light**) indicated here as a percentage of the light striking the surface.

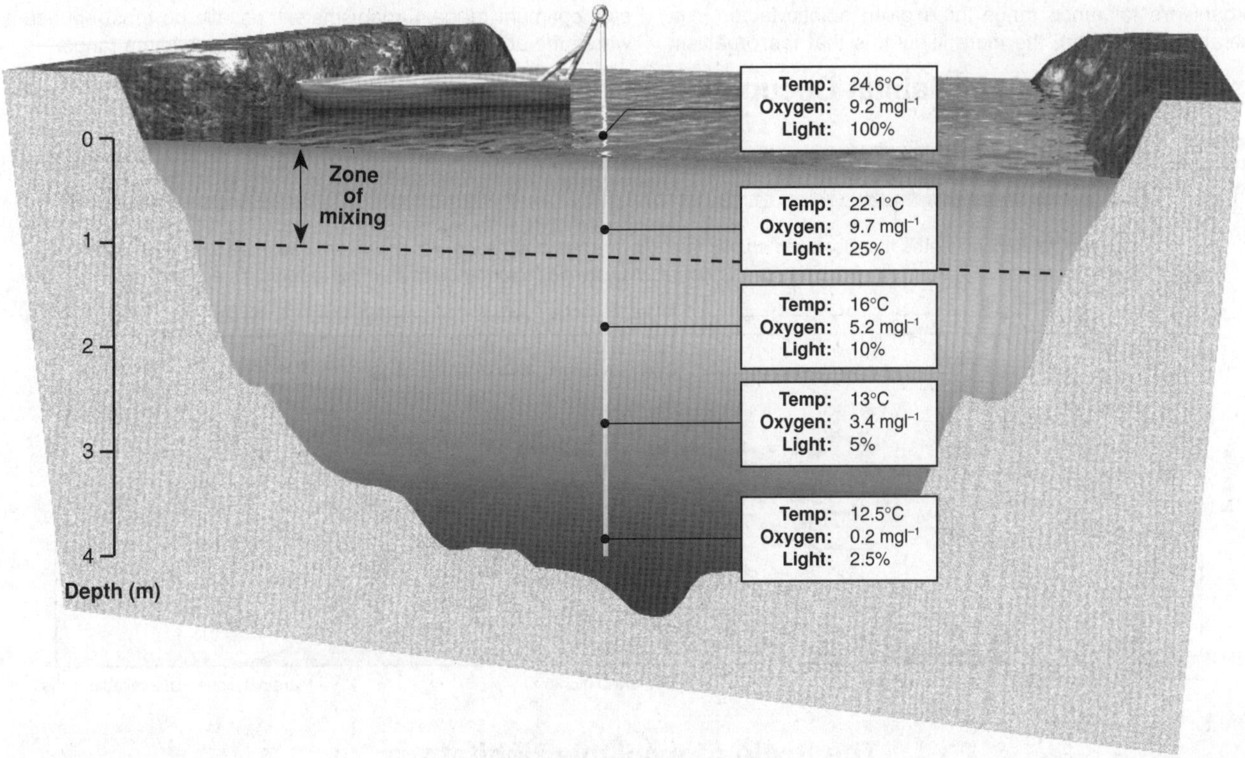

14. With respect to the diagram above, describe the environmental gradient (general trend) from surface to lake bottom for:

(a) Water temperature: ______

(b) Dissolved oxygen: ______

(c) Light penetration: ______

15. During the summer months, the warm surface waters are mixed by gentle wind action. Deeper cool waters are isolated from this surface water. This sudden change in the temperature profile is called a **thermocline** which itself is a further barrier to the mixing of shallow and deeper water.

(a) Explain the effect of the thermocline on the dissolved oxygen at the bottom of the lake: ______

(b) Explain what causes the oxygen level to drop to the low level: ______

16. Many of these shallow lakes can undergo great changes in their salinity (sodium, magnesium, and calcium chlorides):

(a) Name an event that could suddenly reduce the salinity of a small lake: ______

(b) Name a process that can gradually increase the salinity of a small lake: ______

17. Describe the general effect of physical gradients on the distribution of organisms in habitats: ______

Ecosystems

Habitat

The environment in which a species population (or a individual organism) lives (including all the physical and biotic factors) is termed its **habitat**. Within a prescribed habitat, each species population has a range of tolerance to variations in its physical and chemical environment. Within the population, individuals will have slightly different tolerance ranges based on small differences in genetic make-up, age, and health. The wider an organism's tolerance range for a given abiotic factor (e.g. temperature or salinity), the more likely it is that the organism will be able to survive variations in that factor. Species **dispersal** is also strongly influenced by **tolerance range**. The wider the tolerance range of a species, the more widely dispersed the organism is likely to be. As well as a tolerance range, organisms have a narrower **optimum range** within which they function best. This may vary from one stage of an organism's development to another or from one season to another. Every species has its own optimum range. Organisms will usually be most abundant where the abiotic factors are closest to the optimum range.

Habitat Occupation and Tolerance Range

Examples of abiotic factors influencing niche size:

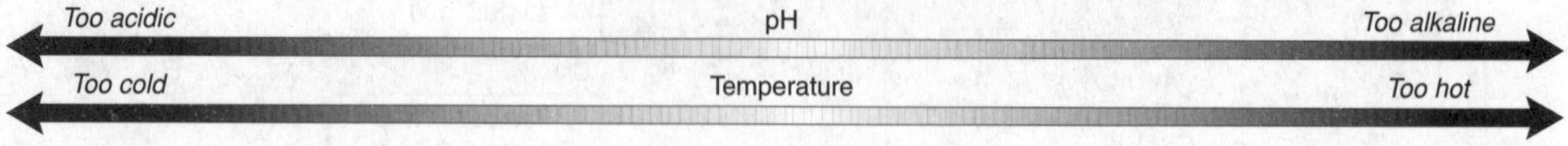

The law of tolerances states that "*for each abiotic factor, a species population (or organism) has toleance range within which it can survive. Toward the extremes of this range, that abiotic factor tends to limit the organism's ability to survive*".

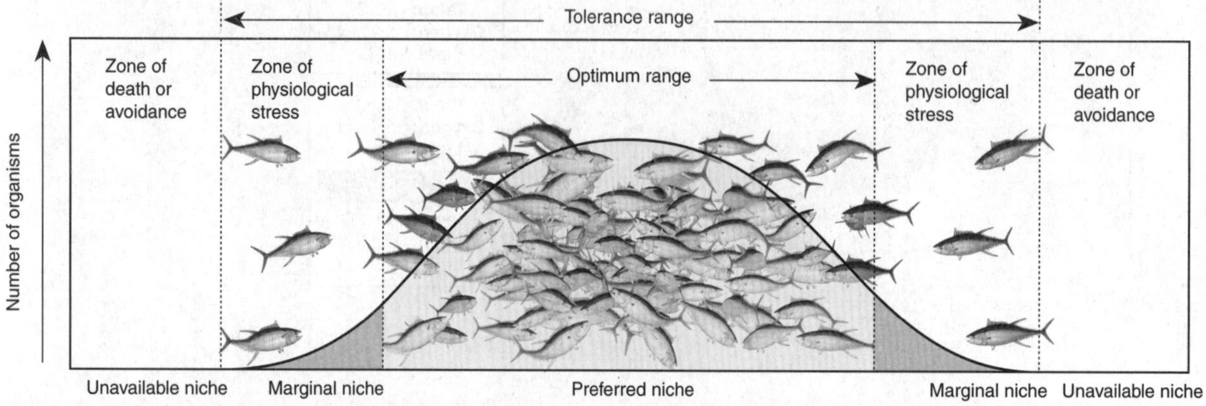

The Scale of Available Habitats

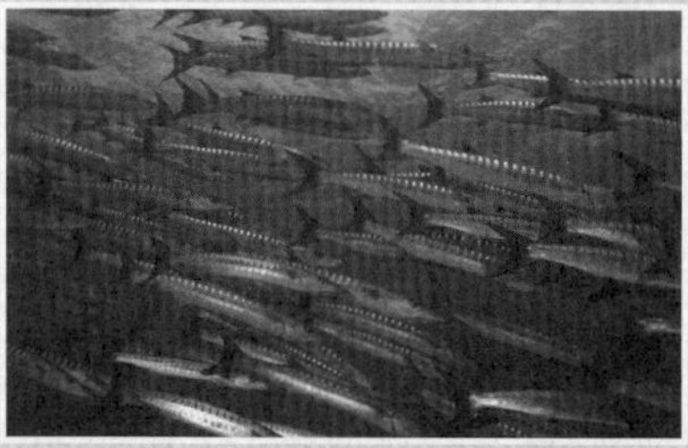

A habitat may be vast and relatively homogeneous, as is the open ocean. Barracuda (above) occur around reefs and in the open ocean where they are aggressive predators.

For non-mobile organisms, such as the fungus above, a suitable habitat may be defined by the particular environment in a relatively tiny area, such as on this decaying log.

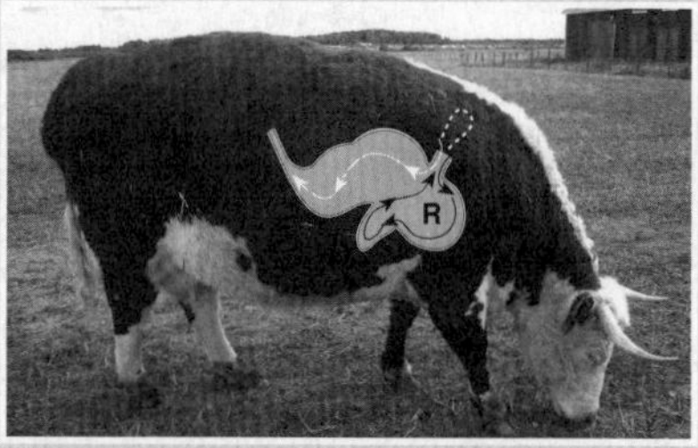

For microbial organisms, such the bacteria and protozoans of the ruminant gut, the habitat is defined by the chemical environment within the rumen (R) of the host animal, in this case, a cow.

1. Explain how an organism's habitat occupation relates to its tolerance range: ______

2. (a) Identify the range in the diagram above in which most of the species population is found. Explain why this is the case: ______

(b) Describe the greatest constraints on an organism's growth and reproduction within this range: ______

3. Describe some probable stresses on an organism forced into a marginal niche: ______

Related activities: Ecological Niche

Ecological Niche

The **ecological niche** describes the functional position of a species in its ecosystem; how it responds to the distribution of resources and how it, in turn, alters those resources for other species. The full range of environmental conditions (biological and physical) under which an organism can exist describes its **fundamental niche.** As a result of direct and indirect interactions with other organisms, species are usually forced to occupy a niche that is narrower than this and to which they are best adapted. This is termed the **realised niche**. From the concept of the niche arose the idea that two species with the same niche requirements could not coexist, because they would compete for the same resources, and one would exclude the other. This is known as **Gause's competitive exclusion principle.** If two species compete for some of the same resources (e.g. food items of a particular size), their resource use curves will overlap (below, right). Within the zone of overlap, competition will be intense.

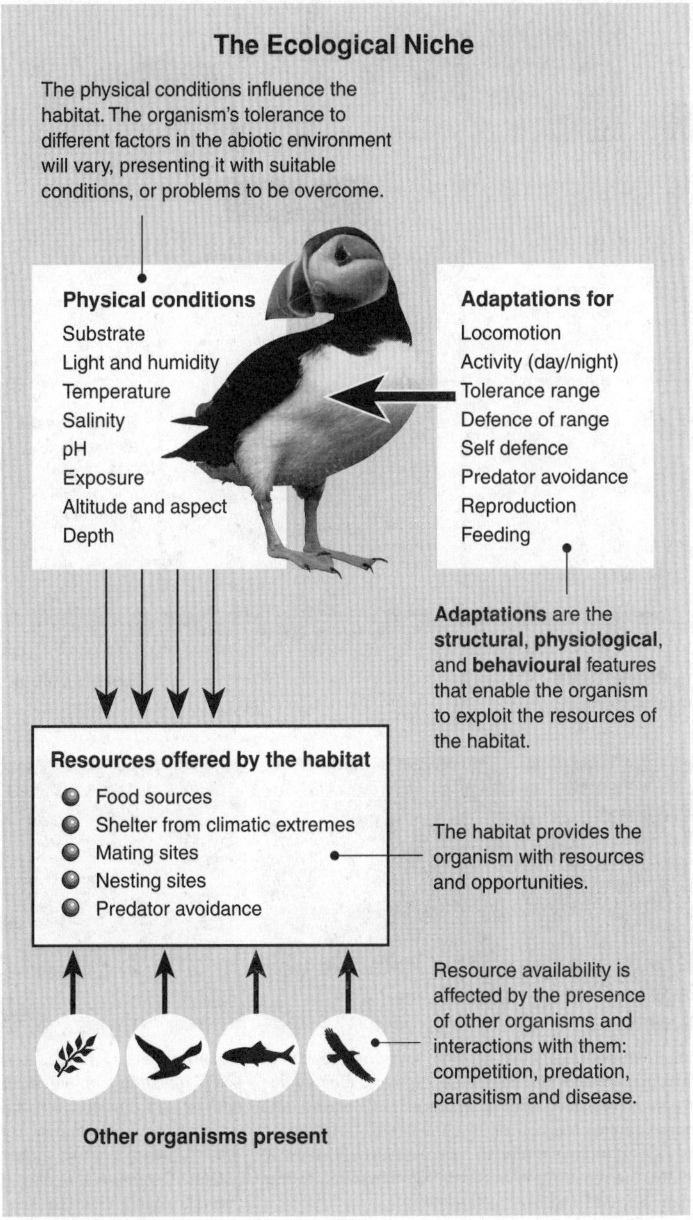

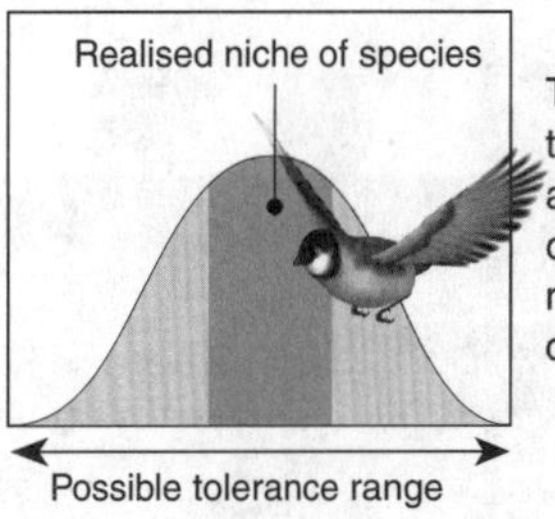

The realised niche

The tolerance range represents the potential (**fundamental**) niche a species could exploit. The actual or **realised** niche of a species is narrower than this because of competition with other species.

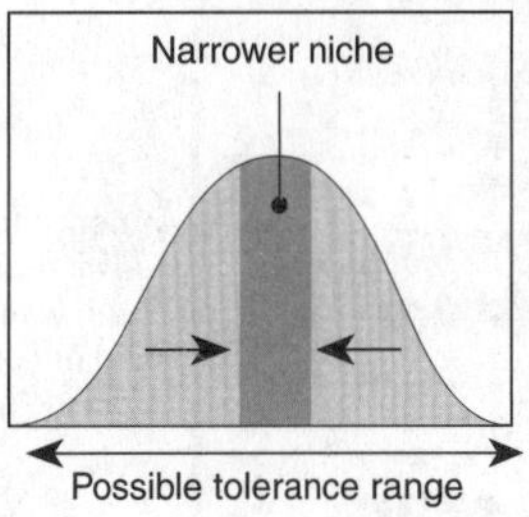

Interspecific competition

If two (or more) species compete for some of the same resources, their resource use curves will overlap (below). Within the zone of overlap, resource competition will be intense and selection will favour specialisation to occupy a narrower niche (left).

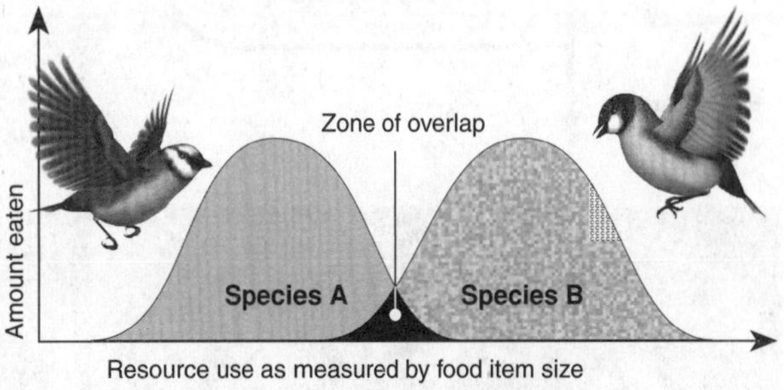

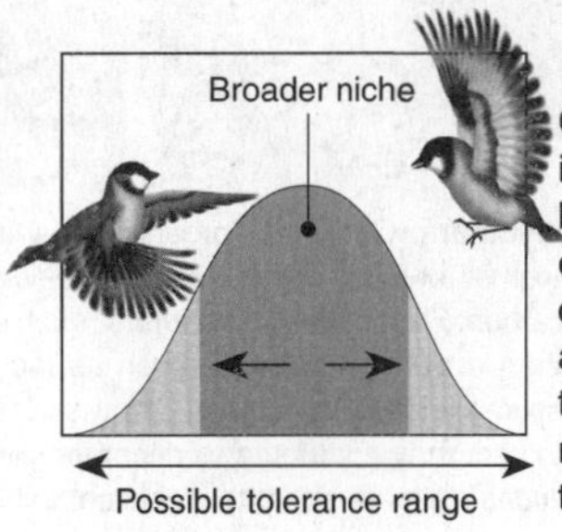

Intraspecific competition

Competition is strongest between individuals of the same species, because their resource needs exactly overlap. When intraspecific competition is intense, individuals are forced to exploit resources in the extremes of their tolerance range. This leads to expansion of the realised niche.

1. (a) Explain in what way the realised niche could be regarded as flexible: ______

(b) Describe factors that might constrain the extent of the realised niche: ______

2. Explain the contrasting effects of interspecific competition and intraspecific competition on niche breadth:

______

Energy Inputs and Outputs

Within ecosystems, organisms are assigned to **trophic** levels based on the way in which they obtain their energy. **Producers** or **autotrophs** manufacture their own food from simple inorganic substances. Most producers utilise sunlight as their energy source for this, but some use simple chemicals. The **consumers** or **heterotrophs** (herbivores, carnivores, omnivores, decomposers, and detritivores), obtain their energy from other organisms. Energy flows through trophic levels rather inefficiently, with only 5-20% of usable energy being transferred to the subsequent level. Energy not used for metabolic processes is lost as heat.

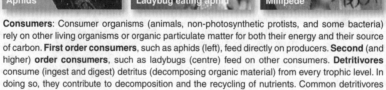

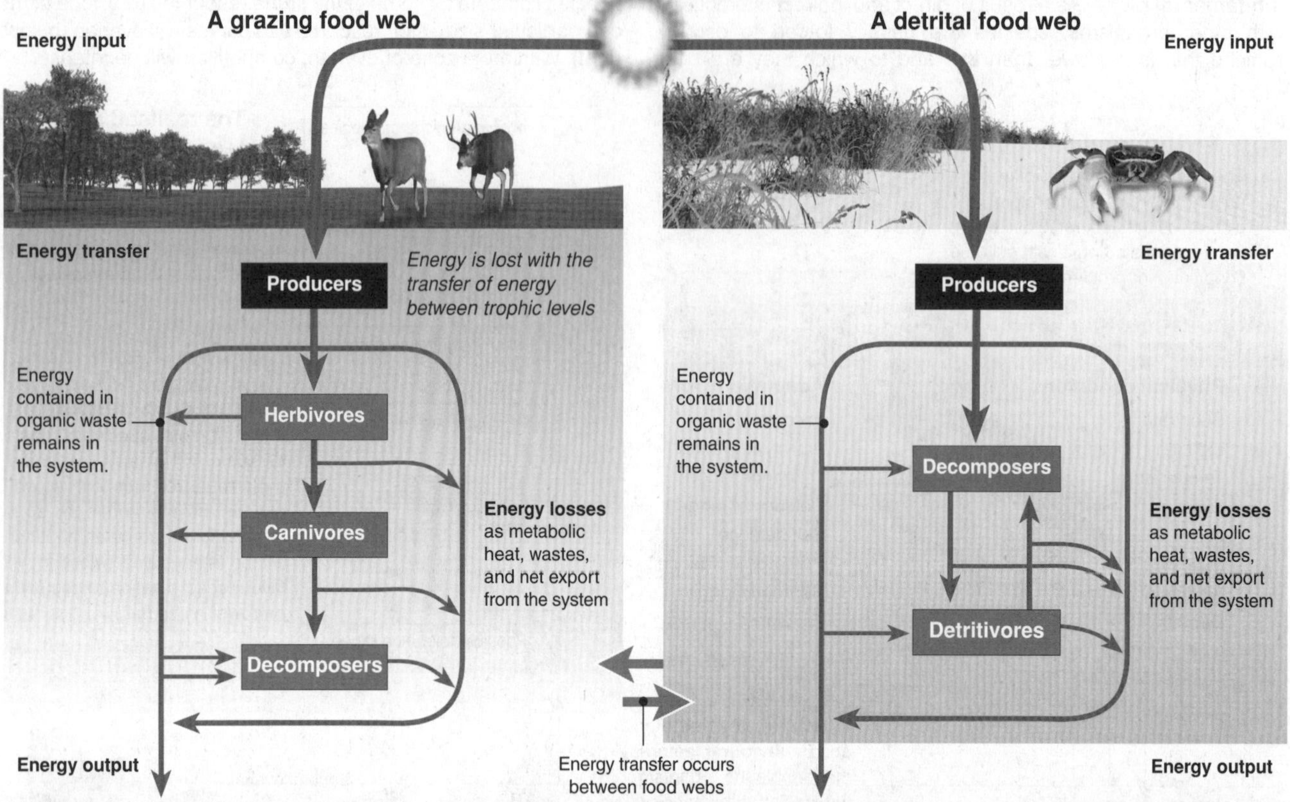

Producers (green plants, algae, and some bacteria) make their own food from simple inorganic carbon sources (e.g. CO_2). Sunlight is the most common energy source for this process.

Consumers: Consumer organisms (animals, non-photosynthetic protists, and some bacteria) rely on other living organisms or organic particulate matter for both their energy and their source of carbon. **First order consumers**, such as aphids (left), feed directly on producers. **Second** (and higher) **order consumers**, such as ladybugs (centre) feed on other consumers. **Detritivores** consume (ingest and digest) detritus (decomposing organic material) from every trophic level. In doing so, they contribute to decomposition and the recycling of nutrients. Common detritivores includes millipedes (right), woodlice, and many terrestrial worms.

Decomposers (fungi and some bacteria) obtain their energy and carbon from the extracellular breakdown of (usually dead) organic matter (DOM). Decomposers play a central role in nutrient cycling.

1. Describe the differences between **producers** and **consumers** with respect to their role in energy transfers:

____

____

2. With respect to energy flow, describe a major difference between a detrital and a grazing food web: ________

____

____

3. Distinguish between detritivores and decomposers with respect to how their contributions to nutrient cycling:

____

____

Food Chains and Webs

Every ecosystem has a **trophic structure**: a hierarchy of feeding relationships which determines the pathways for energy flow and nutrient cycling. Species are assigned to trophic levels on the basis of their sources of nutrition, with the first trophic level (the **producers**), ultimately supporting all other (consumer) levels. Consumers are ranked according to the trophic level they occupy, although some consumers may feed at several different trophic levels. The sequence of organisms, each of which is a source of food for the next, is called a **food chain**. The different food chains in an ecosystem are interconnected to form a complex web of feeding interactions called a **food web**. In the example of a lake ecosystem below, your task is assemble the organisms into a food web in a way that illustrates their trophic status and their relative trophic position(s).

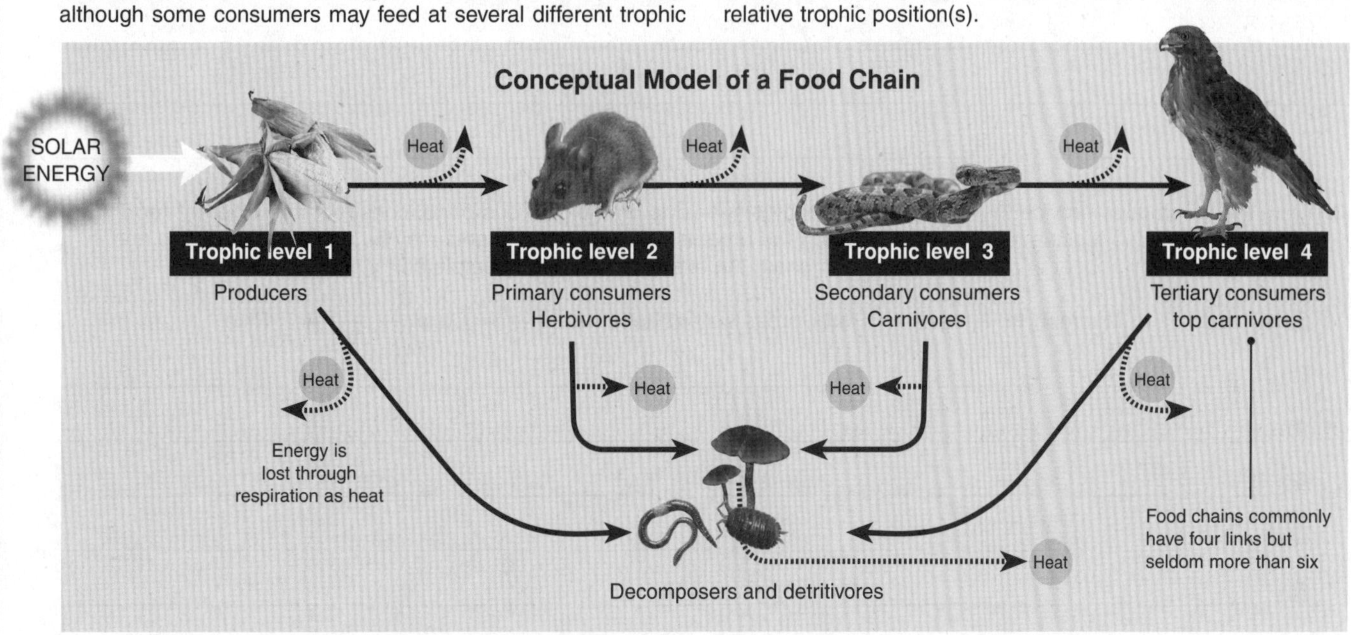

Components of a Simple Lake Ecosystem

Autotrophic protists
e.g. Chlamydomonas One of the genera that form the phytoplankton (or algae).

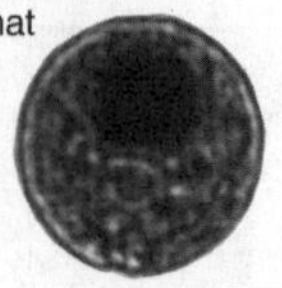

Macrophytes
A variety of species of macroscopic water plants adapted for being submerged, free-floating, or growing at the lake margin.

Protozan (*e.g. Paramecium*)
Ciliated protozoa such as *Paramecium* feed primarily on bacteria and microscopic algae such as *Chlamydomonas.*

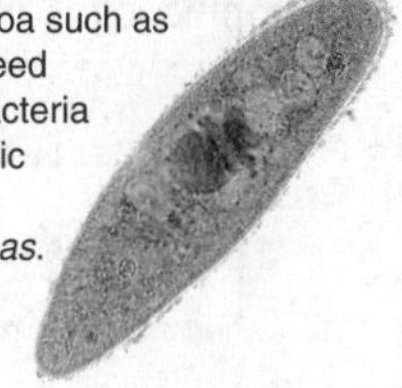

Daphnia
Small freshwater crustacean. It feeds on planktonic algae by filtering them from the water with its limbs.

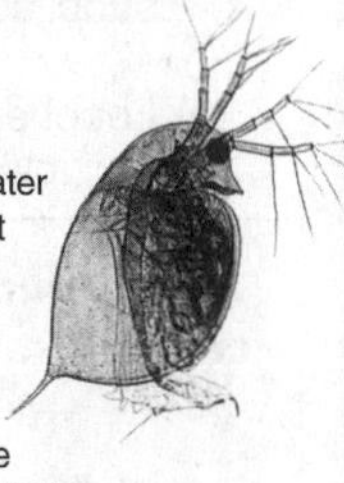

Great pond snail (*Limnaea*)
Omnivorous pond snail, eating both plant and animal material, living or dead, although the main diet is aquatic macrophytes.

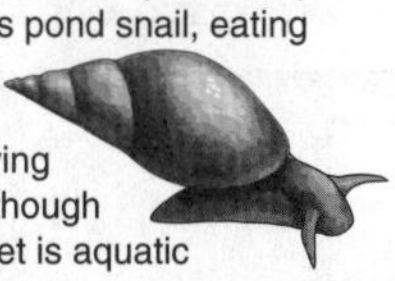

Diving beetle (*Dytiscus*)
Predators of aquatic insect larvae and adult insects blown into the lake. The will also eat organic detritus collected from the bottom mud.

Asplanchna
A large, carnivorous **rotifer** that feeds on protozoa and young zooplankton (e.g. *Daphnia*). Note that most rotifers are small herbivores.

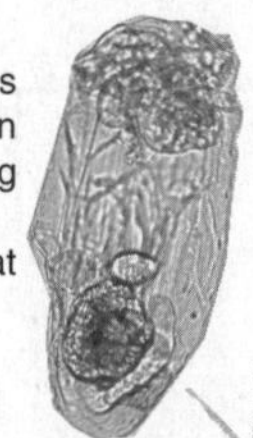

Herbivorous water beetles
(*e.g. Hydrophilus*)
Feed on water plants, although the young beetle larvae are carnivorous, feeding primarily on small pond snails.

Leech (*Glossiphonia*)
Fluid feeding predator of smaller invertebrates, including rotifers, small pond snails, and worms.

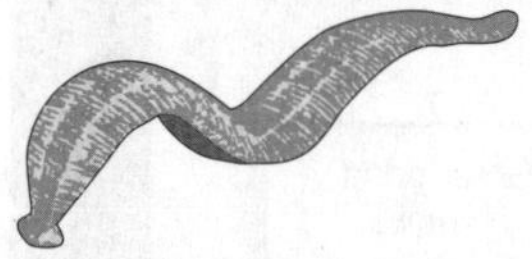

Mosquito larva
The larvae of most mosquito species, e.g. *Culex*, feed on planktonic algae before passing through a pupal stage and undergoing metamorphosis into adult mosquitoes.

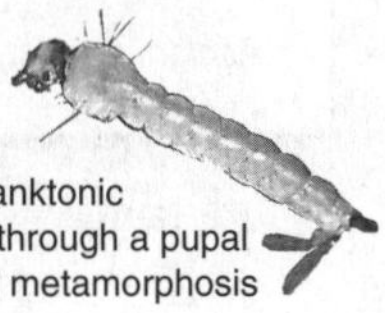

Hydra
A small carnivorous cnidarian that captures small prey items such as small *Daphnia* and insect arvae using its stinging cells on the tentacles.

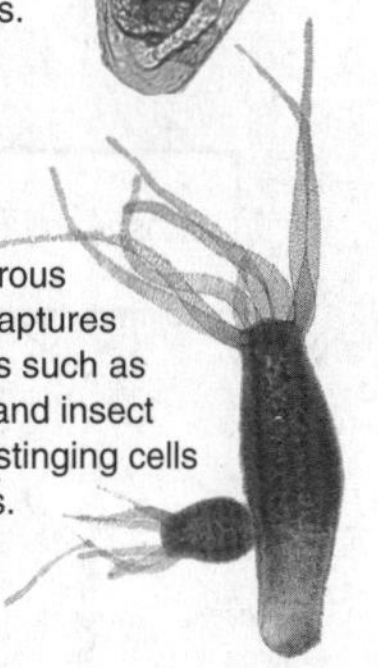

Dragonfly larva
Large aquatic insect larvae that are feed on small invertebrates including *Hydra, Daphnia*, other insect larvae, and leeches.

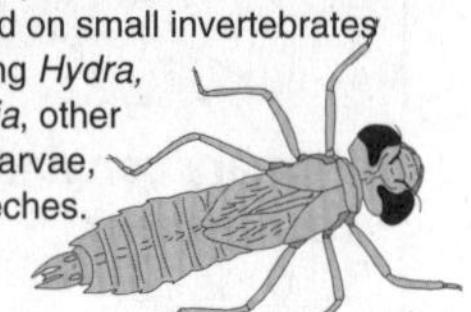

Carp (*Cyprinus*)
A heavy bodied freshwater fish that feeds mainly on bottom living insect larvae and snails, but will also take some plant material (not algae).

Three-spined stickleback (*Gasterosteus*)
A common fish of freshwater ponds and lakes. It feeds mainly on small invertebrates such as *Daphnia* and insect larvae.

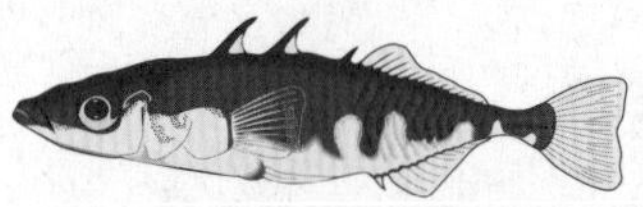

Pike (*Esox lucius*)
A top ambush predator of all smaller fish and amphibians, although they are also opportunistic predators of rodents and small birds.

Detritus
Decaying organic matter from within the lake itself or it may be washed in from the ake margins.

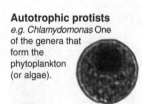

Related activities: Energy Inputs and Outputs, Energy Flow in an Ecosystem

1. (a) Describe what happens to the **amount** of energy available to each successive trophic level in a food chain:

   ____

   (b) Explain why this is the case: ____

   ____

2. Describe the trophic structure of ecosystems, including reference to **food chains** and **trophic** levels:

   ____

   ____

   ____

3. From the information provided for the lake food web components on the previous page, construct **five** different **food chains** to show the feeding relationships between the organisms. Some food chains may be shorter than others and some species will appear in more than one food chain. An example has been completed for you.

   Example 1: Macrophyte ⟶ Herbivorous water beetle ⟶ Carp ⟶ Pike

   (a) ____

   (b) ____

   (c) ____

   (d) ____

   (e) ____

4. (a) Use the food chains created above to help you to draw up a **food web** for this community. Use the information supplied to draw arrows showing the flow of **energy** between species (only energy **from** the detritus is required).

   (b) Label each species to indicate its position in the food web, i.e. its trophic level (**T1, T2, T3, T4, T5**). Where a species occupies more than one trophic level, indicate this, e.g. **T2/3**:

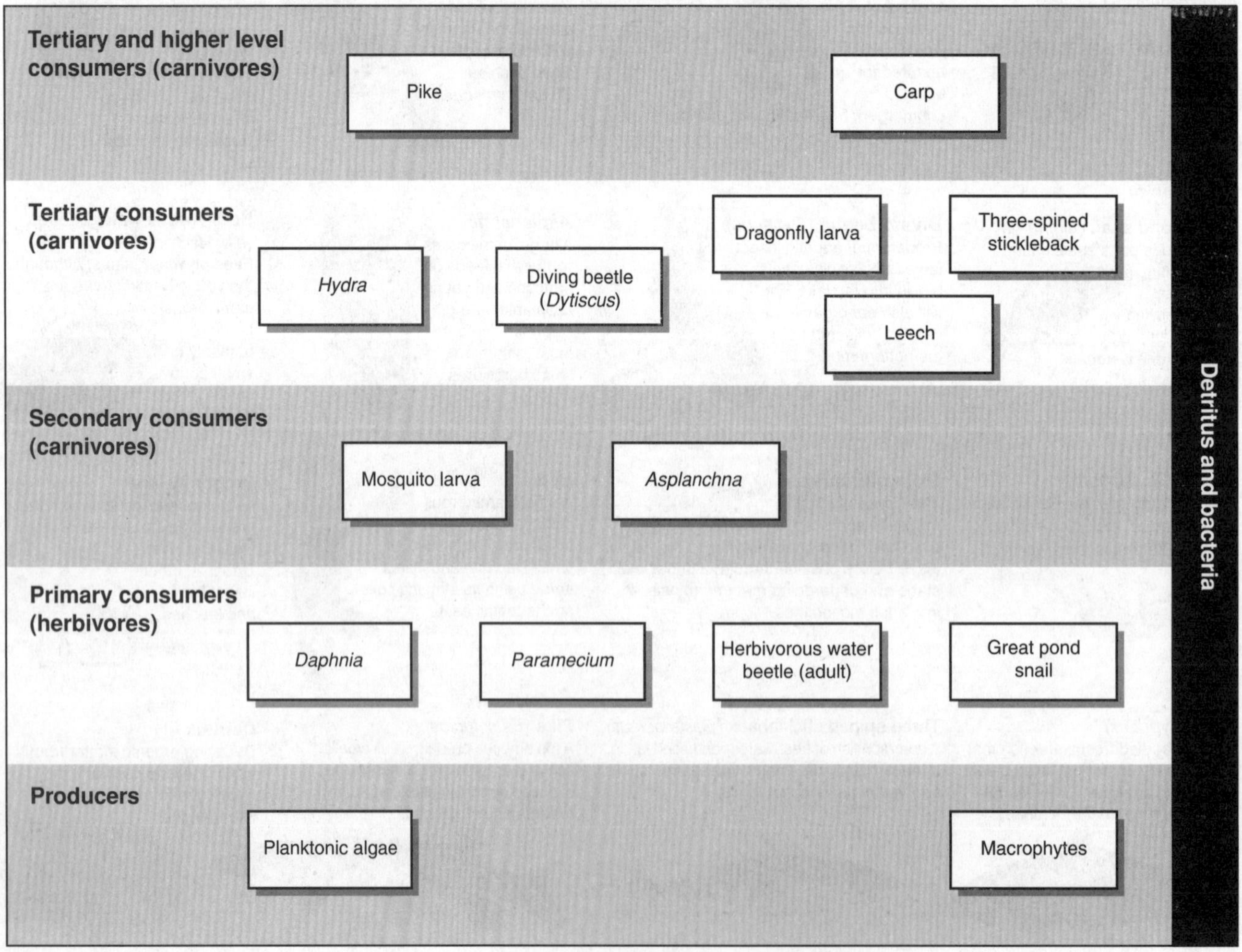

Energy Flow in an Ecosystem

The flow of energy through an ecosystem can be measured and analysed. It provides some idea as to the energy trapped and passed on at each trophic level. Each trophic level in a food chain or web contains a certain amount of biomass: the dry weight of all organic matter contained in its organisms. Energy stored in biomass is transferred from one trophic level to another (by eating, defaecation etc.), with some being lost as low-grade heat energy to the environment in each transfer. Three definitions are useful:

- **Gross primary production**: The total of organic material produced by plants (including that lost to respiration).
- **Net primary production**: The amount of biomass that is available to consumers at subsequent trophic levels.
- **Secondary production**: The amount of biomass at higher trophic levels (consumer production). Production figures are sometimes expressed as rates (productivity).

The percentage of energy transferred from one trophic level to the next varies between 5% and 20% and is called the **ecological efficiency** (efficiency of energy transfer). An average figure of 10% is often used. The path of energy flow in an ecosystem depends on its characteristics. In a tropical forest ecosystem, most of the primary production enters the detrital and decomposer food chains. However, in an ocean ecosystem or an intensively grazed pasture more than half the primary production may enter the grazing food chain.

Energy Flow Through an Ecosystem

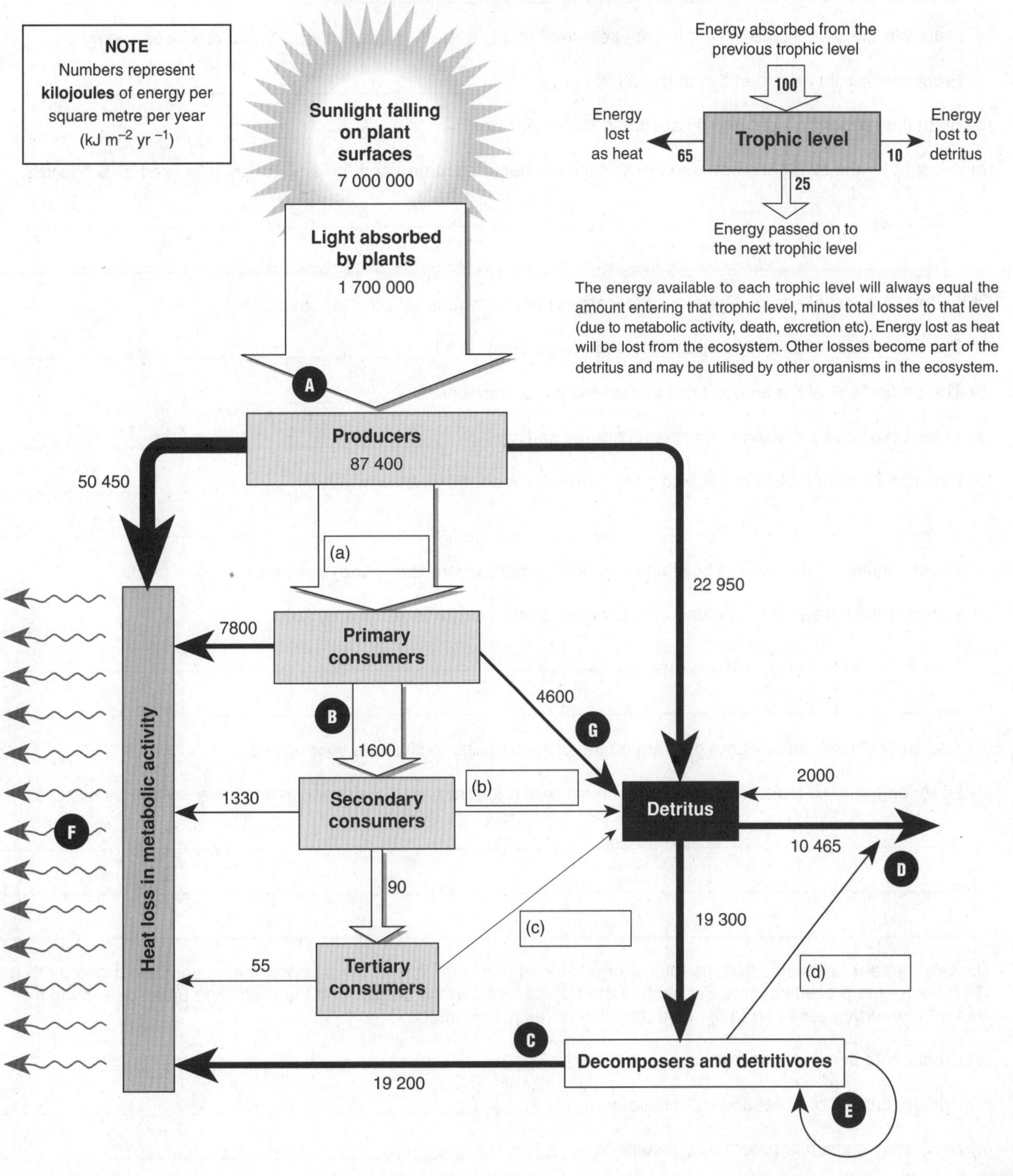

The energy available to each trophic level will always equal the amount entering that trophic level, minus total losses to that level (due to metabolic activity, death, excretion etc). Energy lost as heat will be lost from the ecosystem. Other losses become part of the detritus and may be utilised by other organisms in the ecosystem.

Ecosystems

1. Study the diagram on the previous page illustrating energy flow through a hypothetical ecosystem. Use the example at the top of the page as a guide to calculate the missing values (a)–(d) in the diagram. Note that the sum of the energy inputs always equals the sum of the energy outputs. Place your answers in the spaces provided on the diagram.

2. Describe the original source of energy that powers this ecosystem: ______________________

3. Identify the processes that are occurring at the points labelled **A – G** on the diagram:

A. ______________________ E. ______________________

B. ______________________ F. ______________________

C. ______________________ G. ______________________

D. ______________________

4. (a) Calculate the percentage of light energy falling on the plants that is absorbed at point **A**:

Light absorbed by plants ÷ sunlight falling on plant surfaces x 100 = ______________________

(b) Describe what happens to the light energy that is not absorbed: ______________________

____

5. (a) Calculate the percentage of light energy absorbed that is actually converted (fixed) into producer energy:

Producers ÷ light absorbed by plants x 100 = ______________________

(b) State the **amount** of light energy absorbed that is **not** fixed: ______________________

(c) Account for the difference between the amount of energy absorbed and the amount actually fixed by producers:

____

____

6. Of the total amount of energy **fixed** by producers in this ecosystem (at point **A**) calculate:

(a) The total amount that ended up as metabolic waste heat (in kJ): ______________________

(b) The percentage of the energy fixed that ended up as waste heat: ______________________

7. (a) State the groups for which detritus is an energy source: ______________________

(b) Describe by what means detritus could be removed or added to an ecosystem: ______________________

____

8. In certain conditions, detritus will build up in an environment where few (or no) decomposers can exist.

(a) Describe the consequences of this lack of decomposer activity to the energy flow:

____

____

(b) Add an additional arrow to the diagram on the previous page to illustrate your answer.

(c) Describe three examples of materials that have resulted from a lack of decomposer activity on detrital material:

____

____

____

9. The **ten percent law** states that the total energy content of a trophic level in an ecosystem is only about one-tenth (or 10%) that of the preceding level. For each of the trophic levels in the diagram on the preceding page, determine the amount of energy passed on to the next trophic level as a percentage:

(a) Producer to primary consumer: ______________________

(b) Primary consumer to secondary consumer: ______________________

(c) Secondary consumer to tertiary consumer: ______________________

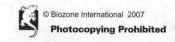

Ecological Pyramids

The trophic levels of any ecosystem can be arranged in pyramid of increasing trophic level. The first trophic level is placed at the bottom and subsequent trophic levels are stacked on top in their 'feeding sequence'. Ecological pyramids can illustrate changes in the numbers, biomass (weight), or energy content of organisms at each level. Each of these three kinds of pyramids tells us something different about the flow of energy and movement of materials between one trophic level and the next. The type of pyramid you choose in order to express information about an ecosystem will depend on what particular features of the ecosystem you are interested in and, of course, the type of data you have collected.

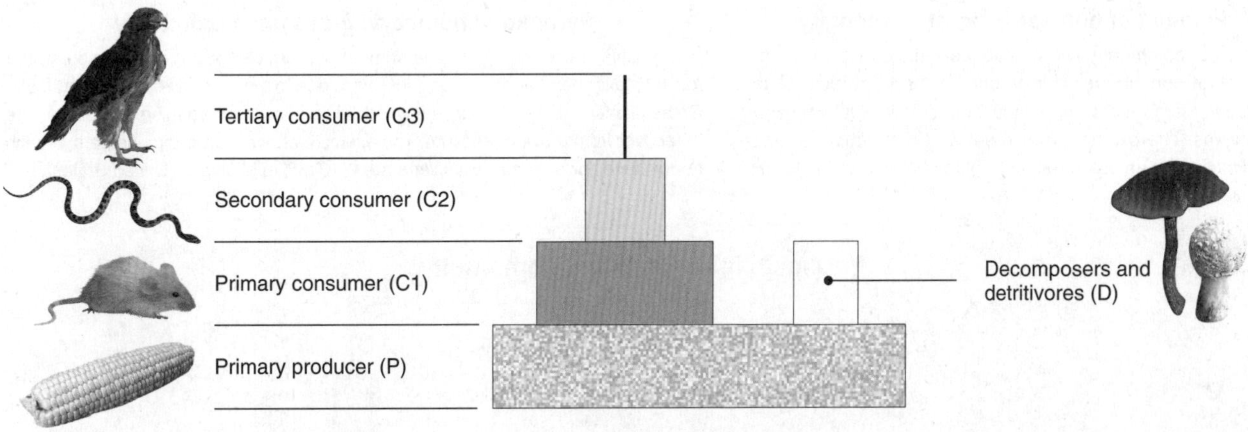

The generalised ecological pyramid pictured above shows a conventional pyramid shape, with a large number (or biomass) of producers forming the base for an increasingly small number (or biomass) of consumers. Decomposers are placed at the level of the primary consumers and off to the side. They may obtain energy from many different trophic levels and so do not fit into the conventional pyramid structure. For any particular ecosystem at any one time (e.g. the forest ecosystem below), the shape of this typical pyramid can vary greatly depending on whether the trophic relationships are expressed as numbers, biomass or energy.

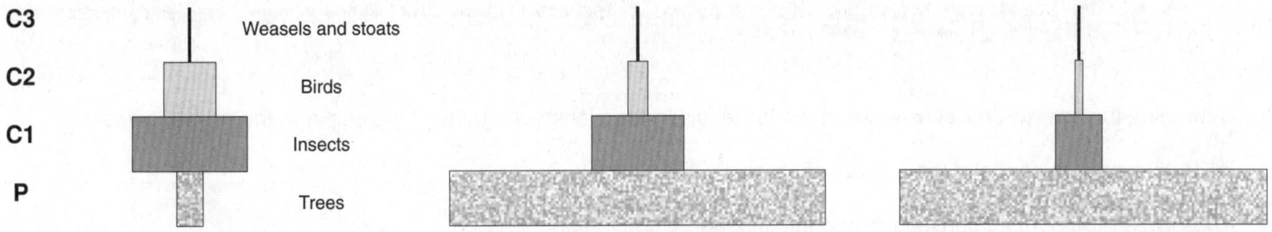

Numbers in a forest community

Pyramids of numbers display the number of individual organisms at each trophic level. The pyramid above has few producers, but they may be of a very large size (e.g. trees). This gives an 'inverted pyramid', although not all pyramids of numbers are like this.

Biomass in a forest community

Biomass pyramids measure the 'weight' of biological material at each trophic level. Water content of organisms varies, so 'dry weight' is often used. Organism size is taken into account, so meaningful comparisons of different trophic levels are possible.

Energy in a forest community

Pyramids of energy are often very similar to biomass pyramids. The energy content at each trophic level is generally comparable to the biomass (i.e. similar amounts of dry biomass tend to have about the same energy content).

1. Describe what the three types of ecological pyramids measure:

   (a) Number pyramid: ______________________

   (b) Biomass pyramid: ______________________

   (c) Energy pyramid: ______________________

2. Explain the advantage of using a biomass or energy pyramid rather than a pyramid of numbers to express the relationship between different trophic levels:

   ____

3. Explain why it is possible for the forest community (on the next page) to have very few producers supporting a large number of consumers:

   ____

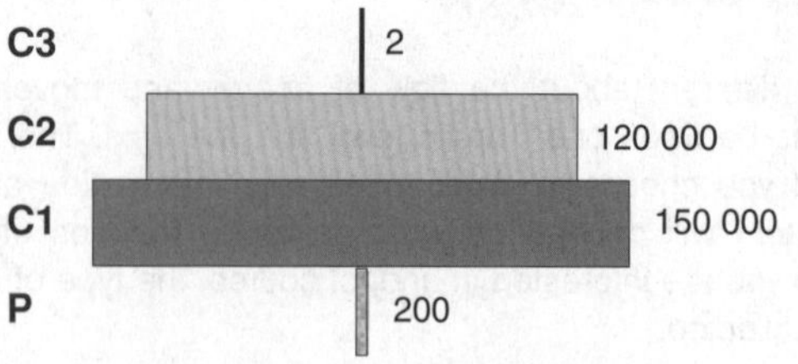

Pyramid of numbers: forest community

In a forest community a few producers may support a large number of consumers. This is due to the large size of the producers; large trees can support many individual consumer organisms. The example above shows the numbers at each trophic level for an oak forest in England, in an area of 10 m^2.

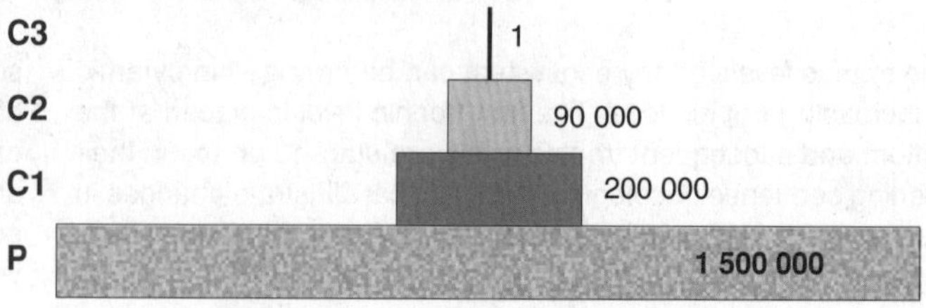

Pyramid of numbers: grassland community

In a grassland community a large number of producers are required to support a much smaller number of consumers. This is due to the small size of the producers. Grass plants can support only a few individual consumer organisms and take time to recover from grazing pressure. The example above shows the numbers at each trophic level for a derelict grassland area (10 m^2) in Michigan, United States.

Pyramids for a Plankton Community

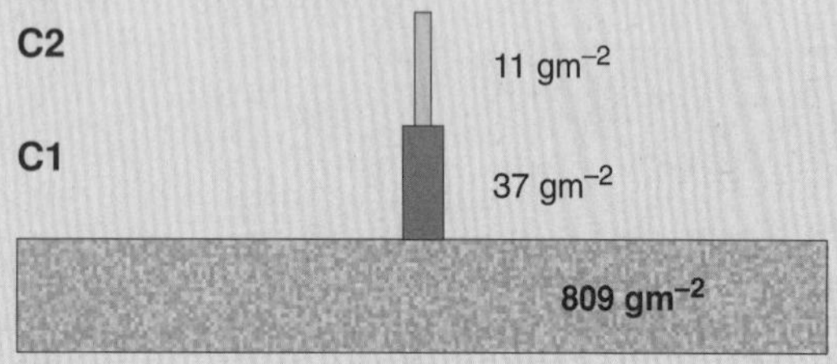

Biomass

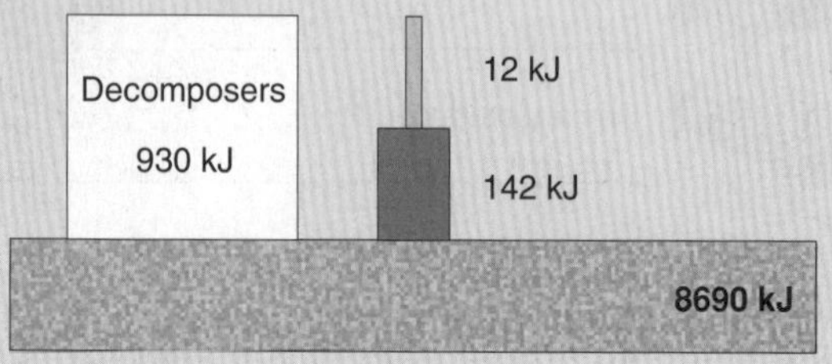

Energy

The pyramids of biomass and energy are virtually identical. The two pyramids illustrated here relate to the same hypothetical plankton community. A large biomass of producers supports a smaller biomass of consumers. The energy at each trophic level is reduced with each progressive stage in the food chain. As a general rule, a maximum of 10% of the energy is passed on to the next level in the food chain. The remaining energy is lost due to respiration, waste, and heat.

4. Determine the **energy transfer** between trophic levels in the plankton community example in the above diagram:

   (a) Between producers and the primary consumers: ____________________

   (b) Between the primary consumers and the secondary consumers: ____________________

   (c) Explain why the energy passed on from the producer to primary consumers is considerably less than the normally expected 10% occurring in most other communities (describe where the rest of the energy was lost to):

   ____

   ____

   (d) After the producers, which trophic group has the greatest energy content: ____________________

   (e) Give a likely explanation why this is the case: ____________________

   ____

   ____

An unusual biomass pyramid

The biomass pyramids of some ecosystems appear rather unusual with an inverted shape. The first trophic level has a lower biomass than the second level. What this pyramid does not show is the rate at which the producers (algae) are reproducing in order to support the larger biomass of consumers.

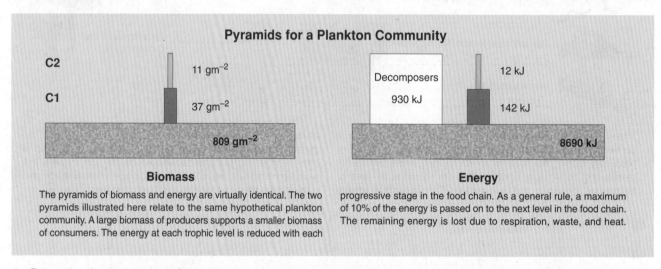

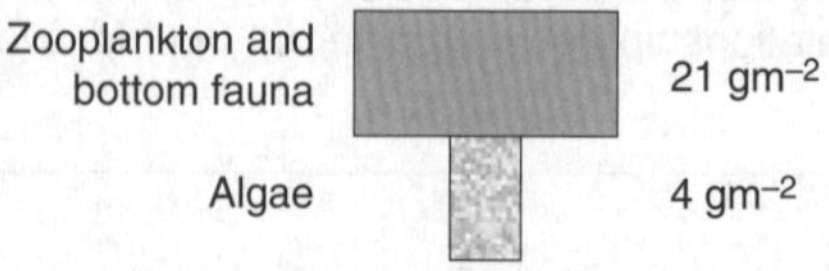

Biomass

5. Give a possible explanation of how a small biomass of producers(algae) can support a larger biomass of consumers (zooplankton):

____

____

____

Plant Productivity

The energy entering ecosystems is fixed by producers in photosynthesis. The rate of photosynthesis is dependent on factors such as temperature and the amount of light, water, and nutrients. The total energy fixed by a plant through photosynthesis is referred to as the **gross primary production** (**GPP**) and is usually expressed as Jm^{-2} (or kJm^{-2}), or as gm^{-2}. However, a portion of this energy is required by the plant for respiration. Subtracting respiration from GPP gives the **net primary production** (**NPP**). The **rate** of biomass production, or **net primary productivity**, is the biomass produced per area per unit time.

Measuring Productivity

Primary productivity of an ecosystem depends on a number of interrelated factors (light intensity, nutrients, temperature, water, and mineral supplies), making its calculation extremely difficult. Globally, the least productive ecosystems are those that are limited by heat energy and water. The most productive ecosystems are systems with high temperatures, plenty of water, and non-limiting supplies of soil nitrogen. The primary productivity of oceans is lower than that of terrestrial ecosystems because the water reflects (or absorbs) much of the light energy before it reaches and is utilised by producers. The table below compares the difference in the net primary productivity of various ecosystems.

Ecosystem Type	Net Primary Productivity kcal m^{-2} y^{-1}	kJ m^{-2} y^{-1}
Tropical rainforest	15 000	63 000
Swamps and marshes	12 000	50 400
Estuaries	9000	37 800
Savanna	3000	12 600
Temperate forest	6000	25 200
Boreal forest	3500	14 700
Temperate grassland	2000	8400
Tundra/cold desert	500	2100
Coastal marine	2500	10 500
Open ocean	800	3360
Desert	< 200	< 840

* Data compiled from a variety of sources.

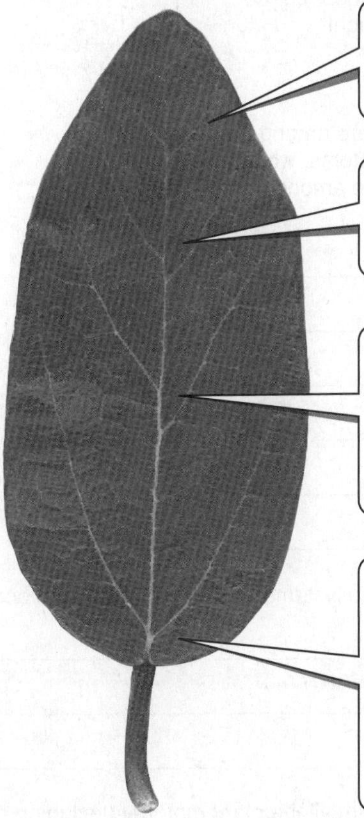

Leaf Area Index (LAI)
Leaf area index is a measure of the total leaf area of a given plant.

Harvestable Dry Biomass
Used for commercial purposes, it is the dry mass of crop available for sale or use.

Relative Growth Rate (R)
Relative growth rate is the gain in mass of plant tissue per unit time.

$$R = \frac{\text{Increase in dry mass in unit time}}{\text{Original dry mass of the plant}}$$

Net Assimilation Rate (NAR)
NAR is the increase in plant weight per unit of leaf area per unit time. Essentially it is the balance between carbon gain from photosynthesis and carbon loss from respiration.

$$NAR = \frac{\text{Increase in dry mass in unit time}}{\text{Leaf area}}$$

Net Primary Productivity of Selected Ecosystems (figures are in kJ m^{-2} y^{-1})

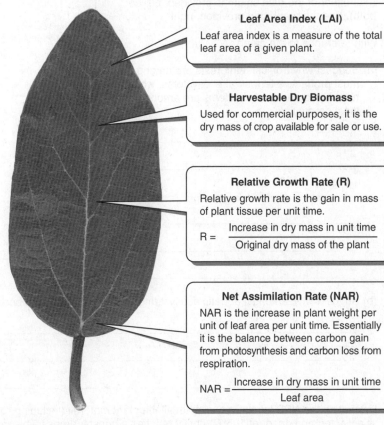

1. Briefly describe three factors that may affect the primary productivity of an ecosystem:

(a) ______________________________

(b) ______________________________

(c) ______________________________

2. Explain the difference between **productivity** and **production** in relation to plants: ______________________________

____

____

Related activities: The Importance of Plants

3. Suggest how the LAI might influence the rate of primary production:

4. Using the data table on the previous page, choose a suitable graph format and plot the differences in the net primary productivity of various ecosystems (use either of the data columns provided, but not both). Use the graph grid provided, right.

5. With reference to the graph:

(a) Suggest why tropical rainforests are among the most productive terrestrial ecosystems, while tundra and desert ecosystems are among the least productive:

(b) Suggest why, amongst aquatic ecosystems, the NPP of the open ocean is low relative to that of coastal systems:

6. Estimating the NPP is relatively simple: all the plant material (including root material) from a measured area (e.g. 1 m^2) is collected and dried (at 105°C) until it reaches a constant mass. This mass, called the **standing crop**, is recorded (in kg m^{-2}). The procedure is repeated after some set time period (e.g. 1 month). The difference between the two calculated masses represents the *estimated* NPP:

(a) Explain why the plant material was dried before weighing:

(b) Define the term **standing crop**:

(c) Suggest why this procedure only provides an estimate of NPP:

(d) State what extra information would be required in order to express the standing crop value in kJ m^{-2}:

(e) Suggest what information would be required in order to calculate the GPP:

7. Intensive horticultural systems achieve very high rates of production (about 10X those of subsistence systems).

(a) Outline the means by which these high rates are achieved:

(b) Comment on the sustainability of these high rates (summary of a group discussion if you wish):

The Carbon Cycle

Carbon is an essential element in living systems, providing the chemical framework to form the molecules that make up living organisms (e.g. proteins, carbohydrates, fats, and nucleic acids). Carbon also makes up approximately 0.03% of the atmosphere as the gas carbon dioxide (CO_2), and it is present in the ocean as carbonate and bicarbonate, and in rocks such as limestone. Carbon cycles between the living (biotic) and non-living (abiotic) environment: it is fixed in the process of photosynthesis and returned to the atmosphere in respiration. Carbon may remain locked up in biotic or abiotic systems for long periods of time as, for example, in the wood of trees or in fossil fuels such as coal or oil. Human activity has disturbed the balance of the carbon cycle (the global carbon budget) through activities such as combustion (e.g. the burning of wood and **fossil fuels**) and deforestation.

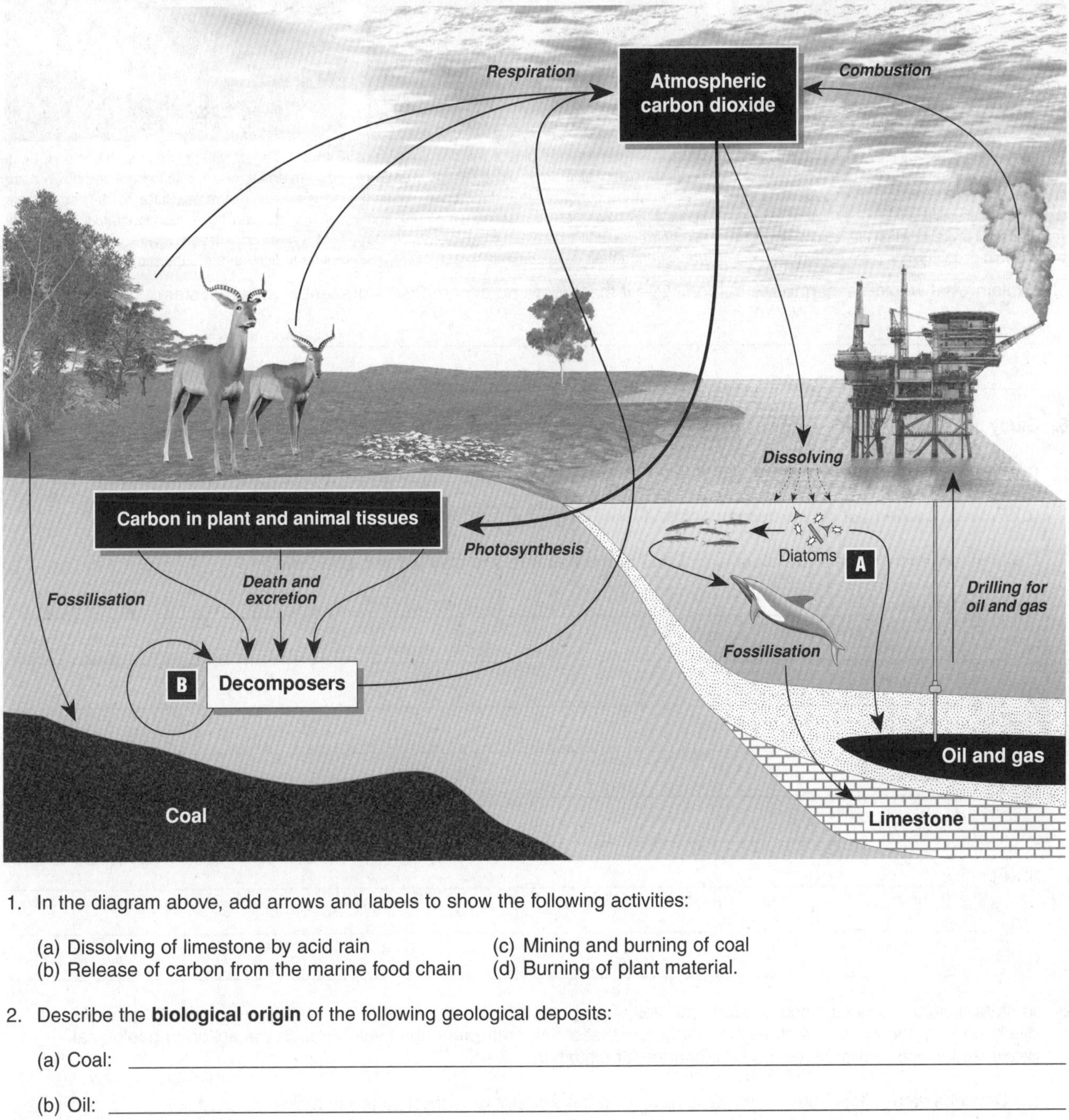

1. In the diagram above, add arrows and labels to show the following activities:

   (a) Dissolving of limestone by acid rain
   (b) Release of carbon from the marine food chain
   (c) Mining and burning of coal
   (d) Burning of plant material.

2. Describe the **biological origin** of the following geological deposits:

   (a) Coal: ______________________

   (b) Oil: ______________________

   (c) Limestone: ______________________

3. Describe the two processes that release carbon into the atmosphere: ______________________

   ____

   ____

4. Name the four geological reservoirs (sinks), in the diagram above, that can act as a source of carbon:

   (a) ______________________ (c) ______________________

   (b) ______________________ (d) ______________________

Termites: These insects play an important role in nutrient recycling. With the aid of symbiotic protozoans and bacteria in their guts, they can digest the tough cellulose of woody tissues in trees. Termites fulfill a vital function in breaking down the endless rain of debris in tropical rainforests.

Dung beetles: Beetles play a major role in the decomposition of animal dung. Some beetles merely eat the dung, but true dung beetles, such as the scarabs and *Geotrupes*, bury the dung and lay their eggs in it to provide food for the beetle grubs during their development.

Fungi: Together with decomposing bacteria, fungi perform an important role in breaking down dead plant matter in the leaf litter of forests. Some mycorrhizal fungi have been found to link up to the root systems of trees where an exchange of nutrients occurs (a mutualistic relationship).

5. Explain what would happen to the carbon cycle if there were no decomposers present in an ecosystem:

6. Study the diagram on the previous page and identify the processes represented at the points labelled [**A**] and [**B**]:

   (a) Process carried out by the diatoms at label **A**:

   (b) Process carried out by the decomposers at label **B**:

7. Explain how each of the three organisms listed below has a role to play in the carbon cycle:

   (a) Dung beetles:

   (b) Termites:

   (c) Fungi:

8. In natural circumstances, accumulated reserves of carbon such as peat, coal and oil represent a **sink** or natural diversion from the cycle. Eventually the carbon in these sinks returns to the cycle through the action of geological processes which return deposits to the surface for oxidation.

   (a) Describe what effect human activity is having on the amount of carbon stored in sinks:

   (b) Explain two global effects arising from this activity:

   (c) Suggest what could be done to prevent or alleviate these effects:

The Nitrogen Cycle

Nitrogen is a crucial element for all living things, forming an essential part of the structure of proteins and nucleic acids. The Earth's atmosphere is about 80% nitrogen gas (N_2), but molecular nitrogen is so stable that it is only rarely available directly to organisms and is often in short supply in biological systems. Bacteria play an important role in transferring nitrogen between the biotic and abiotic environments. Some bacteria are able to fix atmospheric nitrogen, while others convert ammonia to nitrate and thus make it available for incorporation into plant and animal tissues. Nitrogen-fixing bacteria are found living freely in the soil *(Azotobacter)* and living symbiotically with some plants in root nodules *(Rhizobium)*. Lightning discharges also cause the oxidation of nitrogen gas to nitrate which ends up in the soil. Denitrifying bacteria reverse this activity and return fixed nitrogen to the atmosphere. Humans intervene in the nitrogen cycle by producing, and applying to the land, large amounts of nitrogen fertiliser. Some applied fertiliser is from organic sources (e.g. green crops and manures) but much is inorganic, produced from atmospheric nitrogen using an energy-expensive industrial process. Overuse of nitrogen fertilisers may lead to pollution of water supplies, particularly where land clearance increases the amount of leaching and runoff into ground and surface waters.

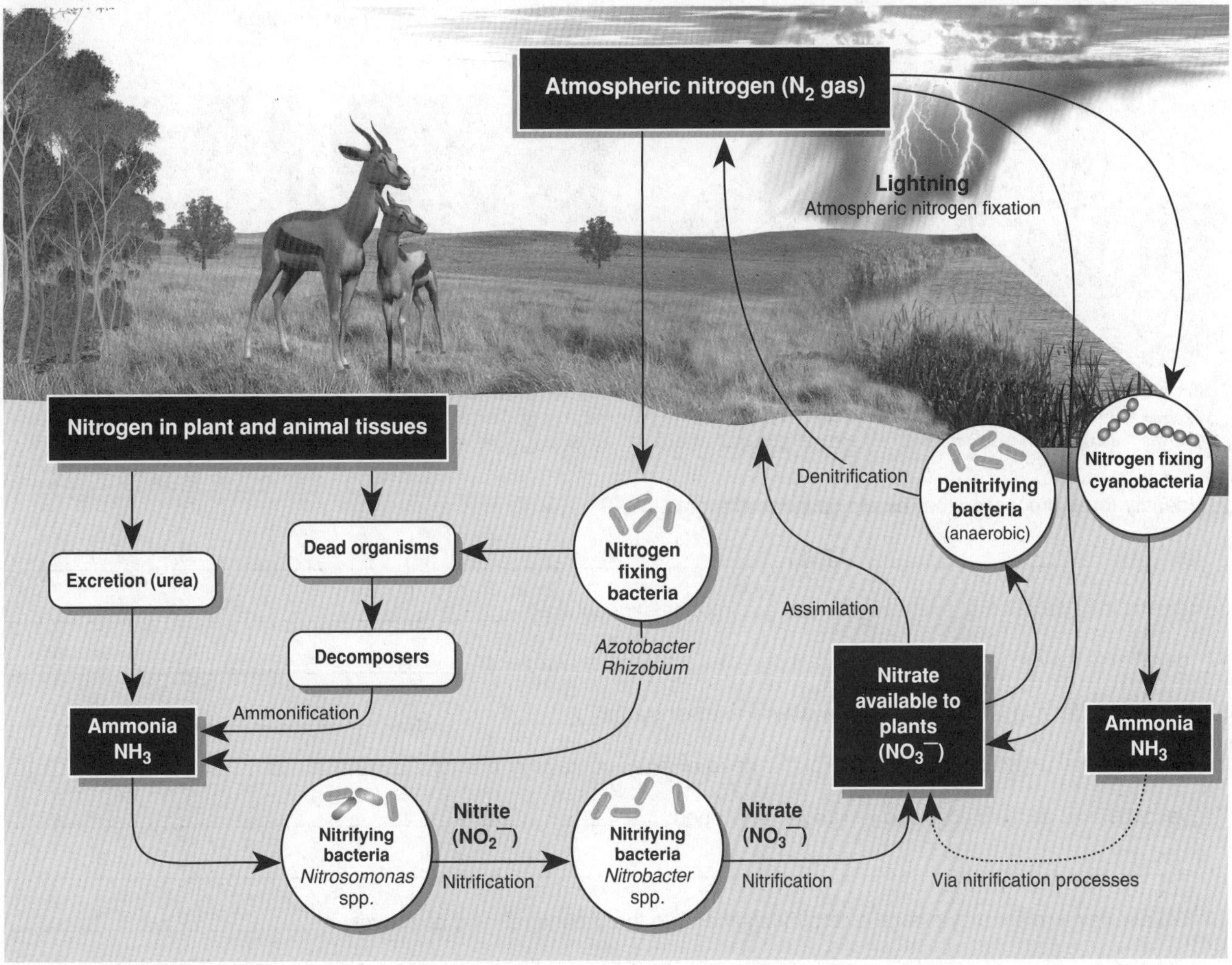

Ecosystems

1. Describe five instances in the nitrogen cycle where **bacterial** action is important. Include the name of each of the processes and the changes to the form of nitrogen involved:

(a) ______________________________

____

(b) ______________________________

____

(c) ______________________________

____

(d) ______________________________

____

(e) ______________________________

____

Human Intervention in the Nitrogen Cycle

Until about sixty years ago, microbial nitrogen fixation was the only mechanism by which nitrogen could be made available to plants. However, during WW II, Fritz Haber developed the **Haber process** whereby nitrogen and hydrogen gas are combined to form gaseous ammonia. The ammonia is converted into ammonium salts and sold as inorganic fertiliser. Its application has revolutionised agriculture by increasing crop yields.

As well as adding nitrogen fertilisers to the land, humans use anaerobic bacteria to break down livestock wastes and release NH_3 into the soil. They also intervene in the nitrogen cycle by discharging **effluent** into waterways. Nitrogen is removed from the land through burning, which releases nitrogen oxides into the atmosphere. It is also lost by mining, harvesting crops, and irrigation, which leaches nitrate ions from the soil.

Crop Rotation

Crop rotation is an agricultural practice in which different crops are cultivated in succession on the same area of land over a period of time. Its purpose is to maintain soil fertility and reduce the adverse effects of pests. Legumes, such as peas and beans, are important in the rotation as they restore nitrogen to the soil. Some crops, like potatoes, suppress weeds and improve soil structure. Other crops that may be included in a typical rotation are wheat, barley, and squash. Different crops have different soil requirements and benefits, so changing crops from year to year minimises deficiencies and allows the soil to replenish.

A typical rotation is of three to five years with plants in different rotations being chosen from different families for their specific contributions to pest management and aspects of soil quality (such as nitrogen content and structure).

Humans may intervene in the nitrogen cycle by applying manure (left), which restores soil nitrogen, and by harvesting crop biomass, which removes material that would potentially rot and replenish soil nitrogen.

Legumes, such as soy beans (above, left) are used in crop rotations to restore soil nitrogen. Alternating between fibrous-rooted and deep-rooted crops (e.g. potatoes) improves soil structure.

2. Identify three processes that **fix** atmospheric nitrogen:

   (a) __________ (b) __________ (c) __________

3. Identify the process that releases nitrogen gas into the atmosphere: __________

4. Identify the main geological reservoir that provides a source of nitrogen: __________

5. Identify the form in which nitrogen is available to most plants: __________

6. Identify a vital organic compound that plants need nitrogen containing ions for: __________

7. Describe how animals acquire the nitrogen they need: __________

8. Explain why farmers may plough a crop of legumes into the ground rather than harvest it: __________

9. Describe five ways in which humans may intervene in the nitrogen cycle and the effects of these interventions:

   (a) __________

   (b) __________

   (c) __________

   (d) __________

   (e) __________

The Water Cycle

The hydrologic cycle (water cycle), collects, purifies, and distributes the Earth's fixed supply of water. The main processes in this water recycling are described below. Besides replenishing inland water supplies, rainwater causes erosion and is a major medium for transporting dissolved nutrients within and among ecosystems. On a global scale, evaporation (conversion of water to gaseous water vapour) exceeds precipitation (rain, snow etc.) over the oceans. This results in a net movement of water vapour (carried by winds) over the land. On land, precipitation exceeds evaporation. Some of this precipitation becomes locked up in snow and ice, for varying lengths of time. Most forms surface and groundwater systems that flow back to the sea, completing the major part of the cycle. Living organisms, particularly plants, participate to varying degrees in the water cycle. Over the sea, most of the water vapour is due to evaporation alone. However on land, about 90% of the vapour results from plant transpiration. Animals (particularly humans) intervene in the cycle by utilising the resource for their own needs.

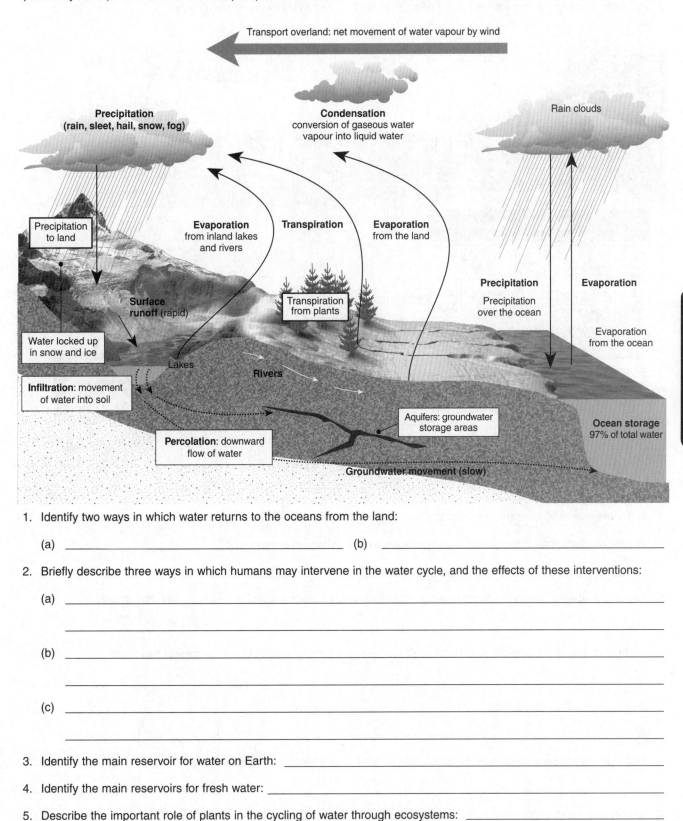

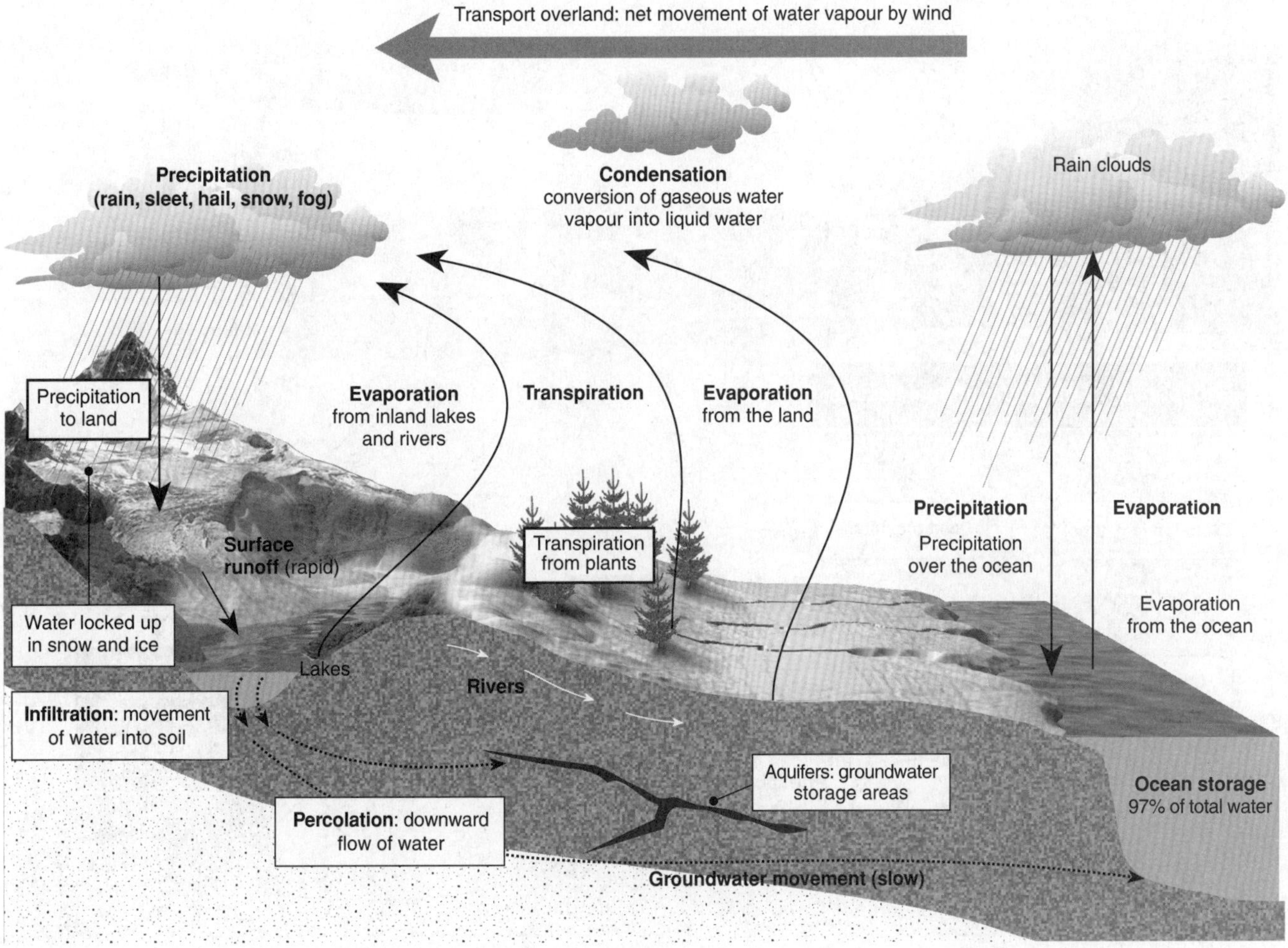

1. Identify two ways in which water returns to the oceans from the land:

   (a) ______________________ (b) ______________________

2. Briefly describe three ways in which humans may intervene in the water cycle, and the effects of these interventions:

   (a) ______________________

   (b) ______________________

   (c) ______________________

3. Identify the main reservoir for water on Earth: ______________________

4. Identify the main reservoirs for fresh water: ______________________

5. Describe the important role of plants in the cycling of water through ecosystems: ______________________

Related activities: Atmosphere and Climate, Global Water Resources, Water Use

The Phosphorus Cycle

Phosphorus is an essential component of nucleic acids and ATP. Unlike carbon, phosphorus has no atmospheric component; cycling of phosphorus is very slow and tends to be local. Small losses from terrestrial systems by leaching are generally balanced by gains from weathering. In aquatic and terrestrial ecosystems, phosphorus is cycled through food webs. Bacterial decomposition breaks down the remains of dead organisms and excreted products. Phosphatising bacteria further break down these products and return phosphates to the soil. Phosphorus is lost from ecosystems through run-off, precipitation, and sedimentation. Sedimentation may lock phosphorus away but, in the much longer term, it can become available again through processes such as geological uplift. Some phosphorus returns to the land as **guano**; phosphate-rich manure (typically of fish eating birds). This return is small though compared with the phosphate transferred to the oceans each year by natural processes and human activity. Excess phosphorus entering water bodies through runoff is a major contributor to **eutrophication** and excessive algal and weed growth, primarily because phosphorus is often limiting in aquatic systems.

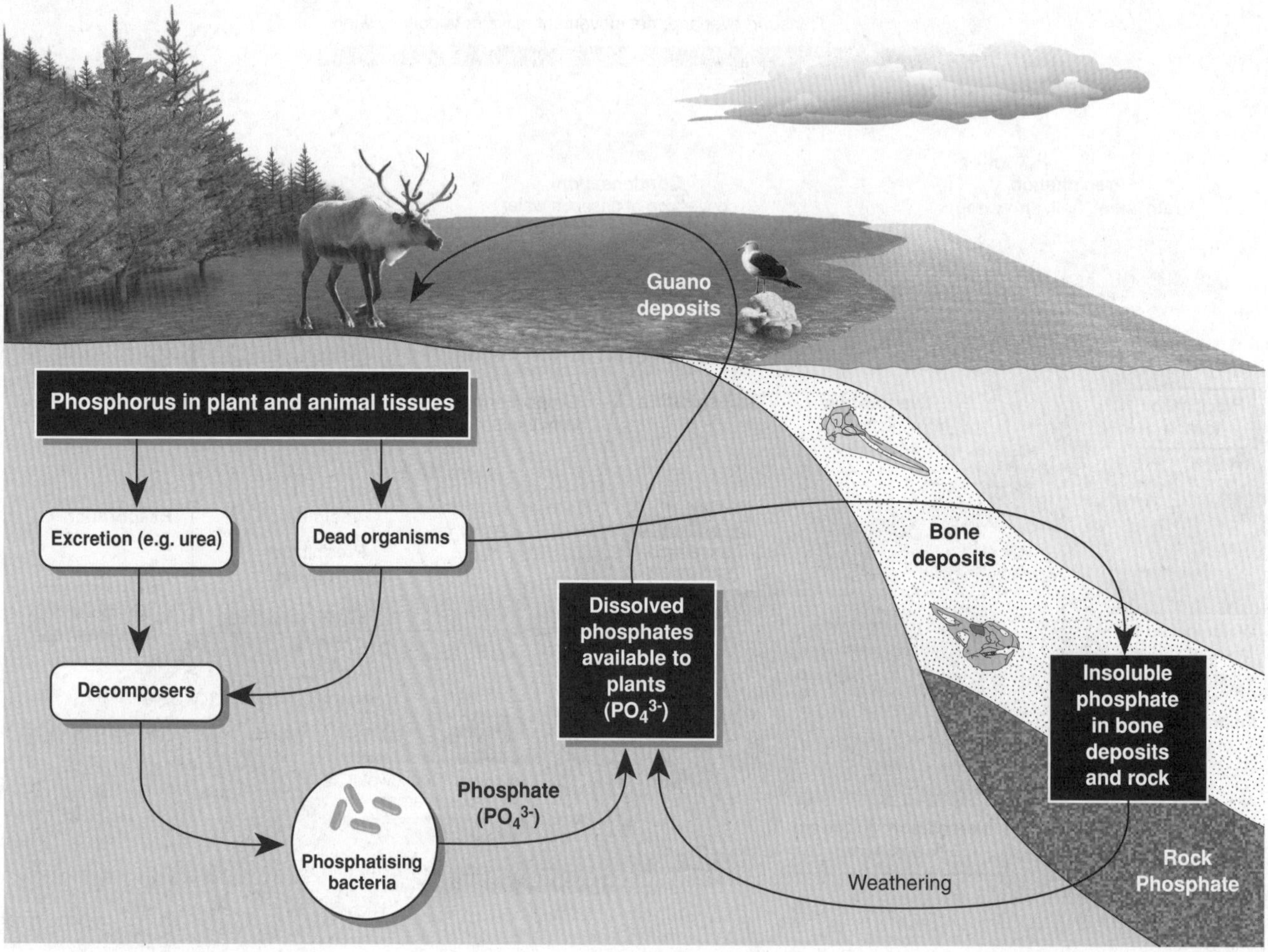

1. In the diagram, add an arrow and label to show where one human activity might intervene in the phosphorus cycle.
2. Identify two instances in the phosphorus cycle where bacterial action is important:

   (a) ______________________ (b) ______________________
3. Name two types of molecules found in living organisms which include phosphorus as a part of their structure:

   (a) ______________________ (b) ______________________
4. Name and describe the origin of three forms of inorganic phosphate making up the geological reservoir:

   (a) ______________________

   (b) ______________________

   (c) ______________________
5. Describe the processes that must occur in order to make rock phosphate available to plants again: ______________________

   ____
6. Identify one major difference between the phosphorus and carbon cycles: ______________________

   ____

Related activities: Water Pollution

The Sulfur Cycle

Although much of the Earth's sulfur is tied up underground in rock and mineral deposits and ocean sediments, it plays a central role in the biosphere. Sulfur is an essential component of proteins and sulfur compounds are important in determining the acidity of precipitation, surface water, and soil. Sulfur circulates through the biosphere in the sulfur cycle, which is complicated because of the many oxidation states of sulfur, including hydrogen sulfide (H_2S), sulfur dioxide (SO_2) sulfate (SO_4^{-2}), and elemental sulfur. Both inorganic processes and living organisms (especially bacteria) are responsible for these transformations. Human activity also releases large quantities of sulfur, primarily through combustion of sulfur-containing coal and oil, but also as a result of refining petroleum, smelting, and other industrial processes. Although SO_2 and sulfate aerosols contribute to air pollution, they also absorb UV radiation and create cloud cover that increases the Earth's albedo (reflectivity) and may offset the effects of rising greenhouse gases. Sulfate aerosols are also produced as a result of the biogenic activity of marine plankton (which release dimethylsulfide or DMS into the atmosphere) and thus may play a natural role in global climate regulation.

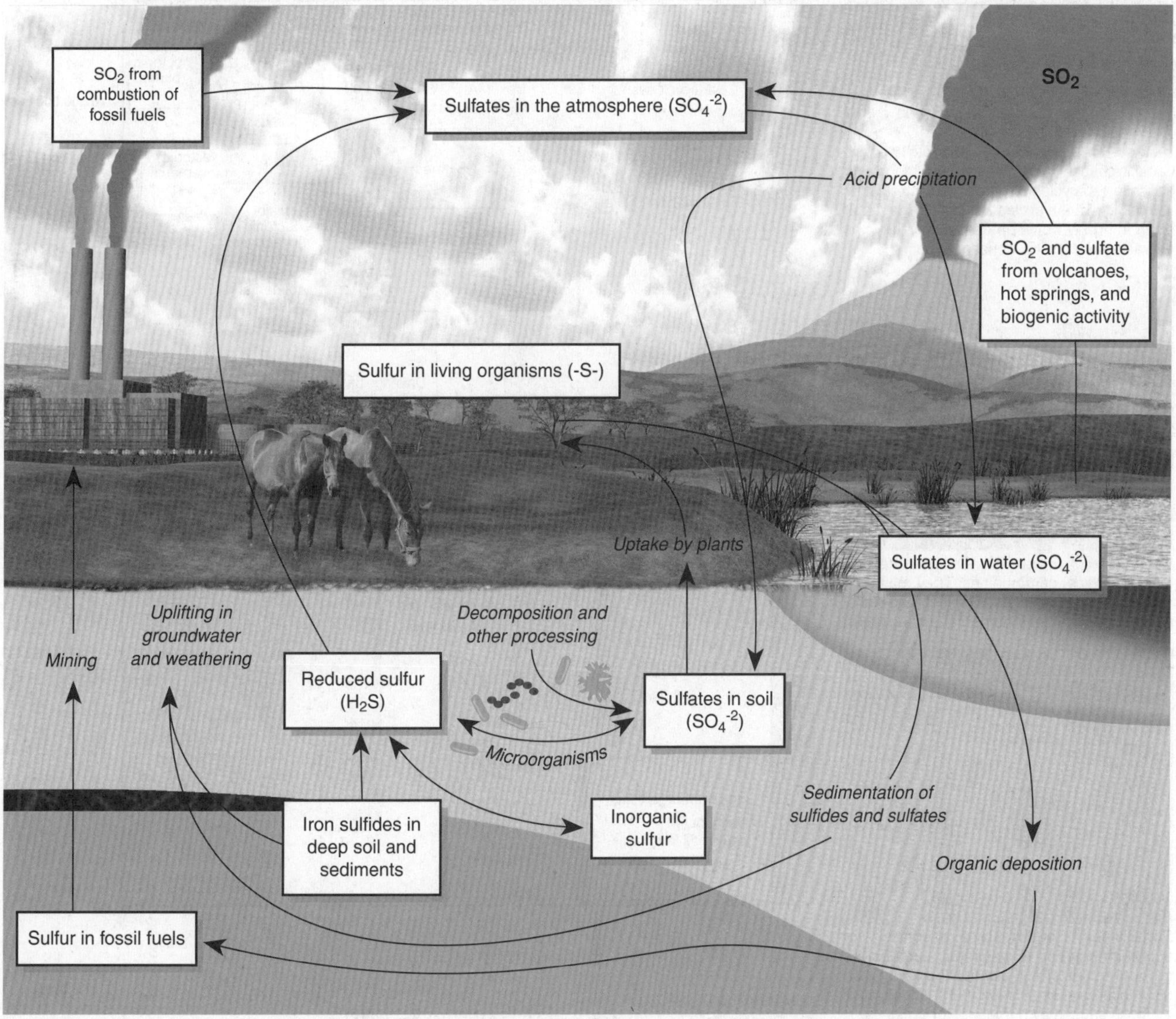

1. Describe two ways in which sulfur can enter the atmosphere from natural sources:

   (a) ______________________ (b) ______________________

2. Describe two ways in which sulfur can enter the atmosphere from as a result of human activity:

   (a) ______________________ (b) ______________________

3. Describe three processes that make sulfur available for uptake by plants:

   (a) ______________________

   (b) ______________________

   (c) ______________________

4. Describe two major roles of sulfur in the biosphere:

   (a) ______________________ (b) ______________________

Environmental Change

Environmental changes come from three sources: the **biosphere** itself, **geological forces** (crustal movements and plate tectonics), and **cosmic forces** (the movement of the moon around the Earth, and the Earth and planets around the sun). All three forces can cause cycles, steady states, and trends (directional changes) in the environment. Environmental trends (such as climate cooling) cause long term changes in communities. Some short term cycles may also influence patterns of behaviour and growth in many species, regulating endogenous cyclical behaviour patterns, called **biological rhythms**.

Climatic change during the last 2-3 million years has involved cycles of glacial and interglacial conditions. These cycles are largely the result of an interplay between astronomical cycles and atmospheric CO_2 concentrations.

Volcanic eruptions may have a large effect on local biological communities. They may also cause prolonged changes to regional and global weather (e.g. Mount Pinatubo eruption, 1989).

Some weather patterns are responsible for subtle changes to ecosystems, such as the gradual onset of a drought. They may also provide large scale and forceful changes, such as those caused by hurricanes or cyclones.

Time scale and geographic extent of environmental change

(Time scale: horizontal axis / geographic extent: vertical axis)

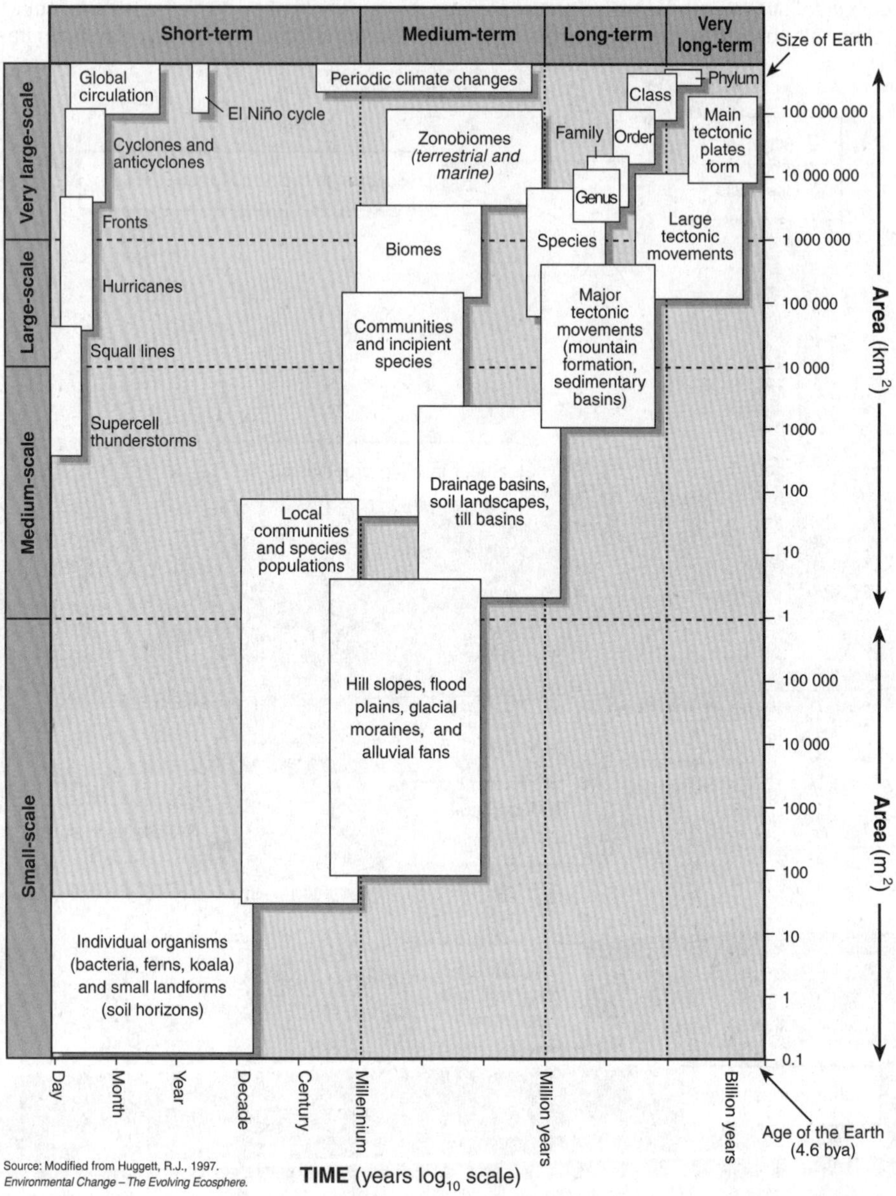

Source: Modified from Huggett, R.J., 1997. *Environmental Change – The Evolving Ecosphere.*

1. Periodic long term changes in the Earth's orbit, a change in the sun's heat output, and continental drift may have been the cause of cycles of climate change in the distant past. These climate changes involved a cooling of the Earth:

   (a) Name these periods of climate cooling: ______________________

   (b) Describe two changes to the landscape that occurred during this period: ______________________

2. On the diagram above, colour-code each of the rectangles to indicate the four themes of environmental change:

   Climatic: __________ Ecological: __________ Tectonic: __________ Evolutionary: __________

3. Explain to what extent these four types of environmental change are interlinked: ______________________

____

____

____

Related activities: Ecosystem Stability, Ecological Succession

Ecosystem Stability

Ecological theory suggests that all species in an ecosystem contribute in some way to ecosystem function. Therefore, species loss past a certain point is likely to have a detrimental effect on the functioning of the ecosystem and on its ability to resist change (its **stability**). Although many species still await discovery, we do know that the rate of species extinction is increasing. Scientists estimate that human destruction of natural habitat is driving up to 100 000 species to extinction every year. This substantial loss of biodiversity has serious implications for the long term stability of many ecosystems.

The Concept of Ecosystem Stability

The stability of an ecosystem refers to its apparently unchanging nature over time. Ecosystem stability has various components, including **inertia** (the ability to resist disturbance) and **resilience** (ability to recover from external disturbances). Ecosystem stability is closely linked to the biodiversity of the system, although it is difficult to predict which factors will stress an ecosystem beyond its range of tolerance. It was once thought that the most stable ecosystems were those with the greatest number of species, since these systems had the greatest number of biotic interactions operating to buffer them against change. This assumption is supported by experimental evidence but there is uncertainty over what level of biodiversity provides an insurance against catastrophe.

Single species crops (monocultures), such as the soy bean crop (above, left), represent low diversity systems that can be vulnerable to disease, pests, and disturbance. In contrast, natural grasslands (above, right) may appear homogeneous, but contain many species which vary in their predominance seasonally. Although they may be easily disturbed (e.g. by burning) they are very resilient and usually recover quickly.

Tropical rainforests (above, left) represent the highest diversity systems on Earth. Whilst these ecosystems are generally resistant to disturbance, once degraded, (above, right) they have little ability to recover. The biodiversity of ecosystems at low latitudes is generally higher than that at high latitudes, where climates are harsher, niches are broader, and systems may be dependent on a small number of key species.

Community Response to Environmental Change

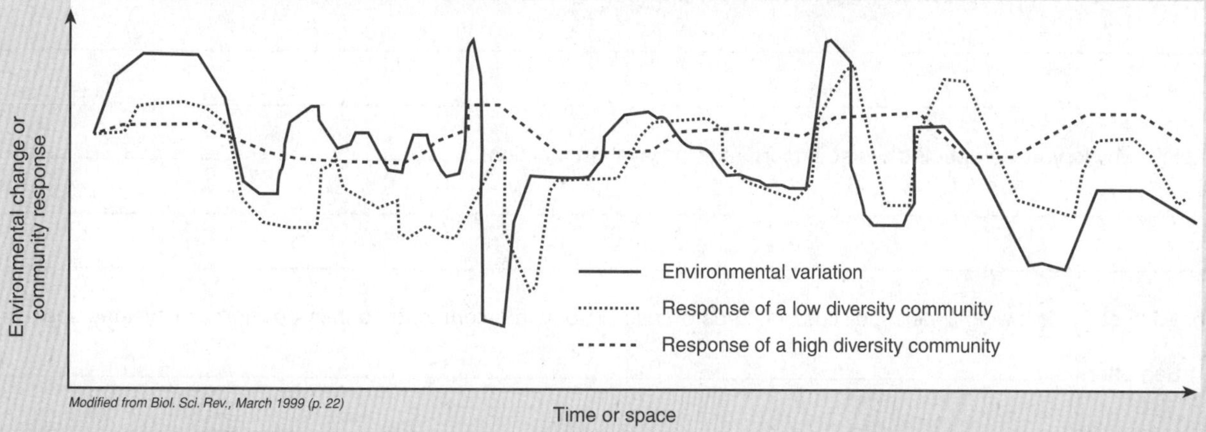

In models of ecosystem function, higher species diversity increases the stability of ecosystem functions such as productivity and nutrient cycling. In the graph above, note how the low diversity system varies more consistently with the environmental variation, whereas the high diversity system is buffered against major fluctuations. In any one ecosystem, some species may be more influential than others in the stability of the system. Such **keystone (key) species** have a disproportionate effect on ecosystem function due to their pivotal role in some ecosystem function such as nutrient recycling or production of plant biomass.

Elephants can change the entire vegetation structure of areas into which they migrate. Their pattern of grazing on taller plant species promotes a predominance of lower growing grasses with small leaves.

Termites are amongst the few larger soil organisms able to break down plant cellulose. They shift large quantities of soil and plant matter and have a profound effect on the rates of nutrient processing in tropical environments.

The starfish *Pisaster* is found along the coasts of North America where it feeds on mussels. If it is removed, the mussels dominate, crowding out most algae and leading to a decrease in the number of herbivore species.

Ecosystems

Related activities: Monitoring Change in an Ecosystem, Loss of Biodiversity

Keystone Species in North America

Gray wolf

Beaver, *Castor canadensis*

Sea otter, *Enhydra lutris*

istockphotos.com

Quaking aspen

JB-BU

Gray or **timber wolves** (*Canis lupus*) are a keystone predator and were once widespread in North American ecosystems. Historically, wolves were eliminated from Yellowstone National Park because of their perceived threat to humans and livestock. As a result, elk populations increased to the point that they adversely affected other flora and fauna. Wolves have since been reintroduced to the park and balance is returning to the ecosystem.

Two smaller mammals are also important keystone species in North America. **Beavers** (top) play a crucial role in biodiversity and many species, including 43% of North America's endangered species, depend partly or entirely on beaver ponds. **Sea otters** are also critical to ecosystem function. When their numbers were decimated by the fur trade, sea urchin populations exploded and the kelp forests, on which many species depend, were destroyed.

Quaking aspen (*Populus tremuloides*) is one of the most widely distributed tree species in North America, and aspen communities are among the most biologically diverse in the region, with a rich understorey flora supporting an abundance of wildlife. Moose, elk, deer, black bear, and snowshoe hare browse its bark, and aspen groves support up to 34 species of birds, including ruffed grouse, which depends heavily on aspen for its winter survival.

1. Suggest one probable reason why high biodiversity promotes greater ecosystem stability: ______________________

____

____

2. Explain why **keystone species** are so important to ecosystem function: ______________________

____

____

3. For each of the following species, discuss features of their biology that contribute to their position as keystone species:

   (a) Sea otter: ____

   ____

   (b) Beaver: ____

   ____

   (c) Gray wolf: ____

   ____

   (d) Quaking aspen: ____

   ____

4. Giving examples, explain how the actions of humans to remove a keystone species might result in ecosystem change:

____

____

____

____

Ecological Succession

Ecological succession is the process by which communities in a particular area change over time. Succession takes place as a result of complex interactions of biotic and abiotic factors. Early communities modify the physical environment. This change results in a change in the biotic community which further alters the physical environment and so on. Each successive community makes the environment more favourable for the establishment of new species. A succession (called a **sere**) proceeds in stages, called seral stages, until the formation of the final climax community, which is stable until further disturbance. Early successional communities are characterised by having a low species diversity, a simple structure, and broad niches. In contrast, community structure in climax communities is complex, with a large number of species interactions. Niches are usually narrow and species diversity is high.

Composition of the community changes with time →

Past community → **Present community** → **Future community**

- **Past community:** Some species in the **past community** were outcompeted, and/or did not tolerate altered abiotic conditions
- **Present community:** The **present community** modifies such abiotic factors as:
  - Light intensity
  - Light quality
  - Wind speed
  - Wind direction
  - Air temperature
  - Soil water
  - Soil composition
  - Humidity
- **Future community:** Changing conditions in the **present community** will allow new species to become established. These will make up the **future community**

TG

Primary Succession

Primary succession refers to colonisation of regions where there is no preexisting community. Examples include the emergence of new volcanic islands, new coral atolls, or islands where the previous community has been extinguished by a volcanic eruption (e.g. the Indonesian island of Krakatau). A classic set colonisation sequence (as depicted below) is relatively uncommon. In reality, the rate at which plants colonise bare ground and the sequence of plant communities that develop are influenced by the local conditions and the dispersal mechanisms of plants in the surrounding region.

Bare rock or volcanic ash | Lichens → Mosses and liverworts → Ferns and grasses → Shrubs, includng nitrogen fixers → Mature trees (**Climax community**)

RA

Secondary Succession: Catastrophe Cycle

(800 years for full richness to develop again)

Events that can cause localised or widespread destruction of forest include cyclone damage, forest fires, and hillside slips. The degree of destruction may vary from the loss of canopy trees in storms, to severe species depletion in the case of fire or landslides. The period of time for the canopy trees to reach their former status is considerable. The photograph on the far left shows a forest fire, while the other shows a large number of tree ferns on a landslide in a mid-successional stage of regeneration.

Cyclone, fire or landslide | Grasses and bracken fern → Abundant saplings of canopy tree species → Dense stand of canopy trees suppresses scrub underneath → Canopy trees increase in height, with natural thinning → Mature trees (**Climax community**)

BH

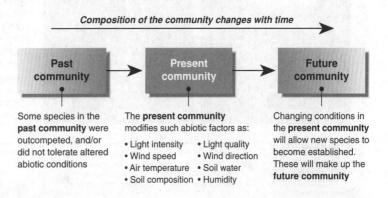

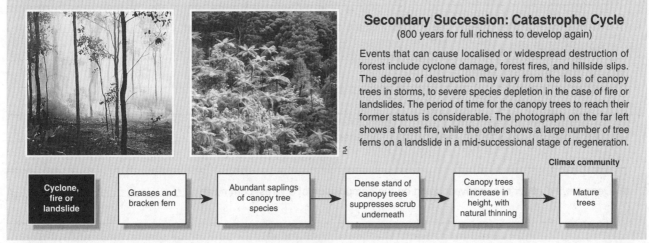

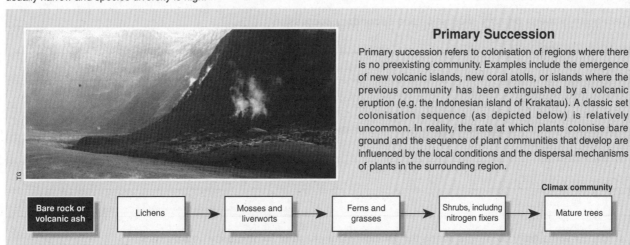

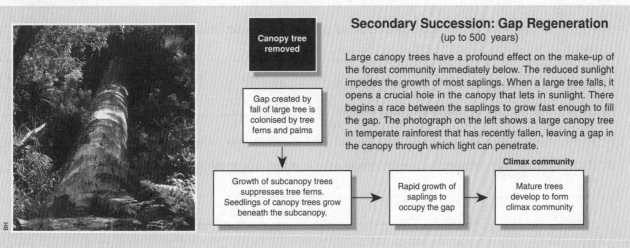

Secondary Succession: Gap Regeneration

(up to 500 years)

Large canopy trees have a profound effect on the make-up of the forest community immediately below. The reduced sunlight impedes the growth of most saplings. When a large tree falls, it opens a crucial hole in the canopy that lets in sunlight. There begins a race between the saplings to grow fast enough to fill the gap. The photograph on the left shows a large canopy tree in temperate rainforest that has recently fallen, leaving a gap in the canopy through which light can penetrate.

Canopy tree removed → Gap created by fall of large tree is colonised by tree ferns and palms → Growth of subcanopy trees suppresses tree ferns. Seedlings of canopy trees grow beneath the subcanopy. → Rapid growth of saplings to occupy the gap → Mature trees develop to form climax community (**Climax community**)

Ecosystems

Related activities: Environmental Change
Web links: Mount St. Helens

Secondary Succession in Cleared Land

(150+ years for mature woodland to develop again)

A secondary succession takes place after a land clearance (e.g. from fire or landslide). Such events do not involve loss of the soil and so tend to be more rapid than primary succession, although the time scale depends on the species involved and on climatic and edaphic (soil) factors. Humans may deflect the natural course of succession (e.g. by mowing) and the climax community that results will differ from the natural community. A climax community arising from a **deflected succession** is called a **plagioclimax**.

	Primarily bare earth →	Open pioneer community (annual grasses) →	Grasses and low growing perennials →	Scrub: shrubs and small trees →	Young broad-leaved woodland →	**Climax community** Mature woodland mainly oak
Time to develop (years)		1–2	3–5	16-30	31-150	150 +

1. Distinguish between **primary** succession and **secondary** succession:

2. Suggest why primary successions rarely follow the classic sequence depicted on the previous page:

3. (a) Identify some early colonisers during the establishment phase of a community on bare rock:

   (b) Describe two important roles of the species that are early colonisers of bare slopes:

4. Describe a possible catastrophic event causing succession in a rainforest ecosystem:

5. (a) Describe the effect of selective logging on the composition of a forest community:

   (b) Suggest why selective logging could be considered preferable (for forest conservation) to clear felling of trees:

6. (a) Explain what is meant by a **deflected succession**:

   (b) Discuss the role that deflected successions might have in maintaining managed habitats:

Populations

Investigating the dynamics of populations

Features of populations, population growth and regulation, population age structure, r & K selection, species interactions

Learning Objectives

☐ 1. Compile your own glossary from the **KEY WORDS** displayed in **bold type** in the learning objectives below.

Features of Populations *(pages 64-65, 67)*

☐ 2. Recall the difference between a **population** and a **community**. Explain what is meant by **population density** and distinguish it from **population size**.

☐ 3. Understand that populations are dynamic and exhibit attributes not shown by individuals themselves. Recognise the following population attributes: **density**, **distribution**, birth rate (**natality**), mean (average) age, death rate (**mortality**), **survivorship**, migration rate, **age structure**, **fecundity** (reproductive potential).

☐ 4. Describe, with examples, the distribution patterns of organisms within their range: **uniform** distribution, **random** distribution, **clumped** distribution. Identify the factors governing each type of distribution.

Population Growth and Size *(pages 66-76)*

☐ 5. Recall that populations are dynamic. Outline how population size can be affected by **births**, **deaths**, and **migration** and express the relationship in an equation.

☐ 6. Recognise the value of **life tables** in providing information of patterns of population birth and mortality. Explain the role of **survivorship curves** in analysing populations. Providing examples, describe the features of Type I, II, and III survivorship curves. If required. distinguish between *r* and K selection.

☐ 7. Describe how the trends in population change can be shown in a **population growth curve** of population numbers (Y axis) against time (X axis).

☐ 8. Describe the factors that affect final population size:
(a) **Carrying capacity** of the environment.
(b) **Environmental resistance**.
(c) **Density dependent factors**, e.g. intraspecific competition, interspecific competition, predation.
(d) **Density independent factors**, e.g. climatic events.
(e) **Limiting factors**, e.g. soil nutrient.

☐ 9. Distinguish between **exponential** and **sigmoidal growth curves**. Create labelled diagrams of these curves, indicating the different phases of growth and the factors regulating population growth at each stage.

☐ 10. Recognise patterns of population growth in colonising, stable, declining, and oscillating populations. Relate these patterns to global human populations.

☐ 11. Describe aspects of human **population dynamics** including projected growth rates and distribution of the world's human population, human **demographics**, and the environmental and economic impacts of human population growth. Comment on the increasing **urbanisation** of the world's population and discuss options for the planned development of cities.

Species Interactions *(pages 77-80)*

☐ 12. Explain the nature of the **intra-** and **interspecific interactions** occurring in communities. Recognise: **competition**, **mutualism**, **commensalism**, **exploitation** (parasitism, predation, herbivory), **amensalism**, and **allelopathy**.

Supplementary Texts

See page 7 for additional details of these texts:

■ Miller, G.T, 2007. **Living in the Environment: Principles, Connections & Solutions**, chpt. 7-9.

■ Raven *et. al.*, 2002. **Environment**, chpt. 8-9.

■ Reiss, M. & J. Chapman, 2000. **Environmental Biology** (Cambridge University Press), pp. 3-5.

■ Smith, R. L. & T.M. Smith, 2001. **Ecology and Field Biology**, reading as required.

Presentation MEDIA to support this topic:
ECOLOGY
• Populations & Interactions
• Human Impact

Periodicals

See page 7 for details of publishers of periodicals:

STUDENT'S REFERENCE

■ **Human Population Grows Up** Scientific American, Sept. 2005, pp. 26-33. *Projections for global population change, the earth's carrying capacity, and the effects of the growing human population on environmental sustainability.*

■ **Predator-Prey Relationships** Biol. Sci. Rev., 10(5) May 1998, pp. 31-35. *Predator-prey relationships, and the defence strategies of prey.*

■ **Inside Story** New Scientist, 29 April 2000, pp. 36-39. *Ecological interactions between fungi and plants and animals: what are the benefits?*

■ **The Future of Red Squirrels in Britain** Biol. Sci. Rev., 16(2) Nov. 2003, pp. 8-11. *A further account of the impact of the grey squirrel on Britain's native red squirrel populations.*

■ **Logarithms and Life** Biol. Sci. Rev., 13(4) March 2001, pp. 13-15. *The basics of logarithmic growth and its application to real populations.*

■ **Time to Rethink Everything** New Scientist, 27 April-18 May 2002 (4 issues). *Globalisation, the impact of humans, & the sustainability of our future.*

■ **Population Bombshell** New Scientist, 11 July 1998 (Inside Science). *Current and predicted growth rates in human populations. This excellent account includes some interesting analyses of population age distributions using pyramids.*

Internet

See pages 4-5 for details of how to access **Bio Links** from our web site: **www.thebiozone.com** From Bio Links, access sites under the topics:

ECOLOGY: > **Populations and Communities**:
• Bull Shoals Lake 1995 report • Communities • Death squared • Competition • Interactions • Intraspecific relations: Cooperation and competition • Population growth and balance • Quantitative population ecology • Species interactions

Features of Populations

Populations have a number of attributes that may be of interest. Usually, biologists wish to determine **population size** (the total number of organisms in the population). It is also useful to know the **population density** (the number of organisms per unit area). The density of a population is often a reflection of the **carrying capacity** of the environment, i.e. how many organisms an environment can support. Populations also have structure; particular ratios of different ages and sexes. These data enable us to determine whether the population is declining or increasing in size. We can also look at the **distribution** of organisms within their environment and so determine what particular aspects of the habitat are favoured over others. One way to retrieve information from populations is to **sample** them. Sampling involves collecting data about features of the population from samples of that population (since populations are usually too large to examine in total). Sampling can be carried out directly (by sampling the population itself using appropriate equipment) or indirectly (e.g. by monitoring calls or looking for droppings or other signs). Some of the population attributes that we can measure or calculate are illustrated on the diagram below.

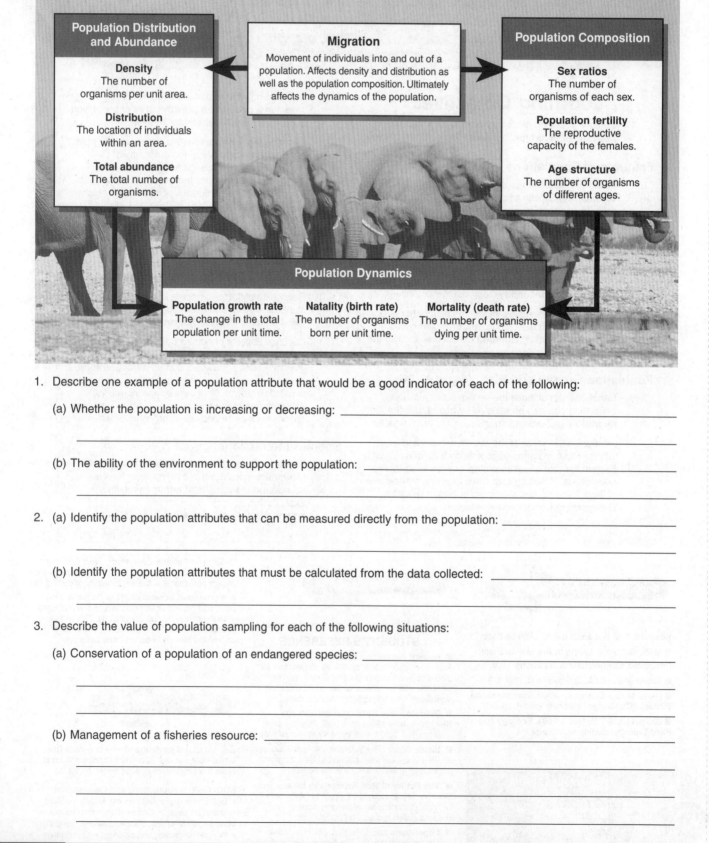

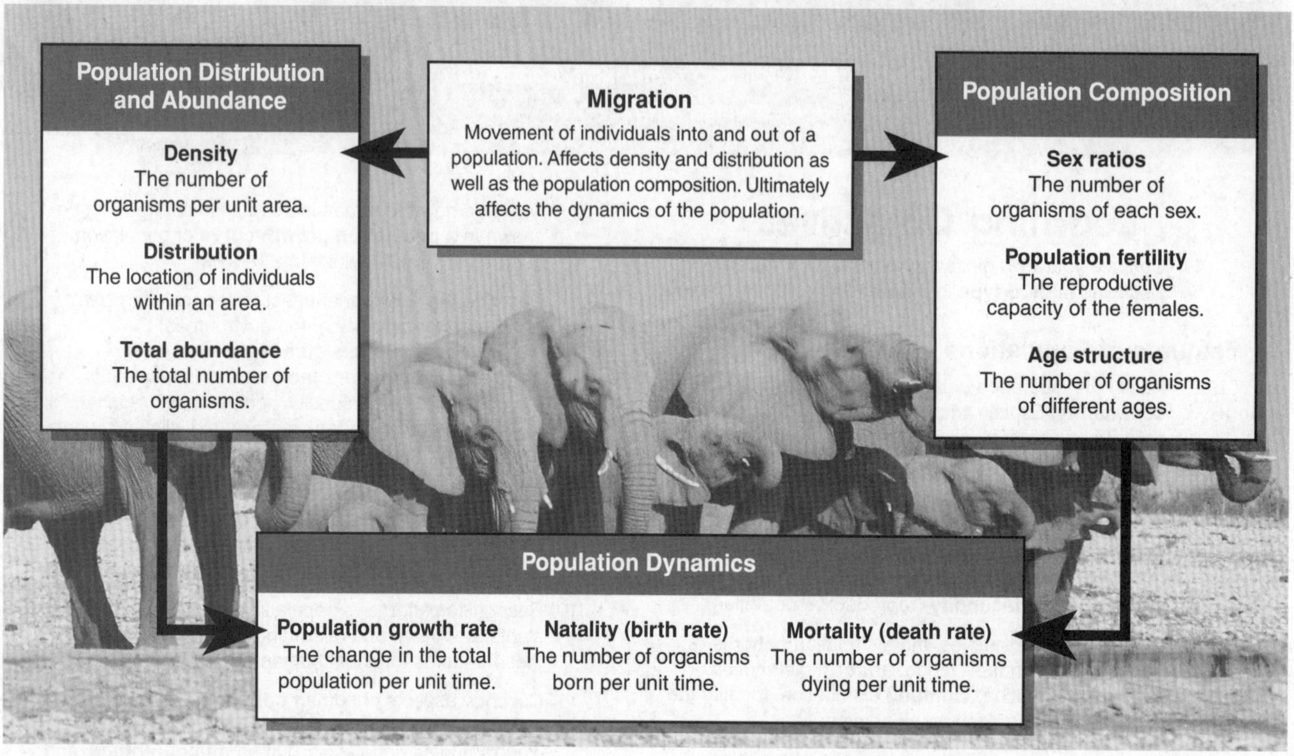

Population Distribution and Abundance
Density
The number of organisms per unit area.
Distribution
The location of individuals within an area.
Total abundance
The total number of organisms.
Migration
Movement of individuals into and out of a population. Affects density and distribution as well as the population composition. Ultimately affects the dynamics of the population.
Population Composition
Sex ratios
The number of organisms of each sex.
Population fertility
The reproductive capacity of the females.
Age structure
The number of organisms of different ages.
Population Dynamics
Population growth rate
The change in the total population per unit time.
Natality (birth rate)
The number of organisms born per unit time.
Mortality (death rate)
The number of organisms dying per unit time.

1. Describe one example of a population attribute that would be a good indicator of each of the following:

   (a) Whether the population is increasing or decreasing: ______

   (b) The ability of the environment to support the population: ______

2. (a) Identify the population attributes that can be measured directly from the population: ______

   (b) Identify the population attributes that must be calculated from the data collected: ______

3. Describe the value of population sampling for each of the following situations:

   (a) Conservation of a population of an endangered species: ______

   (b) Management of a fisheries resource: ______

Related activities: Density and Distribution

Density and Distribution

Distribution and density are two interrelated properties of populations. Population density is the number of individuals per unit area (for land organisms) or volume (for aquatic organisms). Careful observation and precise mapping can determine the distribution patterns for a species. The three basic distribution patterns are: random, clumped and uniform. In the diagram below, the circles represent individuals of the same species. It can also represent populations of different species.

Low Density

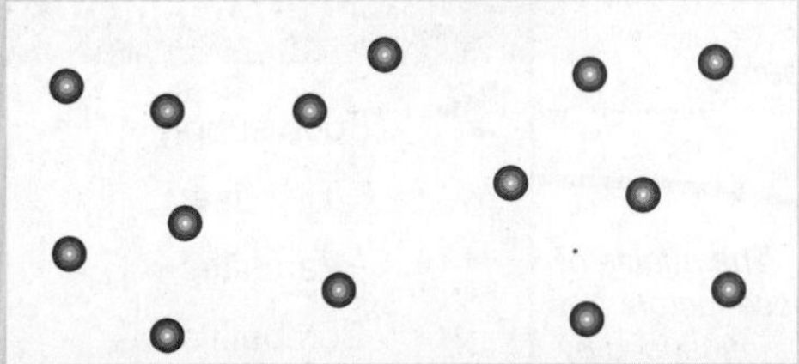

In low density populations, individuals are spaced well apart. There are only a few individuals per unit area or volume (e.g. highly territorial, solitary mammal species).

High Density

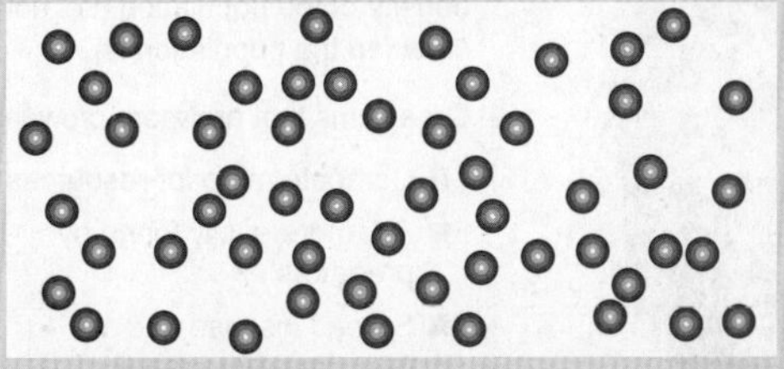

In high density populations, individuals are crowded together. There are many individuals per unit area or volume (e.g. colonial organisms, such as many corals).

Tigers are solitary animals, found at low densities.

Termites form well organised, high density colonies.

Random Distribution

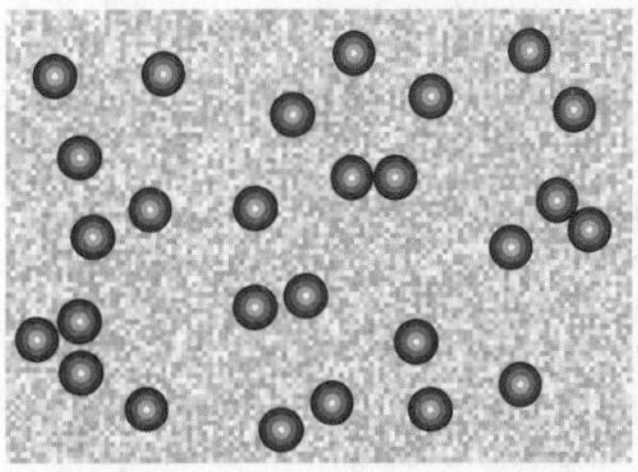

Random distributions occur when the spacing between individuals is irregular. The presence of one individual does not directly affect the location of any other individual. Random distributions are uncommon in animals but are often seen in plants.

Clumped Distribution

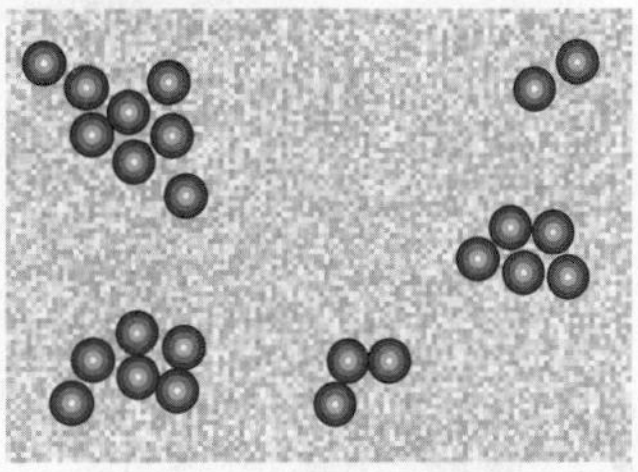

Clumped distributions occur when individuals are grouped in patches (sometimes around a resource). The presence of one individual increases the probability of finding another close by. Such distributions occur in herding and highly social species.

Uniform Distribution

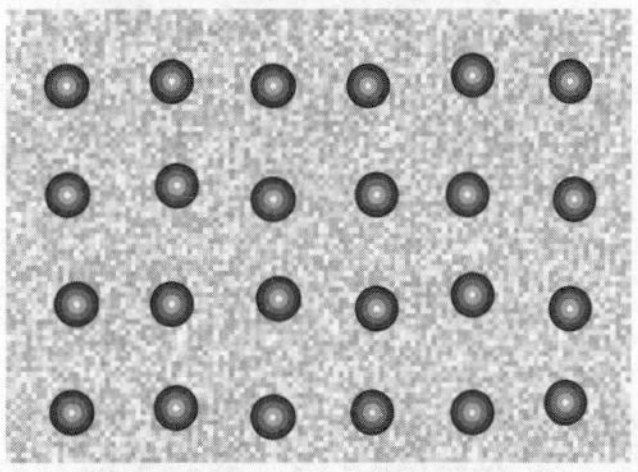

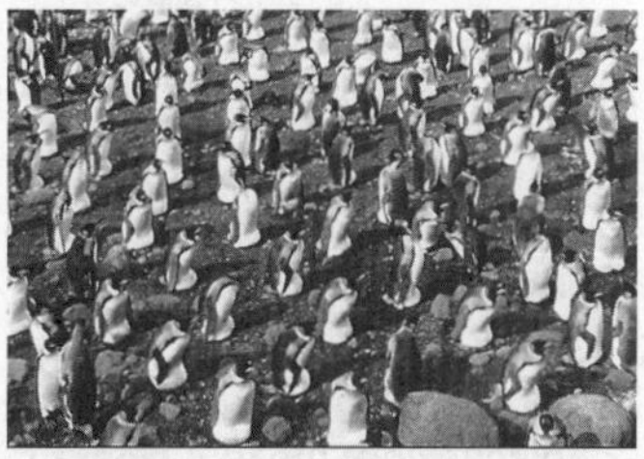

Regular distribution patterns occur when individuals are evenly spaced within the area. The presence of one individual decreases the probability of finding another individual very close by. The penguins illustrated above are also at a high density.

1. Describe why some organisms may exhibit a clumped distribution pattern because of:

   (a) Resources in the environment: ______________________________

   (b) A group social behaviour: ______________________________

2. Describe a social behaviour found in some animals that may encourage a uniform distribution:

   ____

3. Describe the type of environment that would encourage uniform distribution:

   ____

4. Describe an example of each of the following types of distribution pattern:

   (a) Clumped: ______________________________

   (b) Random (more or less): ______________________________

   (c) Uniform (more or less): ______________________________

Population Regulation

Very few species show continued exponential growth. Population size is regulated by factors that limit population growth. The diagram below illustrates how population size can be regulated by environmental factors. **Density independent factors** may affect all individuals in a population equally. Some, however, may be better able to adjust to them. **Density dependent factors** have a greater affect when the population density is higher. They become less important when the population density is low.

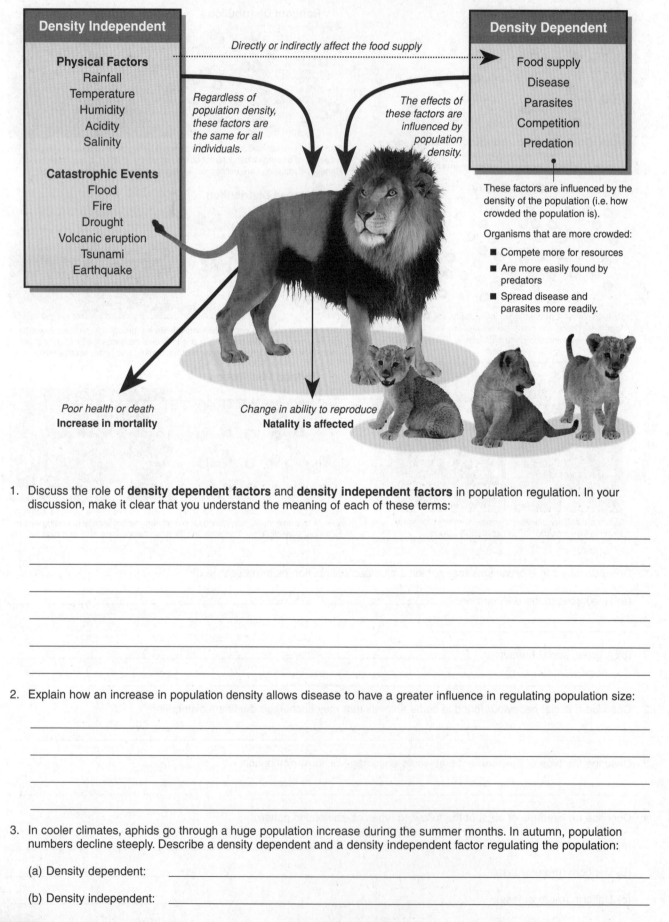

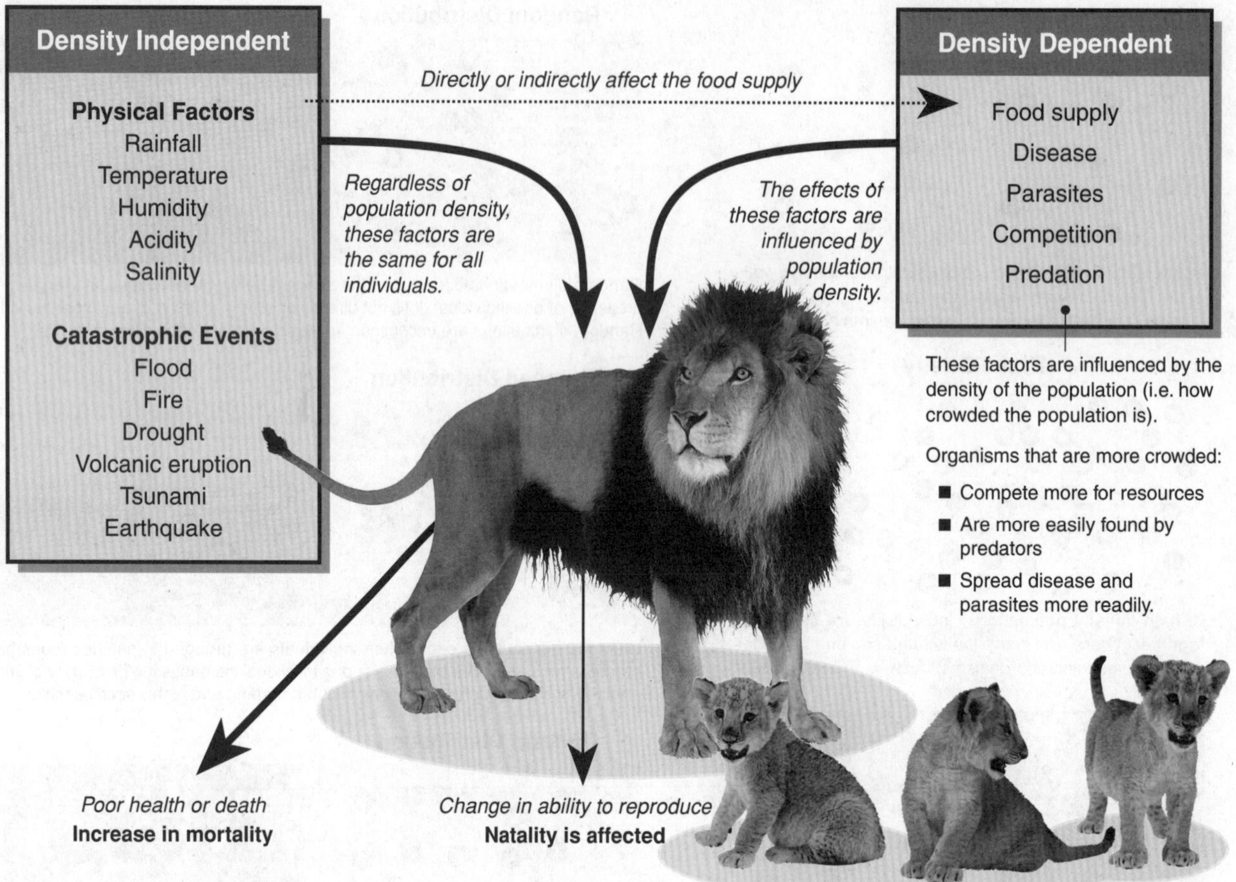

Density Independent
Physical Factors
Rainfall
Temperature
Humidity
Acidity
Salinity
Catastrophic Events
Flood
Fire
Drought
Volcanic eruption
Tsunami
Earthquake
Directly or indirectly affect the food supply
Density Dependent
Food supply
Disease
Parasites
Competition
Predation
Regardless of population density, these factors are the same for all individuals.
The effects of these factors are influenced by population density.
These factors are influenced by the density of the population (i.e. how crowded the population is).
Organisms that are more crowded:
Compete more for resources
Are more easily found by predators
Spread disease and parasites more readily.
Poor health or death
Increase in mortality
Change in ability to reproduce
Natality is affected

1. Discuss the role of **density dependent factors** and **density independent factors** in population regulation. In your discussion, make it clear that you understand the meaning of each of these terms:

2. Explain how an increase in population density allows disease to have a greater influence in regulating population size:

3. In cooler climates, aphids go through a huge population increase during the summer months. In autumn, population numbers decline steeply. Describe a density dependent and a density independent factor regulating the population:

   (a) Density dependent:

   (b) Density independent:

Population Growth

Organisms do not generally live alone. A **population** is a group of organisms of the same species living together in one geographical area. This area may be difficult to define as populations may comprise widely dispersed individuals that come together only infrequently (e.g. for mating). The number of individuals comprising a population may also fluctuate considerably over time. These changes make populations dynamic: populations gain individuals through births or immigration, and lose individuals through deaths and emigration. For a population in **equilibrium**, these factors balance out and there is no net change in the population abundance. When losses exceed gains, the population declines.

Births, *deaths*, *immigrations* (movements into the population) and *emigrations* (movements out of the population) are events that determine the numbers of individuals in a population. Population growth depends on the number of individuals added to the population from births and immigration, minus the number lost through deaths and emigration. This is expressed as:

Population growth =

Births – Deaths + Immigration – Emigration
(B) (D) (I) (E)

The difference between immigration and emigration gives *net migration*. Ecologists usually measure the **rate** of these events. These rates are influenced by environmental factors and by the characteristics of the organisms themselves. Rates in population studies are commonly expressed in one of two ways:

- Numbers per unit time, e.g. 20 150 live births per year.
- Per capita rate (number per head of population), e.g. 122 live births per 1000 individuals per year (12.2%).

Calculating change in population numbers

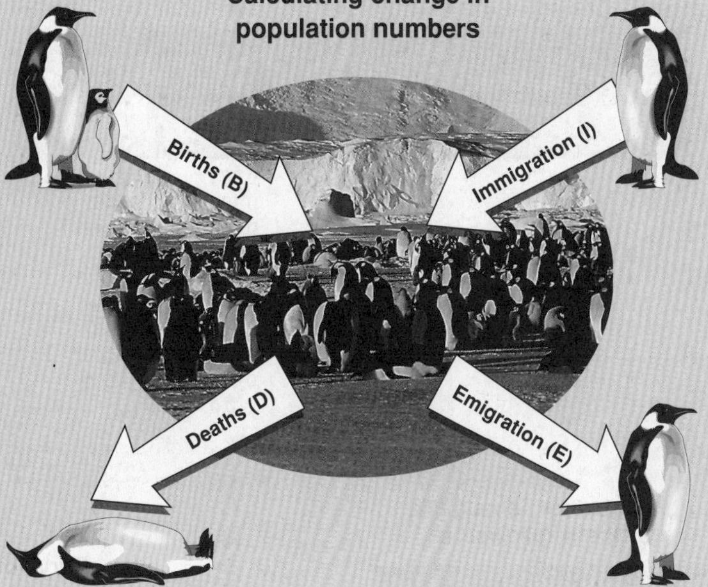

1. Define the following terms used to describe changes in population numbers:

   (a) Death rate (mortality): __________

   (b) Birth rate (natality): __________

   (c) Immigration: __________

   (d) Emigration: __________

   (e) Net migration rate: __________

2. Using the terms, B, D, I, and E (above), construct equations to express the following (the first is completed for you):

   (a) A population in equilibrium: B + I = D + E

   (b) A declining population: __________

   (c) An increasing population: __________

3. The rate of population change can be expressed as the interaction of all these factors:

   Rate of population change = Birth rate – Death rate + Net migration rate (positive or negative)

   Using the formula above, determine the annual rate of population change for Mexico and the United States in 1972:

	USA	Mexico
Birth rate	1.73%	4.3%
Death rate	0.93%	1.0%
Net migration rate	+0.20%	0.0%

   Rate of population change for USA = __________

   Rate of population change for Mexico = __________

4. A population started with a total number of 100 individuals. Over the following year, population data were collected. Calculate birth rates, death rates, net migration rate, and rate of population change for the data below (as percentages):

   (a) Births = 14: Birth rate = __________ (b) Net migration = +2: Net migration rate = __________

   (c) Deaths = 20: Death rate = __________ (d) Rate of population change = __________

   (e) State whether the population is increasing or declining: __________

Populations

Life Tables and Survivorship

Life tables, such as those shown below, provide a summary of mortality for a population (usually for a group of individuals of the same age or **cohort**). The basic data are just the number of individuals remaining alive at successive sampling times (the **survivorship** or **l*x***). Life tables are an important tool when analysing changes in populations over time. They can tell us the ages at which most mortality occurs in a population and can also provide information about life span and population age structure. From basic life table data, biologists derive survivorship curves, based on the l*x* column. Survivorship curves are standardised as the number of survivors per 1000 individuals so that populations of different types can be easily compared.

Life Table and Survivorship Curve for a Population of the Barnacle *Balanus*

Age in years (x)	No. alive each year (N_x)	Proportion surviving at the start of age x (l_x)	Proportion dying between x and x+1 (d_x)	Mortality (q_x)
0	142	1.000	0.563	0.563
1	62	0.437	0.198	0.452
2	34	0.239	0.098	0.412
3	20	0.141	0.035	0.250
4	15	0.106	0.028	0.267
5	11	0.078	0.036	0.454
6	6	0.042	0.028	0.667
7	2	0.014	0.0	0.000
8	2	0.014	0.014	1.000
9	0	0.0	0.0	–

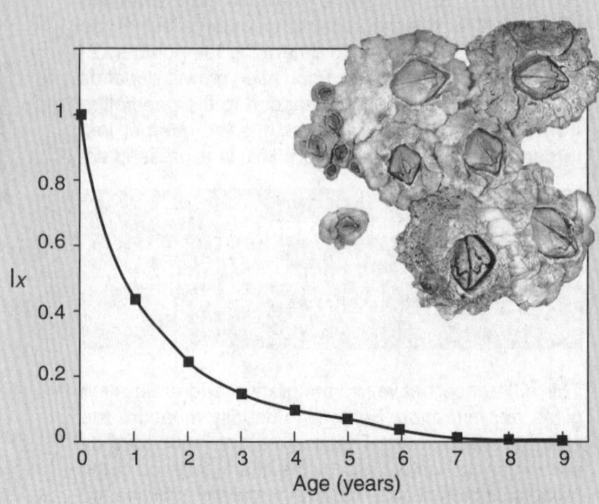

Life Table for Female Elk, Northern Yellowstone

x	l_x	d_x	q_x
0	1000	323	.323
1	677	13	.019
2	664	2	.003
3	662	2	.003
4	660	4	.006
5	656	4	.006
6	652	9	.014
7	643	3	.005
8	640	3	.005
9	637	9	.014
10	628	7	.001
11	621	12	.019
12	609	13	.021
13	596	41	.069
14	555	34	.061
15	521	20	.038
16	501	59	.118
17	442	75	.170
18	367	93	.253
19	274	82	.299
20	192	57	.297
21+	135	135	1.000

Survivorship Curve for Female Elk of Northern Yellowstone National Park

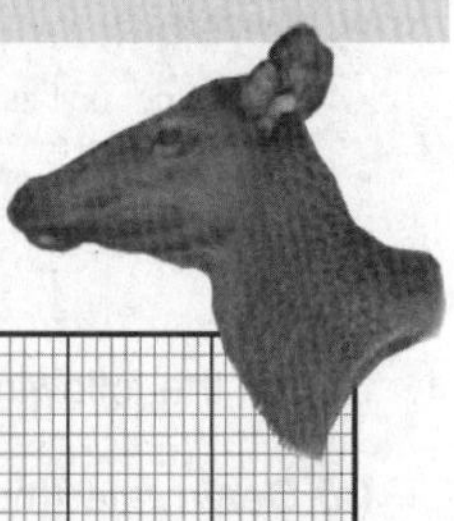

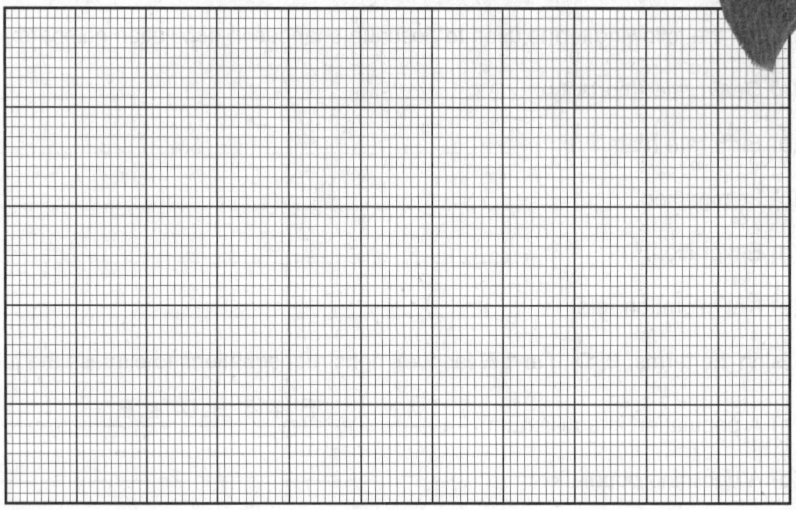

1. (a) In the example of the barnacle *Balanus* above, state when most of the group die: ____________

   (b) Identify the type of survivorship curve is represented by these data (see next page): ____________

2. (a) Using the grid, plot a survivorship curve for elk hinds (above) based on the life table data provided:

   (b) Describe the survivorship curve for these large mammals: ____________

   ____

3. Explain how a biologist might use life table data to manage an endangered population: ____________

   ____

   ____

   ____

Related activities: Survivorship Curves, Population Age Structure, *r* and K Selection

Survivorship Curves

The survivorship curve depicts age-specific mortality. It is obtained by plotting the number of individuals of a particular cohort against time. Survivorship curves are standardised to start at 1000 and, as the population ages, the number of survivors progressively declines. The shape of a survivorship curve thus shows graphically at which life stages the highest mortality occurs. Survivorship curves in many populations fall into one of three hypothetical patterns (below). Wherever the curve becomes steep, there is an increase in mortality. The convex Type I curve is typical of populations whose individuals tend to live out their physiological life span. Such populations usually produce fewer young and show some degree of parental care. Organisms that suffer high losses of the early life stages (a Type III curve) compensate by producing vast numbers of offspring. These curves are conceptual models only, against which real life curves can be compared. Many species exhibit a mix of two of the three basic types. Some birds have a high chick mortality (Type III) but adult mortality is fairly constant (Type II). Some invertebrates (e.g. crabs) have high mortality only when moulting and show a stepped curve.

Hypothetical Survivorship Curves

Type I
Late loss survivorship curve

Mortality (death rate) is very low in the infant and juvenile years, and throughout most of adult life. Mortality increases rapidly in old age. **Examples**: Humans (in developed countries) and many other large mammals (e.g. big cats, elephants).

Type II
Constant loss survivorship curve

Mortality is relatively constant through all life stages (no one age is more susceptible than another). **Examples**: Some invertebrates such as *Hydra*, some birds, some annual plants, some lizards, and many rodents.

Type III
Early loss survivorship curve

Mortality is very high during early life stages, followed by a very low death rate for the few individuals reaching adulthood. **Examples**: Many fish (not mouth brooders) and most marine invertebrates (e.g. oysters, barnacles).

Graph of Survivorship in Relation to Age

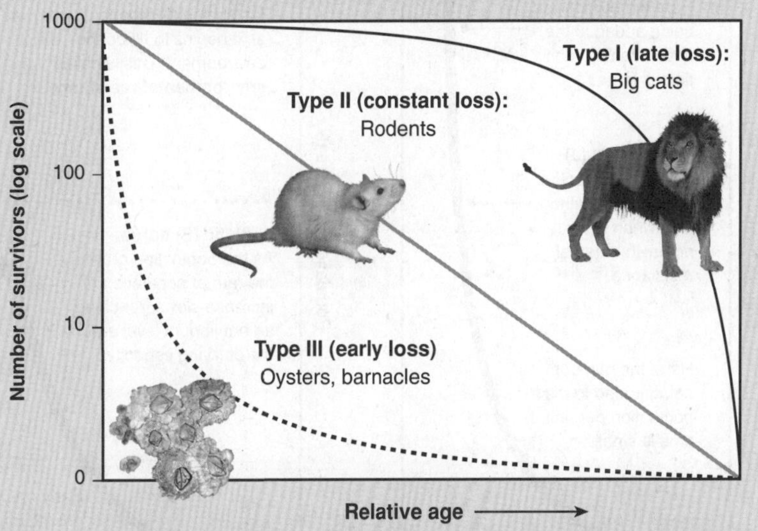

Three basic types of survivorship curves and representative organisms for each type. The vertical axis may be scaled arithmetically or logarithmically.

Elephants have a close matriarchal society and a long period of parental care. Elephants are long-lived and females usually produce just one calf.

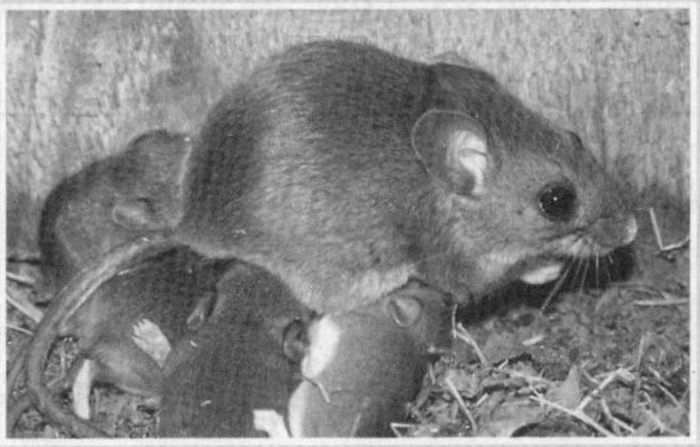

Rodents are well known for their large litters and prolific breeding capacity. Individuals are lost from the population at a more or less constant rate.

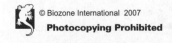

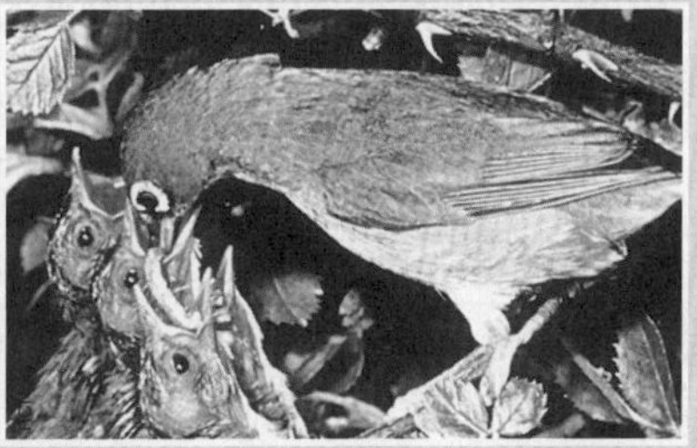

Despite vigilant parental care, many birds suffer high juvenile losses (Type III). For those surviving to adulthood, deaths occur at a constant rate.

1. Explain why human populations might not necessarily show a Type I curve: ____________________

2. Explain how organisms with a Type III survivorship compensate for the high mortality during early life stages:

3. Describe the features of a species with a Type I survivorship that aid in high juvenile survival: ____________________

4. Discuss the following statement: "There is no standard survivorship curve for a given species; the curve depicts the nature of a population at a particular time and place and under certain environmental conditions.":

Populations

Related activities: Life Tables and Survivorship, Population Age Structure, *r* and K Selection

Population Growth Curves

Populations becoming established in a new area for the first time are often termed **colonising populations** (below, left). They may undergo a rapid **exponential** (logarithmic) increase in numbers as there are plenty of resources to allow a high birth rate, while the death rate is often low. Exponential growth produces a J-shaped growth curve that rises steeply as more and more individuals contribute to the population increase. If the resources of the new habitat were endless (inexhaustible) then the population would continue to increase at an **exponential** rate. However, this rarely happens in natural populations. Initially, growth may be exponential (or nearly so), but as the population grows, its increase will slow and it will stabilise at a level that can be supported by the environment (called the carrying capacity or K). This type of growth is called sigmoidal and produces the **logistic growth curve** (below, right). **Established populations** will fluctuate about K, often in a regular way (grey area on the graph below, right). Some species will have populations that vary little from this stable condition, while others may oscillate wildly.

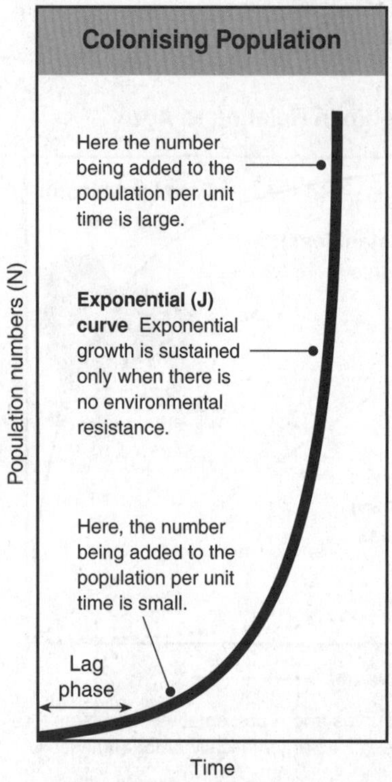

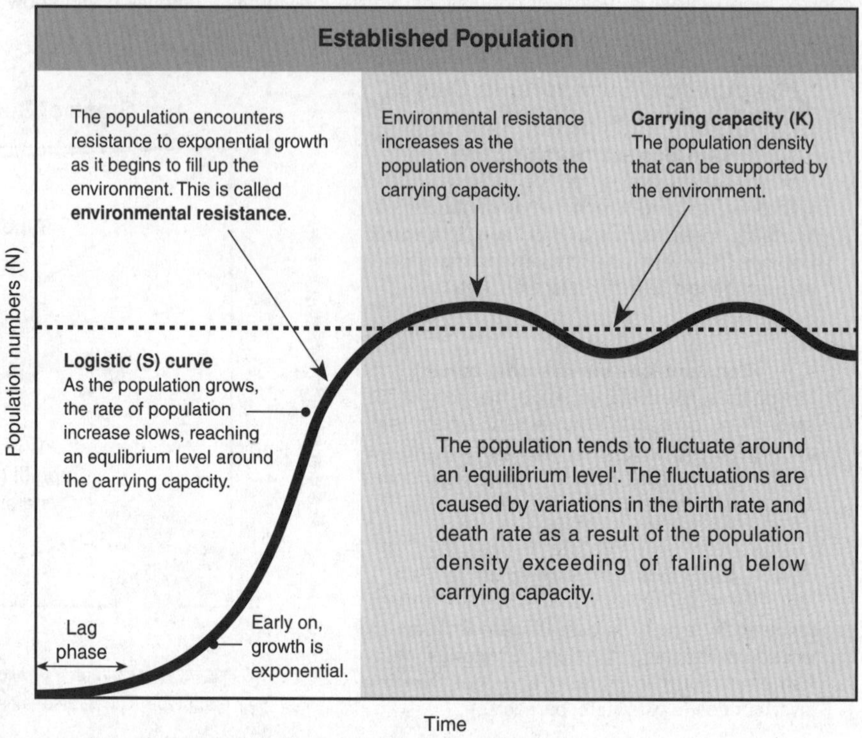

1. Explain why populations tend not to continue to increase exponentially in an environment: ______

2. Explain what is meant by environmental resistance: ______

3. (a) Explain what is meant by carrying capacity: ______

   (b) Explain the importance of **carrying capacity** to the growth and maintenance of population numbers: ______

4. Species that expand into a new area, such as rabbits did in areas of Australia, typically show a period of rapid population growth followed by a slowing of population growth as density dependent factors become more important and the population settles around a level that can be supported by the carrying capacity of the environment.

   (a) Explain why a newly introduced consumer (e.g. rabbit) would initially exhibit a period of exponential population growth:

   (b) Describe a likely outcome for a rabbit population after the initial rapid increase had slowed: ______

5. Describe the effect that introduced grazing species might have on the carrying capacity of the environment:

Related activities: Population Growth, World Population Growth, *r* and K Selection

Population Age Structure

The **age structure** of a population refers to the relative proportion of individuals in each age group in the population. The age structure of populations can be categorised according to specific age categories (such as years or months), but also by other measures such as life stage (egg, larvae, pupae, instars), of size class (height or diameter in plants). Population growth is strongly influenced by age structure; a population with a high proportion of reproductive and prereproductive aged individuals has a much greater potential for population growth than one that is dominated by older individuals. The ratio of young to adults in a relatively stable population of most mammals and birds is approximately 2:1 (below, left). Growing populations in general are characterised by a large and increasing number of young, whereas a population in decline typically has a decreasing number of young. Population age structures are commonly represented as pyramids, in which the proportions of individuals in each age/size class are plotted with the youngest individuals at the pyramid's base. The number of individuals moving from one age class to the next influences the age structure of the population from year to year. The loss of an age class (e.g. through overharvesting) can profoundly influence a population's viability and can even lead to population collapse.

Age Structures in Animal Populations

These theoretical age pyramids, which are especially applicable to birds and mammals, show how growing populations are characterised by a high ratio of young (white bar) to adult age classes (grey bars). Ageing populations with poor production are typically dominated by older individuals.

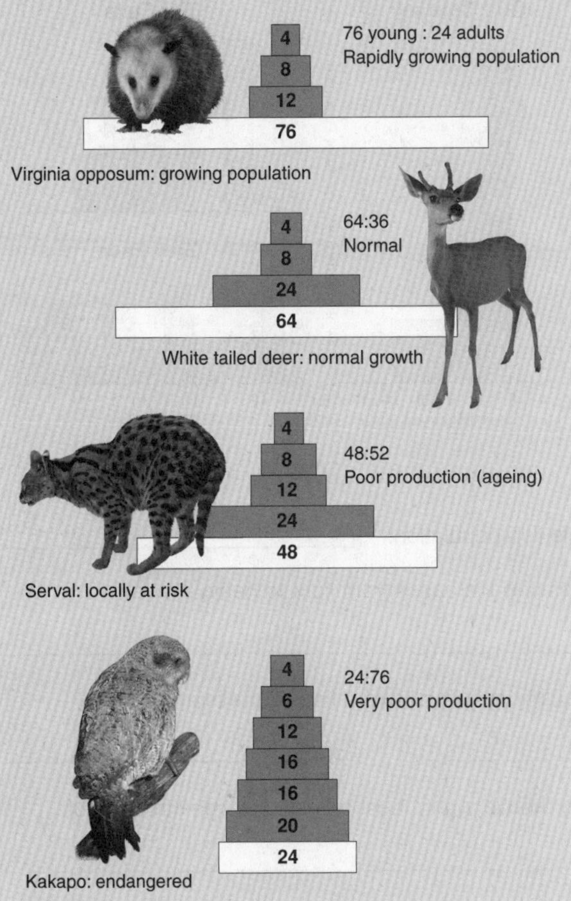

Age Structures in Human Populations

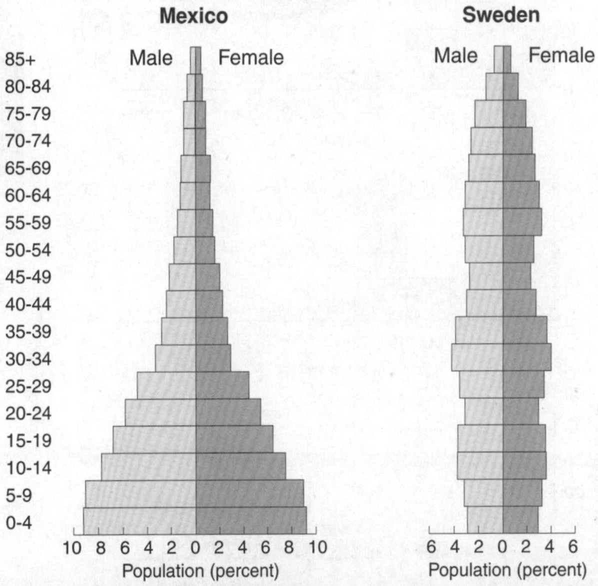

Most of the growth in human populations in recent years has occurred in the developing countries in Africa, Asia, and Central and South America. This is reflected in their age structure; a large proportion of the population comprises individuals younger than 15 years (age pyramid above, left). Even if each has fewer children, the population will continue to increase for many years. The stable age structure of Sweden is shown for comparison.

1. For the theoretical age pyramids above left:

   (a) State the approximate ratio of young to adults in a rapidly increasing population: ____________________

   (b) Suggest why changes in population age structure alone are not necessarily a reliable predictor of population trends:

   ____

   ____

2. Explain why the population of Mexico is likely to continue to increase rapidly even if the rate of population growth slows:

   ____

   ____

Analysis of the age structure of a population can assist in its management because it can indicate where most of the mortality occurs and whether or not reproductive individuals are being replaced. The age structure of both plant and animal populations can be examined; a common method is through an analysis of size which is often related to age in a predictable way.

Managed Fisheries

The graphs below illustrate the age structure of a hypothetical fish population under different fishing pressures. The age structure of the population is determined by analysing the fish catch to determine the frequency of fish in each size (age) class.

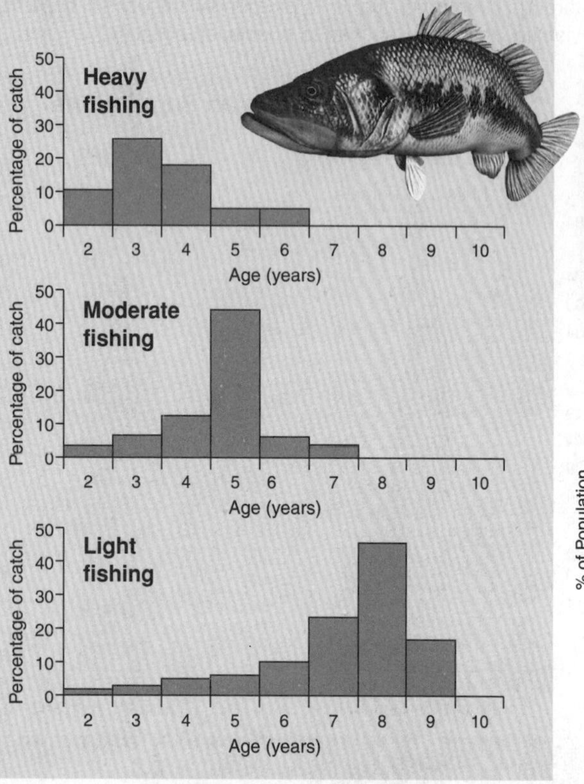

Thatch Palm Populations on Lord Howe Island

Lord Howe Island is a narrow sliver of land approximately 770 km northeast of Sydney. The age structure of populations of the thatch palm *Howea forsteriana* was determined at three locations on the island: the golf course, Grey Face and Far Flats. The height of the stem was used as an indication of age. The differences in age structure between the three sites are mainly due to the extent of grazing at each site.

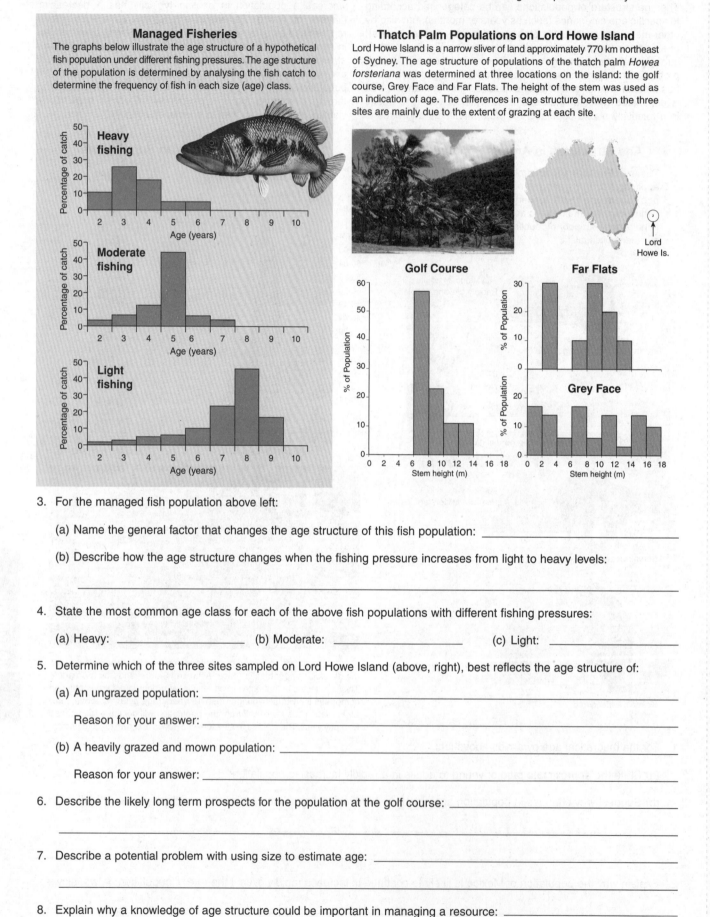

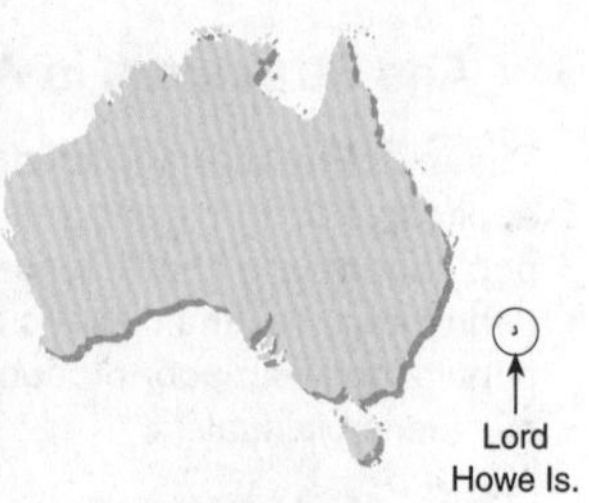

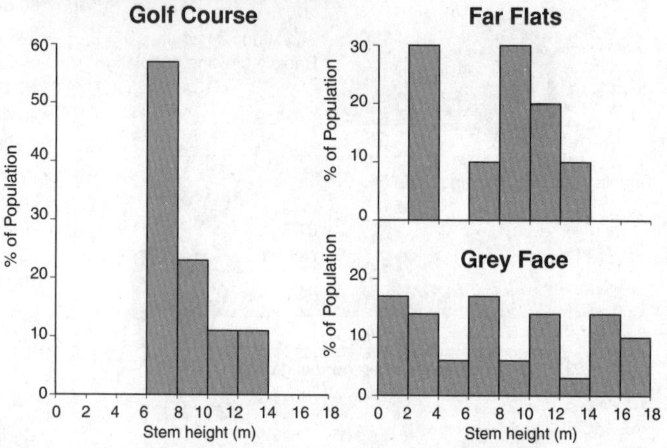

3. For the managed fish population above left:

 (a) Name the general factor that changes the age structure of this fish population: ______________________

 (b) Describe how the age structure changes when the fishing pressure increases from light to heavy levels:

 ____

4. State the most common age class for each of the above fish populations with different fishing pressures:

 (a) Heavy: ____________ (b) Moderate: ____________ (c) Light: ____________

5. Determine which of the three sites sampled on Lord Howe Island (above, right), best reflects the age structure of:

 (a) An ungrazed population: ______________________

 Reason for your answer: ______________________

 (b) A heavily grazed and mown population: ______________________

 Reason for your answer: ______________________

6. Describe the likely long term prospects for the population at the golf course: ______________________

 ____

7. Describe a potential problem with using size to estimate age: ______________________

 ____

8. Explain why a knowledge of age structure could be important in managing a resource: ______________________

 ____

World Population Growth

For most of human history, humans have not been very numerous compared to other species. It took all of human history to reach a population of 1 billion in 1804, but little more than 150 years to reach 3 billion in 1960. The world's population, now at 6.6 billion, is growing at the rate of about 80 million per year. This growth is slower than predicted but the world's population is still expected to increase substantially before stabilising (see Figure 1). World population increase carries important environmental consequences, particularly when it is associated with increasing **urbanisation**. Although the world as a whole still has an average fertility rate of 2.8, growth rates are now lower than at any time since World War II. Continued declines may give human populations time to address some to the major problems posed by the increasing the scope and intensity of human activities.

World Population Density

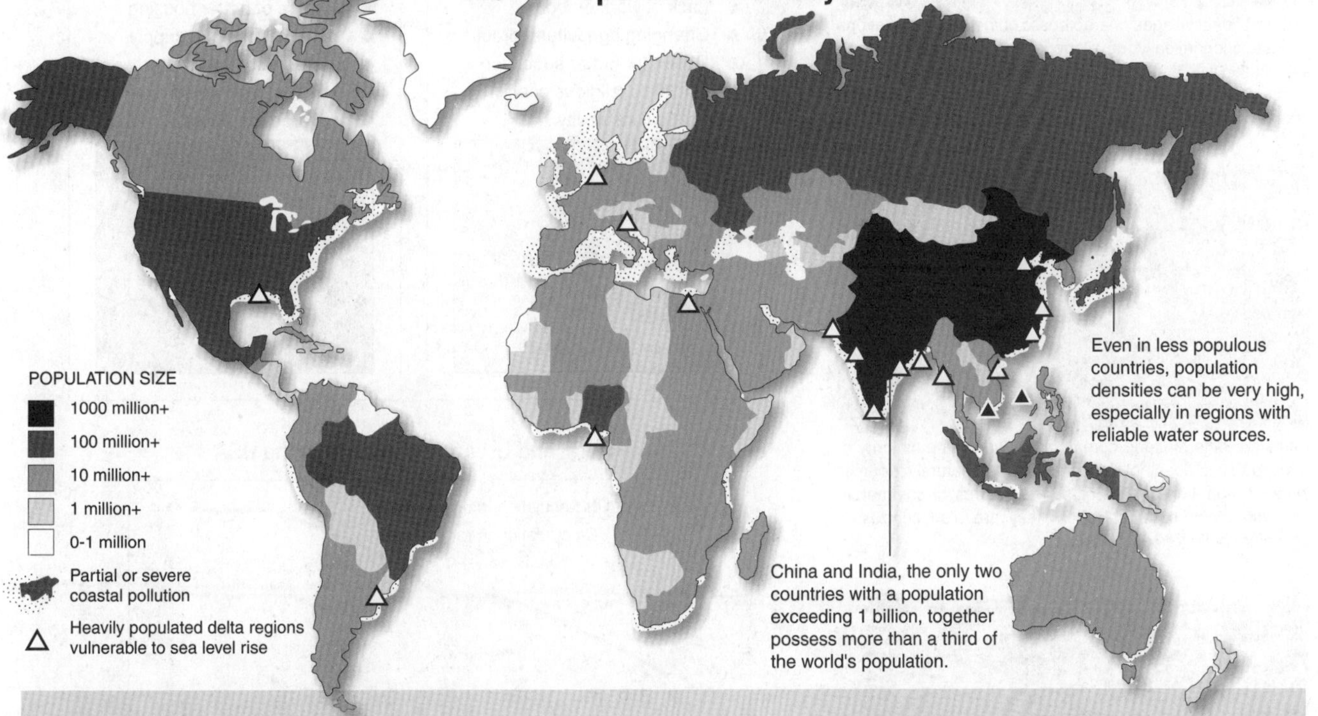

Figure 1: Progress towards population stabilisation

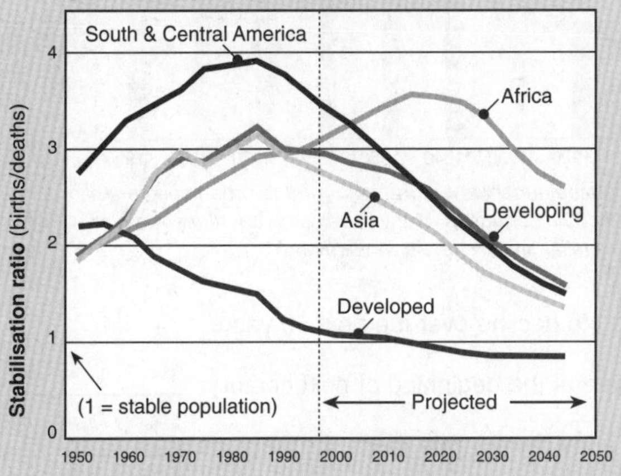

Figure 2: Trends in fertility rates

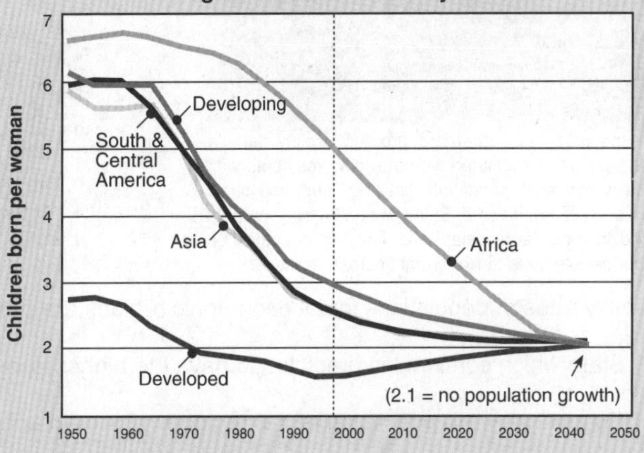

Figure 3: Global population growth

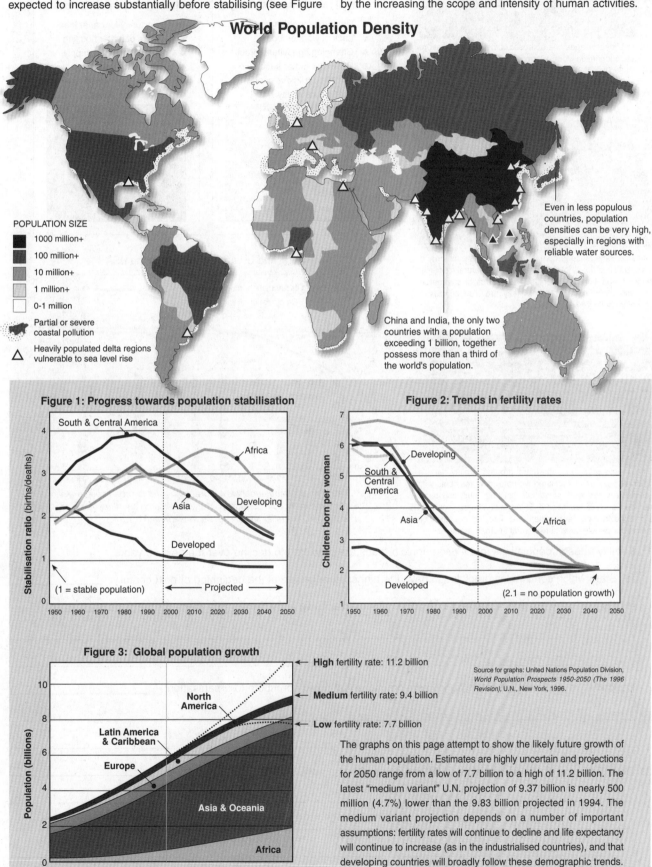

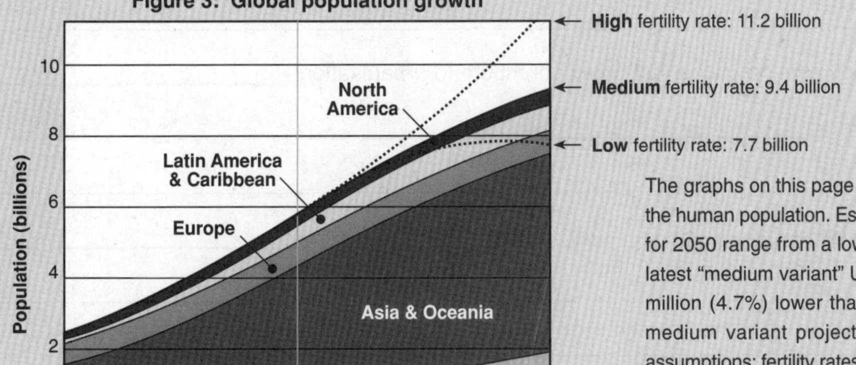

Source for graphs: United Nations Population Division, *World Population Prospects 1950-2050 (The 1996 Revision)*, U.N., New York, 1996.

The graphs on this page attempt to show the likely future growth of the human population. Estimates are highly uncertain and projections for 2050 range from a low of 7.7 billion to a high of 11.2 billion. The latest "medium variant" U.N. projection of 9.37 billion is nearly 500 million (4.7%) lower than the 9.83 billion projected in 1994. The medium variant projection depends on a number of important assumptions: fertility rates will continue to decline and life expectancy will continue to increase (as in the industrialised countries), and that developing countries will broadly follow these demographic trends.

Populations

Related activities: Population Growth, Population Age Structure
Web links: Human Impact: Overpopulation

The Shift to Urban Living

The traditional villages characteristic of the rural populations of less economically developed countries have a close association with the land. The households depend directly on agriculture or harvesting natural resources for their livelihood and are linked through family ties, culture, and economics.

Cities are differentiated communities, where the majority of the population does not depend directly on natural resource-based occupations. While cities are centres of commerce, education, and communication, they are also centres of crowding, pollution, and disease.

Cities, especially those that are growing rapidly, face a range of problems associated with providing residents with adequate water, food, sanitation, housing, jobs, and basic services, such as health care. Slums or squatter settlements are found in most large cities in developing countries as more poor people migrate from rural to urban areas.

The redistribution of people from rural to urban environments, or **urbanisation**, has been an important characteristic of human societies. Almost half of the people in the world already live in urban areas and by the end of the 21st century, this figure is predicted to increase to 80-90%. Urban populations can grow through natural increase (i.e. more births than deaths) or by **immigration**. Immigration is driven both by **push factors** that encourage people to leave their rural environment and **pull factors** that draw them to the cities.

Immigration push factors

- Rural overpopulation
- Lack of work or food
- Changing agricultural practices
- Desire for better education
- Racial or religious conflict
- Political instability

Immigration pull factors

- Opportunity for better jobs
- Chance of better housing
- More reliable food supply
- Opportunity for greater wealth
- Freedom from village traditions
- Government policy

Rural and Urban Populations in the USA

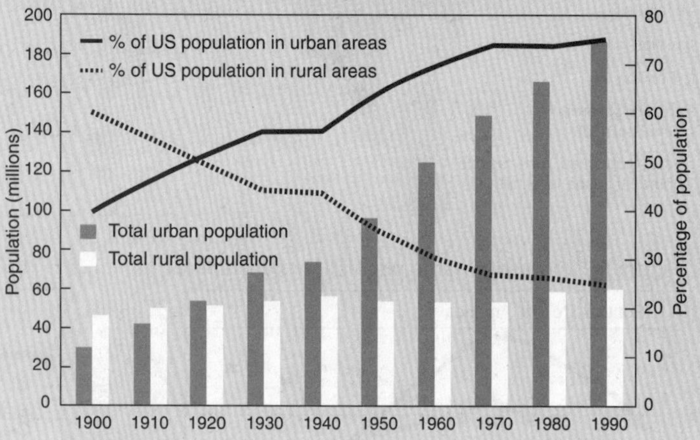

The United States underwent a dramatic rural to urban shift in the 19th and early 20th centturies. Many developing countries are now experiencing similar shifts. Graph compiled from UN data

1. Fertility rates of populations for all geographic regions are predicted to decline over the next 50 years.

   (a) State which continent is predicted to have the highest fertility rate at the beginning of next century: ______________

   (b) Suggest why the population of this region is slower to achieve a low fertility rate than other regions: ______________

   ____

   ____

2. Describe the kinds of changes in agricultural practices that could contribute to urbanisation: ______________

   ____

   ____

3. Describe some of the detrimental effects of urbanisation: ______________

   ____

   ____

   ____

   ____

Humans and Resources

The expanding human population puts increasing strain on the world's resources, creates problems of pollution and waste disposal, and often places other species at risk. Even when resources are potentially sufficient to meet demands, problems with distribution and supply create inequalities between regions. This page outlines some of the problems associated with the availability and use of resources. Although the world's situation might seem bleak, progress is being made towards a more sustainable future. Houses and cities are now being designed with greater energy efficiency in mind, steps are being taken to reduce greenhouse gas emissions and other forms of pollution, and sustainable agricultural practices are being increasingly encouraged. Advancements in technology combined with a political commitment to sustainability will add to these gains.

Air pollution contributes to global warming, ozone depletion, and acid rain, and is set to increase markedly in the next 30 years.

Global water consumption is rising rapidly. The availability of water is likely to become one of the most pressing resource issues of the 21st century.

Combustion engine emissions are increasing rapidly in Asia as their economies develop and their populations become more affluent.

A dwindling supply of fossil fuels provides about 85% of the world's commercial energy. Most of this is consumed by the richest countries.

Global climate change as a result of the greenhouse effect will cause a rise in sea levels and threaten coastal populations such as those in Bangladesh (above).

In industrialised societies, one person consumes many tonnes of raw materials each year, which must be extracted, processed, and then disposed of as waste.

Aquatic environments, such as coral reefs and freshwater habitats in lakes, rivers, and wetlands are at risk from population pressure (58% of the worlds reefs and 34% of all fish species).

Threats to biodiversity from all sources are rapidly reaching a critical level. Current extinction rates have increased 100 to 1000 times due to human impact on natural environments.

Consumption of natural resources, including fuels, water, and minerals, by modern industrial economies remains very high (in the range of 45 to 85 tonnes per person annually).

Forest fires and logging continue to cause shrinkage of the world's tropical and temperate forests. Deforestation in the Amazon doubled from 1994 to 1995 before declining in 1996.

The unsustainable fishing of the world's fish stocks has occurred in many fishing grounds (e.g. cod fishing in the North Atlantic). Many of these fish populations are unlikely to recover.

Although the world's food production is theoretically adequate to meet human needs there are problems with distribution. Some 800 million people remain undernourished.

1. Describe some of the predicted impacts of shortages of fossil fuel and freshwater on world agriculture:

2. "*If the world's resources were distributed more equably, there would be enough for all.*" Discuss this statement:

r and K Selection

Two parameters govern the logistic growth of populations: the intrinsic rate of natural increase or biotic potential (this is the maximum reproductive potential of an organism, symbolised by an italicised ***r***), and the carrying capacity (saturation density) of the environment (represented by the letter **K**). Species can be characterised by the relative importance of *r* and K in their life cycles. Species with a high intrinsic capacity for population increase are called ***r*-selected species**, and include algae, bacteria, rodents, many insects, and most annual plants. These species show life history features associated with rapid growth in disturbed environments. To survive, they must continually invade new areas to compensate for being replaced by more competitive species. In contrast, **K-selected** species, which include most large mammals, birds of prey, and large, long-lived plants, exist near the carrying capacity of their environments and are pushed in competitive environments to use resources more efficiently. These species have fewer offspring and longer lives, and put their energy into nuturing their young to reproductive age. Most organisms have reproductive patterns between these two extremes. Both *r*-selected species (crops) and K-selected species (livestock) are found in agriculture.

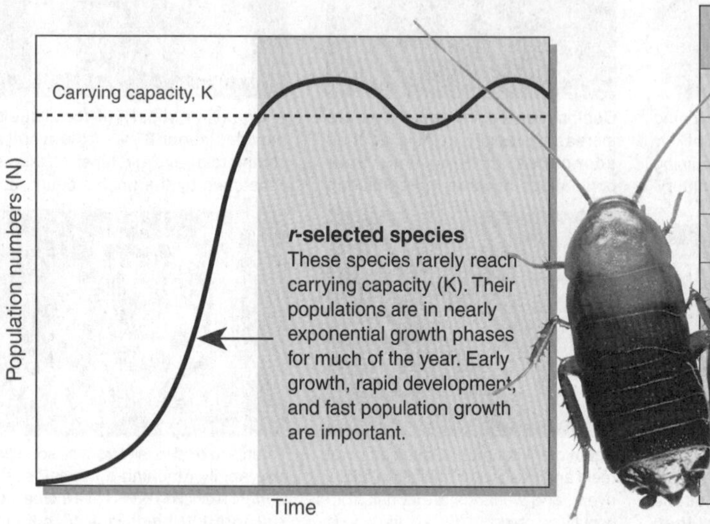
Carrying capacity, K
Population numbers (N)
Time
r-selected species
These species rarely reach carrying capacity (K). Their populations are in nearly exponential growth phases for much of the year. Early growth, rapid development, and fast population growth are important.

Correlates of *r*-Selected Species	
Climate	Variable and/or unpredictable
Mortality	Density independent
Survivorship	Often Type III (early loss)
Population size	Fluctuates wildly. Often below K.
Competition	Variable, often lax. Generalist niche.
Selection favours	Rapid development, high r_m, early reproduction, small body size, single reproduction (annual).
Length of life	Short (usually less than one year)
Leads to:	Productivity

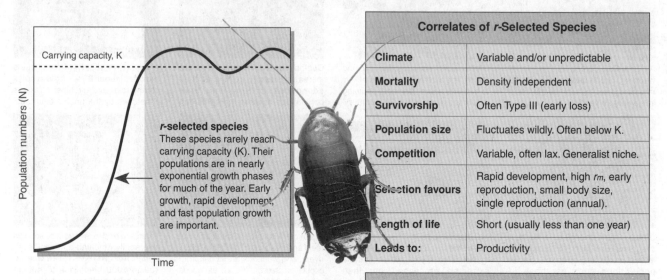

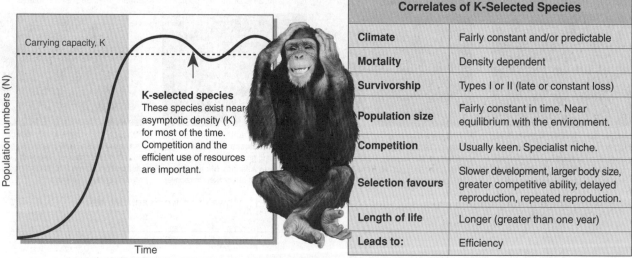

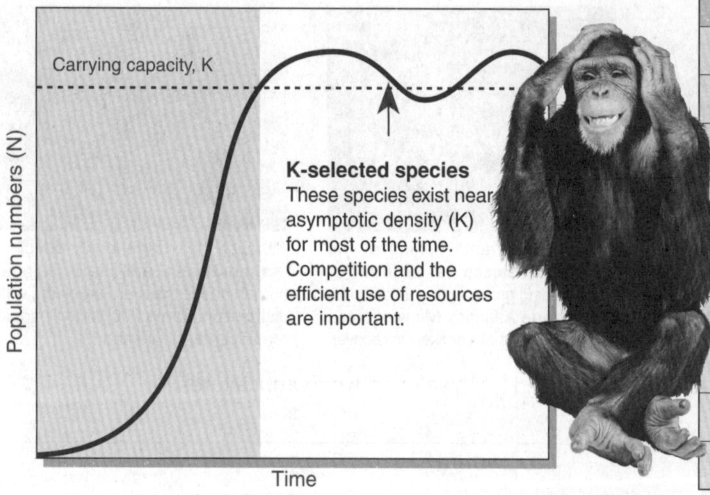
Carrying capacity, K
Population numbers (N)
Time
K-selected species
These species exist near asymptotic density (K) for most of the time. Competition and the efficient use of resources are important.

Correlates of K-Selected Species	
Climate	Fairly constant and/or predictable
Mortality	Density dependent
Survivorship	Types I or II (late or constant loss)
Population size	Fairly constant in time. Near equilibrium with the environment.
Competition	Usually keen. Specialist niche.
Selection favours	Slower development, larger body size, greater competitive ability, delayed reproduction, repeated reproduction.
Length of life	Longer (greater than one year)
Leads to:	Efficiency

1. Explain the significance of the *r* and the K notation when referring to ***r*** and **K selection**: ______________________________

2. Giving an example, explain why *r*-selected species tend to be **opportunists**: ______________________________

3. Explain why K-selected species are also called **competitor species**: ______________________________

4. Suggest why many K-selected species are often vulnerable to extinction: ______________________________

Related activities: Life tables and Survivorshp, Survivorship Curves, Population Growth Curves

Species Interactions

No organism exists in isolation. Each takes part in many interactions, both with other organisms and with the non-living components of the environment. Species interactions may involve only occasional or indirect contact (predation or competition) or they may involve close association or **symbiosis**. Symbiosis is a term that encompasses a variety of interactions involving close species contact. There are three types of symbiosis: **parasitism** (a form of exploitation), **mutualism**, and **commensalism**. Species interactions affect population densities and are important in determining community structure and composition. Some interactions, such as allelopathy, may even determine species presence or absence in an area.

Examples of Species Interactions

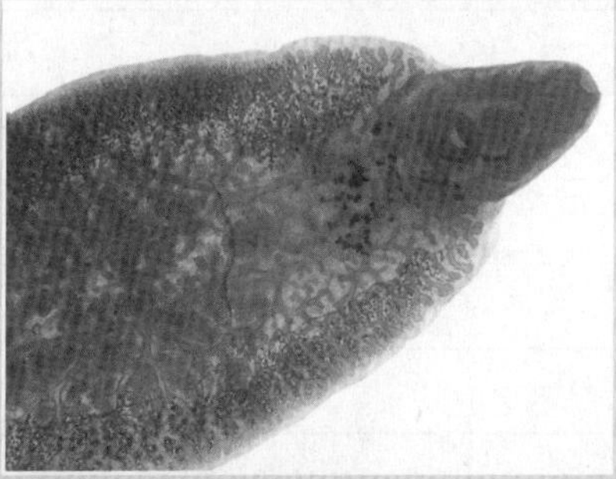

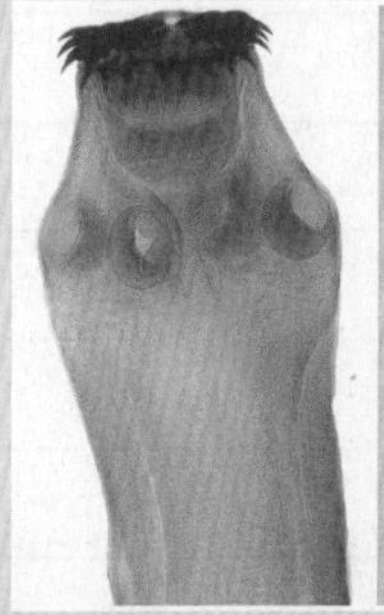

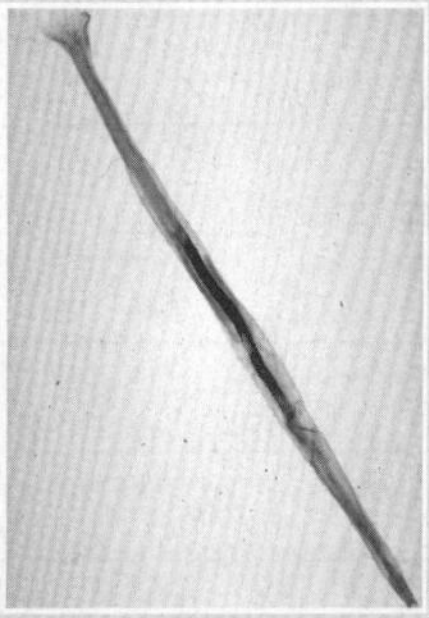

Parasitism is a common exploitative relationship in plants and animals. A parasite exploits the resources of its host (e.g. for food, shelter, warmth) to its own benefit. The host is harmed, but usually not killed. **Endoparasites**, such as liver flukes (left), tapeworms (centre) and nematodes (right)), are highly specialised to live inside their hosts, attached by hooks or suckers to the host's tissues.

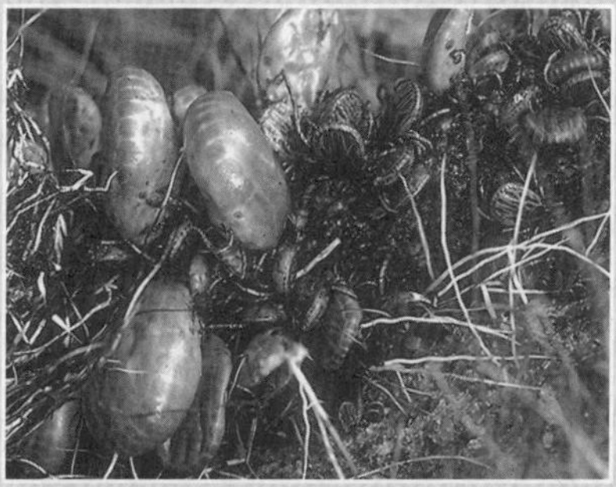

Ectoparasites, such as ticks (above), mites, and fleas, live attached to the outside of the host, where they suck body fluids, cause irritation, and may act as vectors for disease causing microorganisms.

Mutualism involves an intimate association between two species that offers advantage to both. **Lichens** (above) are the result of a mutualism between a fungus and an alga (or cyanobacterium).

Termites have a mutualistic relationship with the cellulose digesting bacteria in their guts. A similar mutualistic relationship exists between ruminants and their gut microflora of bacteria and ciliates.

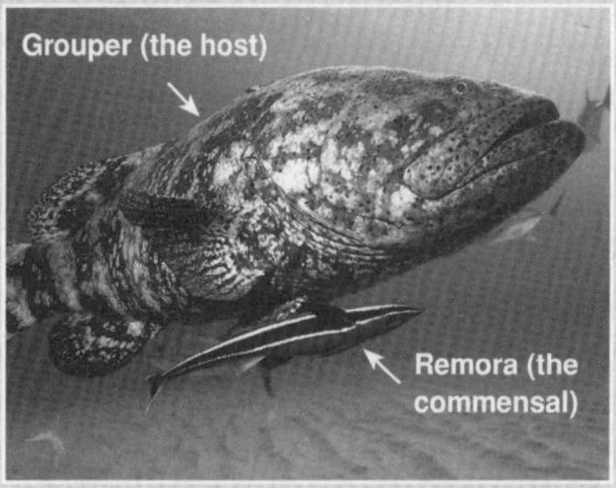

In **commensal** relationships, such as between this large grouper and a remora, two species form an association where one organism, the commensal, benefits and the other is neither harmed or helped.

Many species of decapod crustaceans, such as this anemone shrimp, are commensal with sea anemones. The shrimp gains by being protected from predators by the anemone's tentacles.

Interactions involving **competition** for the same food resources are dominated by the largest, most aggressive species. Here, hyaenas compete for a carcass with vultures and maribou storks.

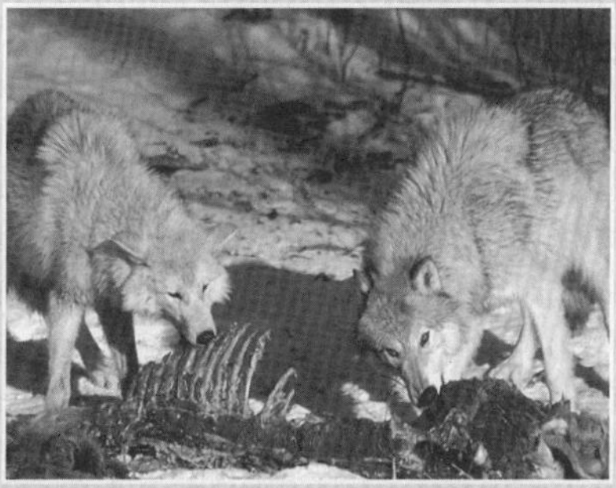

Predation is an easily identified relationship, as one species kills and eats another (above). Herbivory is similar type of exploitation, except that the plant is usually not killed by the herbivore.

1. Discuss each of the following interspecific relationships, including reference to the species involved, their role in the interaction, and the specific characteristics of the relationship:

   (a) **Mutualism** between ruminant herbivores and their gut microflora: ______________________________

   ____

   ____

   ____

   ____

Related activities: Interspecific Competition, Intraspecific Competition
Web links: Ecological Interactions from EcoLibrary, Nearctica Ecology

(b) **Commensalism** between a shark and a remora: ___

(c) **Parasitism** between a tapeworm and its human host: ___

(d) **Parasitism** between a cat flea and its host: ___

2. Summarise your knowledge of species interactions by completing the following, entering a (+), (–), or (0) for species B, and writing a brief description of each term. Codes: (+): species benefits, (–): species is harmed, (0): species is unaffected.

Interaction	Species A	Species B	Description of relationship
(a) Mutualism	+		
(b) Commensalism	+		
(c) Parasitism	–		
(d) Amensalism	0		
(e) Predation	–		
(f) Competition	–		
(g) Herbivory	+		
(h) Antibiosis	+ / 0		

3. For each of the interactions between two species described below, choose the correct term to describe the interaction and assign a +, – or 0 for each species involved in the space supplied. Use the completed table above to help you:

Description	Term	Species A	Species B
(a) A tiny cleaner fish picking decaying food from the teeth of a much larger fish (e.g. grouper).	Mutualism	Cleaner fish +	Grouper +
(b) Ringworm fungus growing on the skin of a young child.		Ringworm	Child
(c) Human effluent containing poisonous substances killing fish in a river downstream of discharge.		Humans	Fish
(d) Humans planting cabbages to eat only to find that the cabbages are being eaten by slugs.		Humans	Slugs
(e) A shrimp that gets food scraps and protection from sea anemones, which appear to be unaffected.		Shrimp	Anemone
(f) Birds follow a herd of antelopes to feed off disturbed insects, antelopes alerted to danger by the birds.		Birds	Antelope

Interspecific Competition

In naturally occurring populations, direct competition between different species (**interspecific competition**) is usually less intense than intraspecific competition because coexisting species have evolved slight differences in their realised niches, even though their fundamental niches may overlap (a phenomenon termed **niche differentiation**). This has been well documented in barnacle species (below). However, when two species with very similar niche requirements are brought into direct competition through the introduction of a foreign species, one usually benefits at the expense of the other, which is excluded (the **competitive exclusion principle**). In Australia, the introduction of foreign, ecologically aggressive species is implicated in the competitive displacement and decline of many native species (see the examples of mallard ducks and mosquito fish, below). Displacement of native species by introduced ones is more likely if the introduced competitor is also adaptable and hardy. It can be difficult to provide evidence of decline in a species as a direct result of competition, but it can often be inferred if the range of the native species contracts and that of the introduced competitor shows a corresponding increase.

Competitive Exclusion in Barnacles

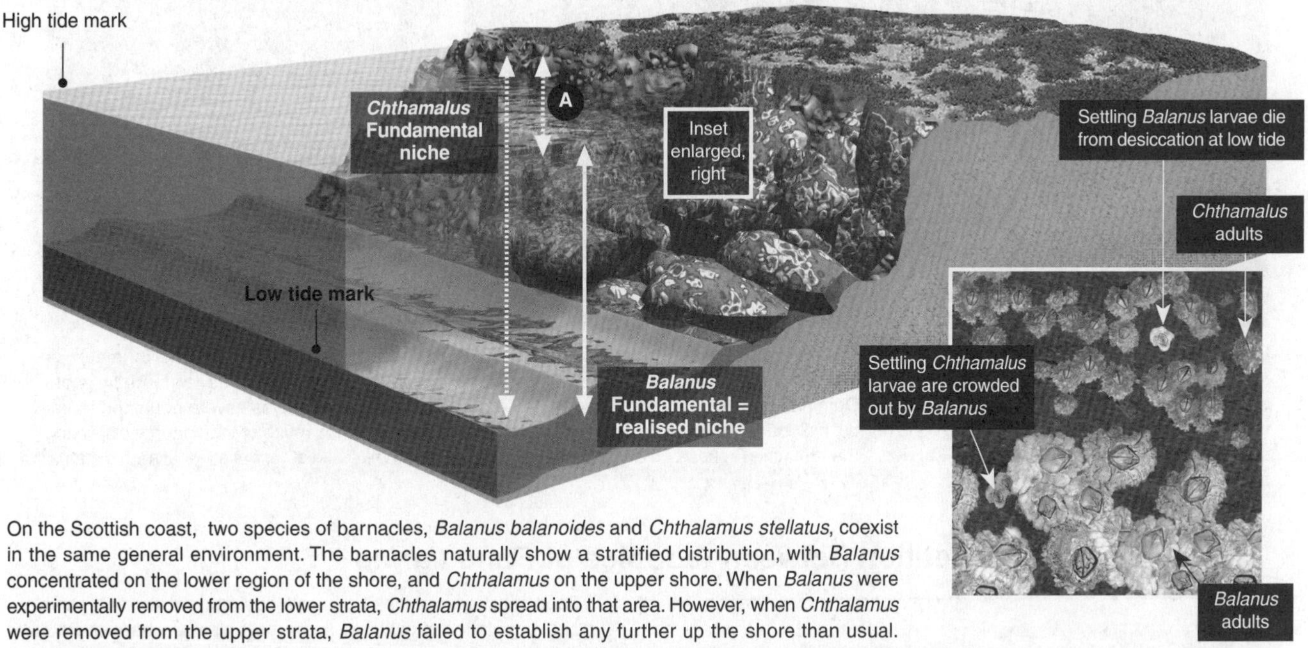

On the Scottish coast, two species of barnacles, *Balanus balanoides* and *Chthalamus stellatus*, coexist in the same general environment. The barnacles naturally show a stratified distribution, with *Balanus* concentrated on the lower region of the shore, and *Chthalamus* on the upper shore. When *Balanus* were experimentally removed from the lower strata, *Chthalamus* spread into that area. However, when *Chthalamus* were removed from the upper strata, *Balanus* failed to establish any further up the shore than usual.

In Australia, the adaptable mallard duck, *Anas platyrhynchos* (left) is at least partly responsible for the decline in native Pacific black ducks (*Anas superciliosa*). Mallards are bigger than the native ducks and can physically bully them in competing for food. They also breed more profusely, with an average clutch size of 11 eggs (compared with 8 in blacks). Mallard males are also very sexually aggressive and will interbreed with blacks to form fertile hybrids so that the pool of "pure bred" blacks is diminished.

Throughout the world, the introduction of mosquito fish, *Gambusia affinis* (left) is implicated in the decline of endemic fish and amphibian species. *Gambusia* is an aggressive, opportunistic species, with a strong competitive advantage. In Australia, evidence suggests that the presence of *Gambusia* results in a decline in native fish and frog populations (e.g. the green and golden bell frog). *Gambusia* are prolific breeders, compete for food and habitat, and prey on the immature stages of native species.

1. (a) In the example of the barnacles (above), describe what is represented by the zone labelled with the arrow **A**:

   (b) Explain the evidence for the barnacle distribution being the result of competitive exclusion:

2. Describe two aspects of the biology of a named introduced species that have helped its success as an invading competitor:

   (a)

   (b)

Intraspecific Competition

Some of the most intense competition occurs between individuals of the same species (**intraspecific competition**). Most populations have the capacity to grow rapidly, but their numbers cannot increase indefinitely because environmental resources are finite. Every ecosystem has a **carrying capacity** (K), defined as the number of individuals in a population that the environment can support. Intraspecific competition for resources increases with increasing population size and, at carrying capacity, it reduces the per capita growth rate to zero. When the demand for a particular resource (e.g. food, water, nesting sites, nutrients, or light) exceeds supply, that resource becomes a **limiting factor**. Populations respond to resource limitation by reducing their population growth rate (e.g. through lower birth rates or higher mortality). The response of individuals to limited resources varies depending on the organism. In many invertebrates and some vertebrates such as frogs, individuals reduce their growth rate and mature at a smaller size. In some vertebrates, territoriality spaces individuals apart so that only those with adequate resources can breed. When resources are very limited, the number of available territories will decline.

Intraspecific Competition

Scramble competition in caterpillars

Direct competition for available food between members of the same species is called **scramble competition**. In some situations where scramble competition is intense, none of the competitors gets enough food to survive.

Contest competition in wolves

In some cases, competition is limited by hierarchies existing within a social group. Dominant individuals receive adequate food, but individuals low in the hierarchy must **contest** the remaining resources and may miss out.

Display of a male anole

Intraspecific competition may be for mates or breeding sites, as well as for food. In anole lizards (above), males have a bright red throat pouch and use much of their energy displaying to compete with other males for available mates.

Competition Between Tadpoles of *Rana tigrina*

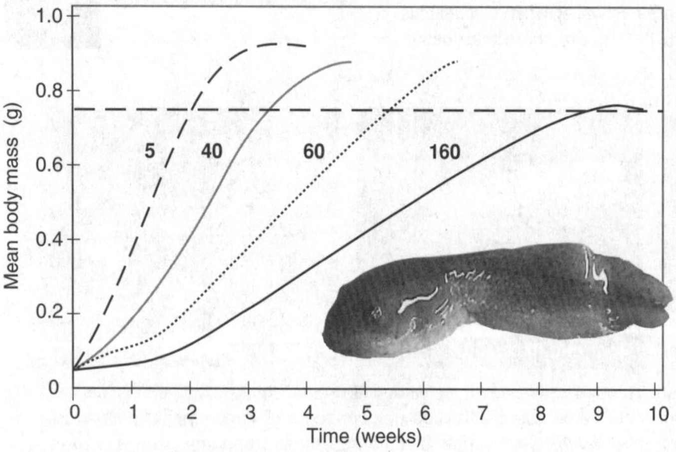

Both figures: Smith & Smith: Ecology and field biology, 2001

Food shortage reduces both individual growth rate and survival, and population growth. In some organisms, where there is a metamorphosis or a series of moults before adulthood (e.g. frogs, crustacean zooplankton, and butterflies), individuals may die before they mature.

The graph (left) shows how the growth rate of tadpoles (*Rana tigrina*) declines as the density increases from 5 to 160 individuals (in the same sized space).

- At high densities, tadpoles grow more slowly, taking longer to reach the minimum size for metamorphosis (0.75 g), and decreasing their chances of successfully metamorphosing from tadpoles into frogs.
- Tadpoles held at lower densities grow faster, to a larger size, metamorphosing at an average size of 0.889 g.
- In some species, such as frogs and butterflies, the adults and juveniles reduce the intensity of intraspecific competition by exploiting different food resources.

1. Using an example, predict the likely effects of **intraspecific competition** on each of the following:

   (a) Individual growth rate: ______________________

   (b) Population growth rate: ______________________

   (c) Final population size: ______________________

2. In the tank experiment with *Rana* (above), the tadpoles were contained in a fixed volume with a set amount of food:

   (a) Describe how *Rana* tadpoles respond to resource limitation: ______________________

   ____

   (b) Categorise the effect on the tadpoles as density-dependent / density-independent (delete one).

   (c) Comment on how much the results of this experiment are likely to represent what happens in a natural population:

   ____

Investigating Ecosystems

Practical investigation of ecosystem structure and function

Field studies: sampling populations and measuring abiotic factors.

Learning Objectives

☐ 1. Compile your own glossary from the **KEY WORDS** displayed in **bold type** in the learning objectives below.

Sampling Populations *(pages 82, 85-96)*

☐ 2. Identify the type of information that can be gained from population studies (e.g. **abundance**, **density**, **age structure**, **distribution**). Explain why we **sample** populations and describe the advantages and drawbacks involved.

☐ 3. A field study should enable you to test a hypothesis about a certain aspect of a population. You should provide an outline of your study including reference to the type of data you will collect and the methods you will use, the size of your sampling unit (e.g. quadrat size) and the number of samples you will take, the assumptions of your investigation, and controls.

☐ 4. Explain how and why sample size affects the accuracy of population estimates. Explain how you would decide on a suitable sample size. Discuss the compromise between sampling accuracy and sampling effort.

☐ 5. Describe the following techniques used to study aspects of populations (e.g. **distribution**, **abundance**, **density**). Identify the advantages and limitations of each method with respect to sampling time, cost, and the suitability to the organism and specific habitat type:
(a) **Direct counts**
(b) **Frame** and/or **point quadrats**
(c) **Belt** and/or **line transects**
(d) **Mark and recapture** and the **Lincoln index**
(e) **Netting** and **trapping**

☐ 6. Recognise the value to population studies of **radio-tracking** and **indirect methods** of sampling such as counting nests, and recording calls and droppings.

☐ 7. Describe the methods used to ensure **random sampling**, and appreciate why this is important.

☐ 8. Describe **qualitative methods** for investigating the distribution of organisms in specific habitats.

☐ 9. Recognise appropriate ways in which different types of data may be recorded, analysed, and presented.

☐ 10. Demonstrate an ability to calculate and use simple statistics (**mean** and **standard deviation**) for the analysis and comparison of population data.

☐ 11. Calculate simple statistical tests, such as the chi-squared and student's *t* test, and apply them appropriately to the analysis and comparison of population data. Recognise that the design of any field study will determine how the data can be analysed.

☐ 12. Describe how the study of ecosystems (both biotic and physical factors) can be used in a practical way to assess ecosystem change.

Measuring Abiotic Factors *(pages 83-84)*

☐ 13. Describe methods to measure abiotic factors in a habitat. Include reference to the following (as appropriate): pH, light, temperature, dissolved oxygen, current speed, total dissolved solids, and conductivity.

☐ 14. Appreciate the influence of abiotic factors on the distribution and abundance of organisms in a habitat.

Supplementary Texts

See page 7 for additional details of these texts:

■ Miller, G.T, 2007. **Living in the Environment: Principles, Connections and Solutions**, see various appendices.

■ Reiss, M. & J. Chapman, 2000. **Environmental Biology** (Cambridge Uni. Press), pp. 6-14, 22-25.

■ Smith, R. L. & T.M. Smith, 2001. **Ecology and Field Biology**, reading as required.

Presentation MEDIA to support this topic:

ECOLOGY
• Practical Ecology

Periodicals

See page 7 for details of publishers of periodicals:

STUDENT'S REFERENCE

■ **Ecological Projects** Biol. Sci. Rev., 8(5) May 1996, pp. 24-26. *A thorough guide to planning and carrying out a field-based project.*

■ **Fieldwork - Sampling Animals** Biol. Sci. Rev., 10(4) March 1998, pp. 23-25. *Appropriate methodology for collecting animals in the field.*

■ **Fieldwork Sampling - Plants** Biol. Sci. Rev., 10(5) May 1998, pp. 6-8. *Excellent article covering the methodology for sampling plant communities.*

■ **Bird Ringing** Biol. Sci. Rev., 14(3) Feb. 2002, pp. 14-19. *Techniques used in investigating populations of highly mobile organisms: mark and recapture, ringing techniques, and application of diversity indices.*

■ **British Butterflies in Decline** Biol. Sci. Rev., 14(4) April 2002, pp. 10-13. *Documented changes in the distribution of British butterfly species. This account includes a description of the techniques used to monitor changes in population numbers.*

■ **Bowels of the Beasts** New Scientist, 22 August 1998, pp. 36-39. *Analyses of the faeces of animals can reveal much about the make-up, size, and genetic diversity of a population.*

TEACHER'S REFERENCE

■ **Ecology Fieldwork in 16 to 19 Biology** SSR, 84(307) December 2002, pp. 87. *Examines the fieldwork opportunities provided to 16-19 students and suggests how these could be enhanced and firmly established for all students studying biology.*

Internet

See pages 4-5 for details of how to access **Bio Links** from our web site: **www.thebiozone.com** From Bio Links, access sites under the topics:

ECOLOGY: > **Environmental Monitoring**: • Remote sensing and monitoring ... *and others* > **Populations and Communities**: • Communities • Quantitative population ecology • Sirtracking for wildlife research

Sampling Populations

Information about the populations of rare organisms in isolated populations may, in some instances, be collected by direct measure (direct counts and measurements of all the individuals in the population). However, in most cases, populations are too large to be examined directly and they must be sampled in a way that still provides information about them. Most practical exercises in population ecology involve the collection or census of living organisms, with a view to identifying the species and quantifying their abundance and other population features of interest. Sampling techniques must be appropriate to the community being studied and the information you wish to obtain. Some of the common strategies used in ecological sampling, and the situations for which they are best suited, are outlined in the table below. It provides an overview of points to consider when choosing a sampling regime. One must always consider the time and equipment available, the organisms involved, and the impact of the sampling method on the environment. For example, if the organisms involved are very mobile, sampling frames are not appropriate. If it is important not to disturb the organisms, observation alone must be used to gain information.

Method	Equipment and procedure	Information provided and considerations for use
Point sampling 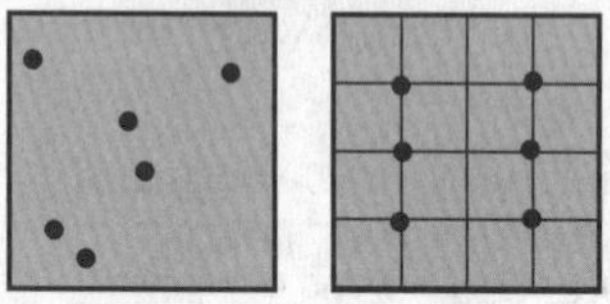 Random; Systematic (grid)	Individual points are chosen on a map (using a grid reference or random numbers applied to a map grid) and the organisms are sampled at those points. Mobile organisms may be sampled using traps, nets etc.	**Useful for:** Determining species abundance and community composition. If samples are large enough, population characteristics (e.g. age structure, reproductive parameters) can be determined. **Considerations**: Time efficient. Suitable for most organisms. Depending on method, environmental disturbance is minimal. Species occurring in low abundance may be missed.
Transect sampling 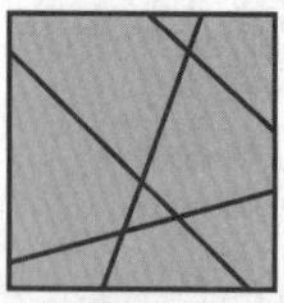 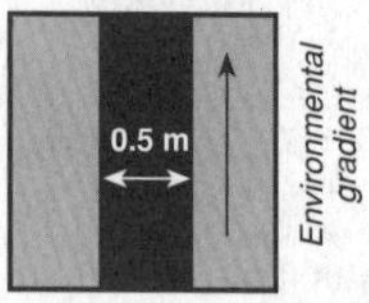	Lines are drawn across a map and organisms occurring along the line are sampled. **Line transects**: Tape or rope marks the line. The species occurring on the line are recorded (all along the line or, more usually, at regular intervals). Lines can be chosen randomly (left) or may follow an environmental gradient. **Belt transects**: A measured strip is located across the study area to highlight any transitions. Quadrats are used to sample the plants and animals at regular intervals along the belt. Plants and immobile animals are easily recorded. Mobile or cryptic animals need to be trapped or recorded using appropriate methods.	**Useful for**: Well suited to determining changes in community composition along an environmental gradient. When placed randomly, they provide a quick measure of species occurrence. **Considerations for line transects**: Time efficient. Most suitable for plants and immobile or easily caught animals. Disturbance to the environment can be minimised. Species occurring in low abundance may be missed. **Considerations for belt transects**: Time consuming to do well. Most suitable for plants and immobile or easily caught animals. Good chance of recording most or all species. Efforts should be made to minimise disturbance to the environment.
Quadrat sampling 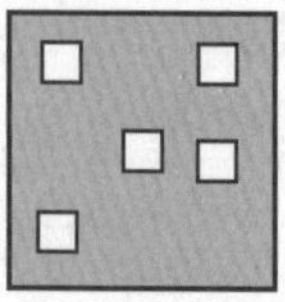	Sampling units or quadrats are placed randomly or in a grid pattern on the sample area. The occurrence of organisms in these squares is noted. Plants and slow moving animals are easily recorded. Rapidly moving or cryptic animals need to be trapped or recorded using appropriate methods.	**Useful for**: Well suited to determining community composition and features of population abundance: species density, frequency of occurrence, percentage cover, and biomass (if harvested). **Considerations**: Time consuming to do well. Most suitable for plants and immobile or easily caught animals. Quadrat size must be appropriate for the organisms being sampled and the information required. Disturbing if organisms are removed.
Mark and recapture (capture-recapture) 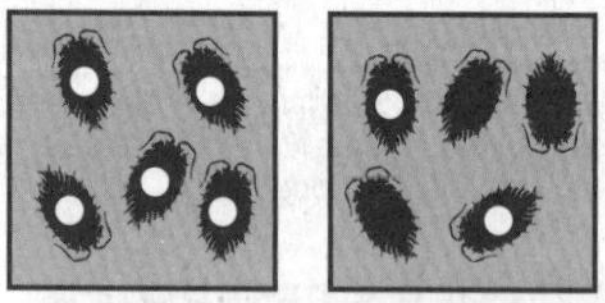 First sample: marked; Second sample: proportion recaptured	Animals are captured, marked, and then released. After a suitable time period, the population is resampled. The number of marked animals recaptured in a second sample is recorded as a proportion of the total.	**Useful for**: Determining total population density for highly mobile species in a certain area (e.g. butterflies). Movements of individuals in the population can be tracked (especially when used in conjunction with electronic tracking devices). **Considerations**: Time consuming to do well. Not suitable for immobile species. Population should have a finite boundary. Period between samplings must allow for redistribution of marked animals in the population. Marking should present little disturbance and should not affect behaviour.

1. Explain why we **sample** populations: ______________________________

____

2. Describe a sampling technique that would be appropriate for determining each of the following:

(a) The percentage cover of a plant species in pasture: ______________________________

(b) The density and age structure of a plankton population: ______________________________

(c) Change in community composition from low to high altitude on a mountain: ______________________________

Monitoring Physical Factors

Most ecological studies require us to measure the physical factors (parameters) in the environment that may influence the abundance and distribution of organisms. In recent years there have been substantial advances in the development of portable, light-weight meters and dataloggers. These enable easy collection and storage of data in the field.

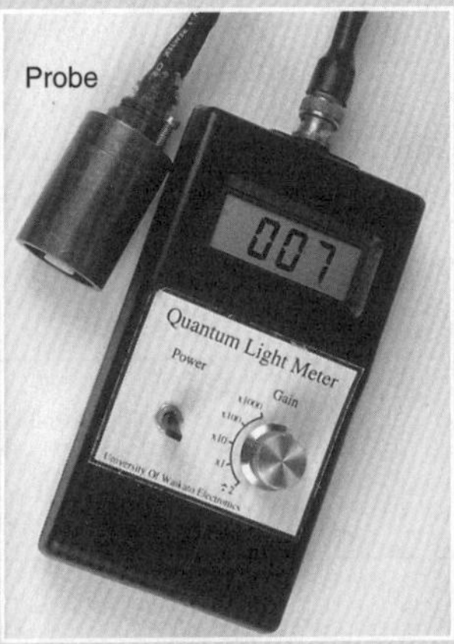

Quantum light meter: Measures light intensity levels. It is not capable of measuring light quality (wavelength).

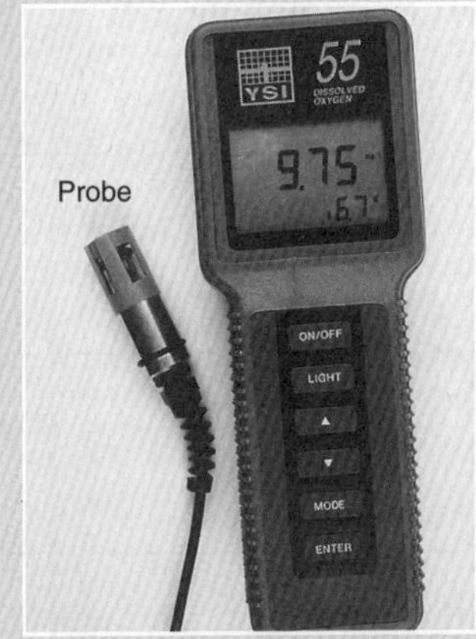

Dissolved oxygen meter: Measures the amount of oxygen dissolved in water (expressed as mgl^{-1}).

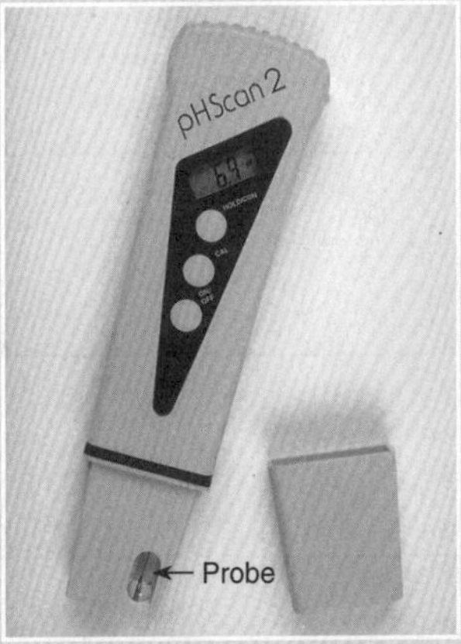

pH meter: Measures the acidity of water or soil, if it is first dissolved in pure water (pH scale 0 to 14).

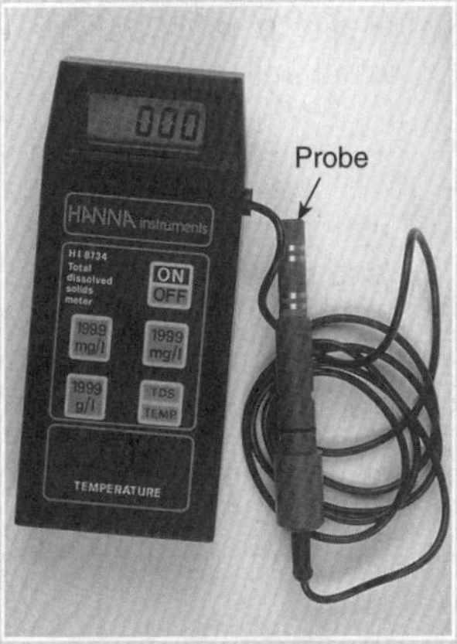

Total dissolved solids (TDS) meter: Measures content of dissolved solids (as ions) in water in mgl^{-1}.

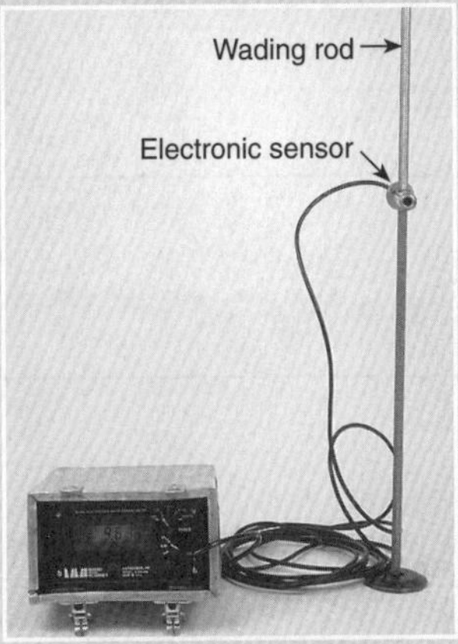

Current meter: The electronic sensor is positioned at set depths in a stream or river on the calibrated wading rod as current readings are taken.

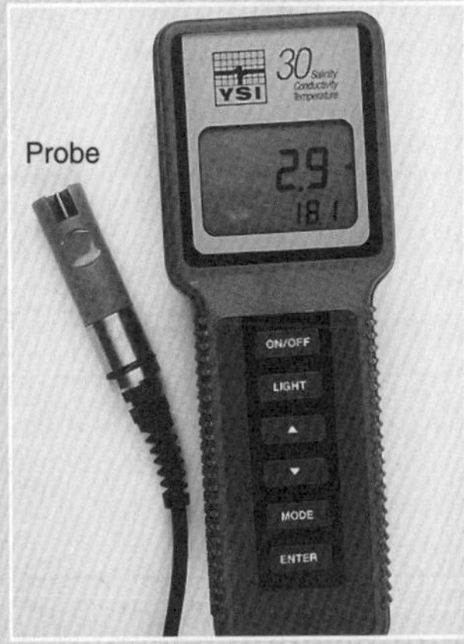

Multipurpose meter: This is a multi-functional meter, which can measure salinity, conductivity and temperature simply by pushing the MODE button.

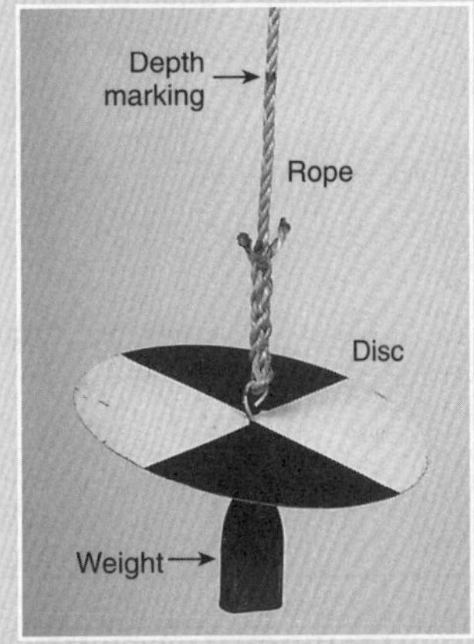

Secchi disc: This simple device is used to provide a crude measure of water clarity (the maximum depth at which the disc can just be seen).

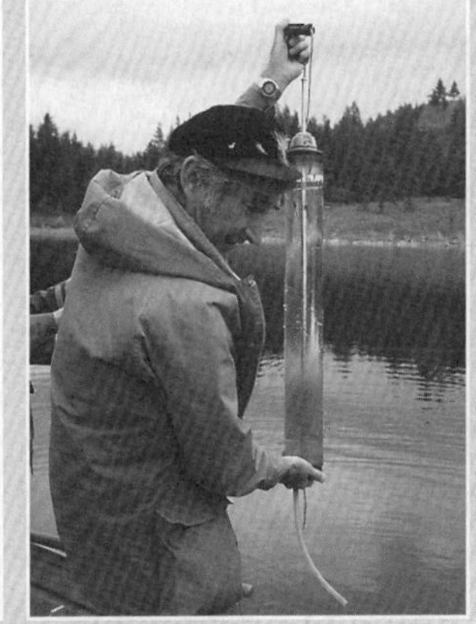

Collecting a water sample: A Nansen bottle is used to collect water samples from a lake for lab analysis, testing for nutrients, oxygen and pH.

This photo JDG. Others, Campus photography, University of Waikato

Dataloggers and Environmental Sensors

Dataloggers are electronic instruments that record measurements over time. They are equipped with a microprocessor, data storage facility, and sensor. Different sensors are employed to measure a range of variables in water (photos A and B) or air (photos C and D), as well as make physiological measurements. The datalogger is connected to a computer, and software is used to set the limits of operation (e.g. the sampling interval) and initiate the logger. The logger is then disconnected and used remotely to record and store data. When reconnected to the computer, the data are downloaded, viewed, and plotted. Dataloggers, such as those pictured here from PASCO, are being increasingly used in professional and school research. They make data collection quick and accurate, and they enable prompt data analysis.

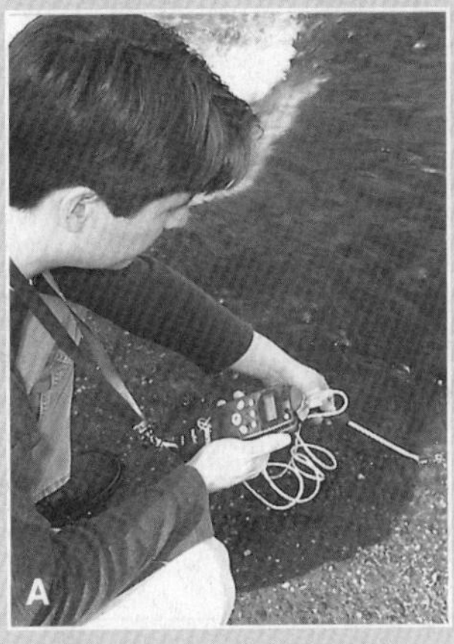

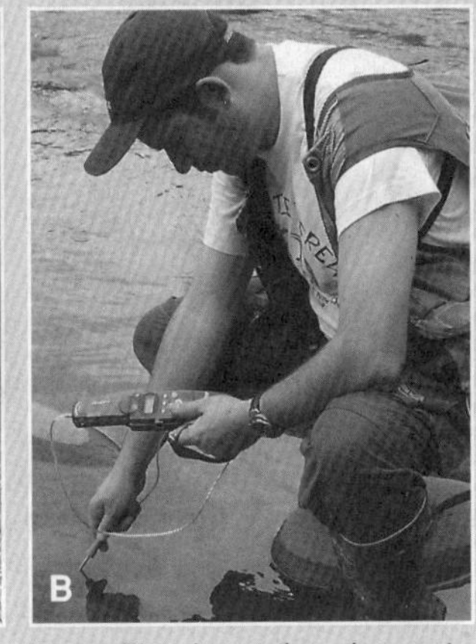

Dataloggers are now widely used to monitor conditions in aquatic environments. Different variables such as pH, temperature, conductivity, and dissolved oxygen can be measured by changing the sensor attached to the logger.

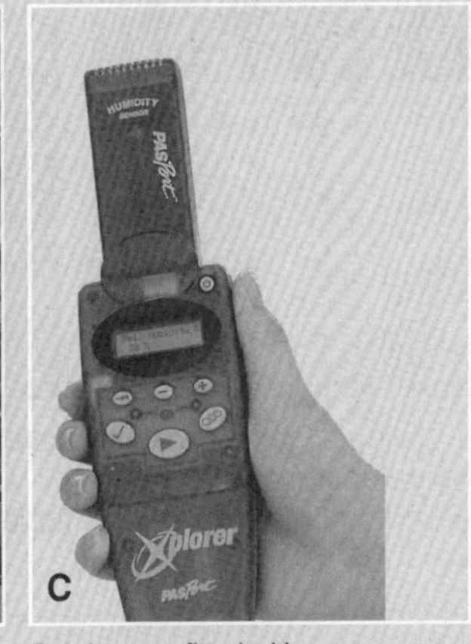

Dataloggers fitted with sensors are portable and easy to use in a wide range of terrestrial environments. They are used to measure variables such as air temperature and pressure, relative humidity, light, and carbon dioxide gas.

Datalogger photos courtesy of PASCO

1. The physical factors of an exposed rocky shore and a sheltered estuarine mudflat differ markedly. For each of the factors listed in the table below, briefly describe how they may differ (if at all):

Environmental parameter	Exposed rocky coastline	Estuarine mudflat
Severity of wave action		
Light intensity and quality		
Salinity/ conductivity		
Temperature change (diurnal)		
Substrate/ sediment type		
Oxygen concentration		
Exposure time to air (tide out)		

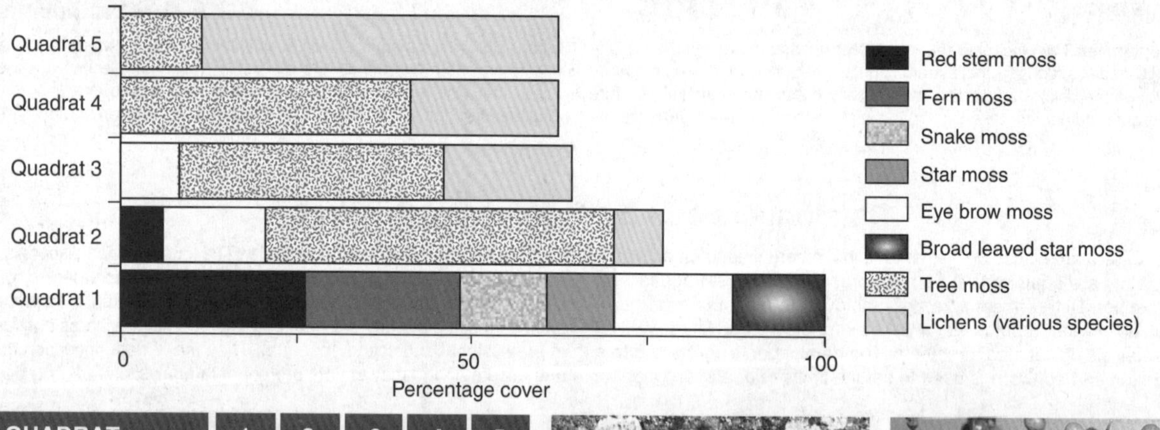

QUADRAT	1	2	3	4	5
Height (m)	0.4	0.8	1.2	1.6	2.0
Light (arbitrary units)	40	56	68	72	72
Humidity (percent)	99	88	80	76	78
Temperature (°C)	12.1	12.2	13	14.3	14.2

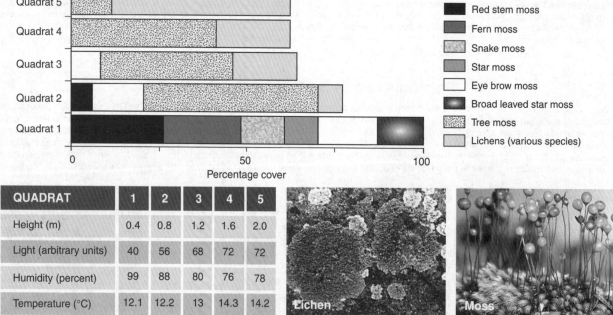

2. The figure (above) shows the changes in vegetation cover along a 2 m vertical transect up the trunk of an oak tree (*Quercus*). Changes in the physical factors light, humidity, and temperature along the same transect were also recorded. From what you know about the ecology of mosses and lichens, account for the observed vegetation distribution:

Quadrat Sampling

Quadrat sampling is a method by which organisms in a certain proportion (sample) of the habitat are counted directly. As with all sampling methods, it is used to estimate population parameters when the organisms present are too numerous to count in total. It can be used to estimate population **abundance** (number), **density**, **frequency of occurrence**, and **distribution**. Quadrats may be used without a transect when studying a relatively uniform habitat. In this case, the quadrat positions are chosen randomly using a random number table.

The general procedure is to count all the individuals (or estimate their percentage cover) in a number of quadrats of known size and to use this information to work out the abundance or percentage cover value for the whole area. The number of quadrats used and their size should be appropriate to the type of organism involved (e.g. grass vs tree).

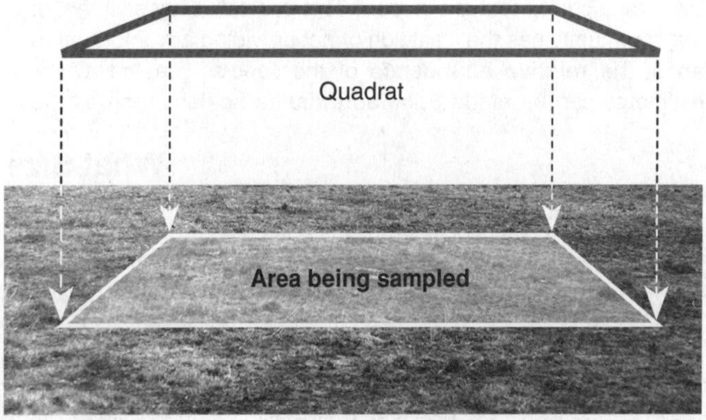
Quadrat
Area being sampled

$$\text{Estimated average density} = \frac{\text{Total number of individuals counted}}{\text{Number of quadrats} \times \text{area of each quadrat}}$$

Guidelines for Quadrat Use:

1. The **area of each quadrat** must be known exactly and ideally quadrats should be the same shape. The quadrat does not have to be square (it may be rectangular, hexagonal etc.).
2. **Enough quadrat samples** must be taken to provide results that are representative of the total population.
3. The **population of each quadrat** must be known exactly. Species must be distinguishable from each other, even if they have to be identified at a later date. It has to be decided beforehand what the count procedure will be and how organisms over the quadrat boundary will be counted.
4. The size of the quadrat should be appropriate to the organisms and habitat, e.g. a large size quadrat for trees.
5. The quadrats must be **representative of the whole area**. This is usually achieved by **random sampling** (right).

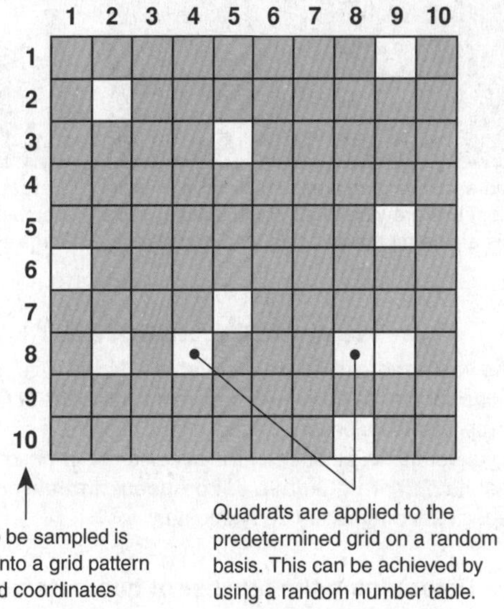
1 2 3 4 5 6 7 8 9 10
1 2 3 4 5 6 7 8 9 10
The area to be sampled is divided up into a grid pattern with indexed coordinates
Quadrats are applied to the predetermined grid on a random basis. This can be achieved by using a random number table.

Sampling a centipede population

A researcher by the name of Lloyd (1967) sampled centipedes in Wytham Woods, near Oxford in England. A total of 37 hexagon–shaped quadrats were used, each with a diameter of 30 cm (see diagram on right). These were arranged in a pattern so that they were all touching each other. Use the data in the diagram to answer the following questions.

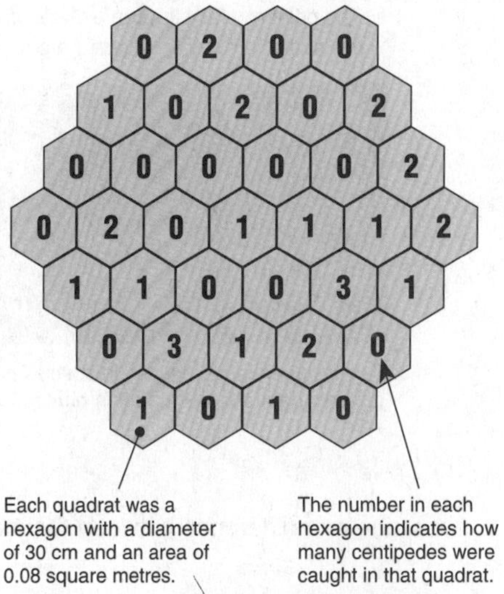
0 2 0 0
1 0 2 0 2
0 0 0 0 0 2
0 2 0 1 1 1 2
1 1 0 0 3 1
0 3 1 2 0
1 0 1 0
Each quadrat was a hexagon with a diameter of 30 cm and an area of 0.08 square metres.
The number in each hexagon indicates how many centipedes were caught in that quadrat.

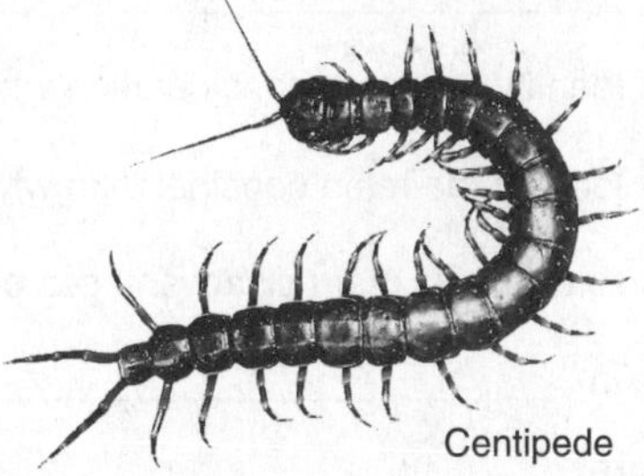
Centipede

1. Determine the average number of centipedes captured per quadrat:

2. Calculate the estimated average density of centipedes per square metre (remember that each quadrat is 0.08 square metres in area):

3. Looking at the data for individual quadrats, describe in general terms the distribution of the centipedes in the sample area:

4. Describe one factor that might account for the distribution pattern:

Quadrat-Based Estimates

The simplest description of a plant community in a habitat is a list of the species that are present. This qualitative assessment of the community has the limitation of not providing any information about the **relative abundance** of the species present. Quick estimates can be made using **abundance scales**, such as the ACFOR scale described below. Estimates of percentage cover provide similar information. These methods require the use of **quadrats**. Quadrats are used extensively in plant ecology. This activity outlines some of the common considerations when using quadrats to sample plant communities.

What Size Quadrat?

Quadrats are usually square, and cover 0.25 m^2 (0.5 m x 0.5 m) or 1 m^2, but they can be of any size or shape, even a single point. The quadrats used to sample plant communities are often 0.25 m^2. This size is ideal for low-growing vegetation, but quadrat size needs to be adjusted to habitat type. The quadrat must be large enough to be representative of the community, but not so large as to take a very long time to use.

A quadrat covering an area of 0.25 m^2 is suitable for most low growing plant communities, such as this alpine meadow, fields, and grasslands.

Larger quadrats (e.g. 1 m^2) are needed for communities with shrubs and trees. Quadrats as large as 4 m x 4 m may be needed in woodlands.

Small quadrats (0.01 m^2 or 100 mm x 100 mm) are appropriate for lichens and mosses on rock faces and tree trunks.

How Many Quadrats?

As well as deciding on a suitable quadrat size, the other consideration is how many quadrats to take (the sample size). In species-poor or very homogeneous habitats, a small number of quadrats will be sufficient. In species-rich or heterogeneous habitats, more quadrats will be needed to ensure that all species are represented adequately.

Determining the number of quadrats needed

- Plot the cumulative number of species recorded (on the *y* axis) against the number of quadrats already taken (on the *x* axis).
- The point at which the curve levels off indicates the suitable number of quadrats required.

Fewer quadrats are needed in species-poor or very uniform habitats, such as this bluebell woodland.

Describing Vegetation

Density (number of individuals per unit area) is a useful measure of abundance for animal populations, but can be problematic in plant communities where it can be difficult to determine where one plant ends and another begins. For this reason, plant abundance is often assessed using **percentage cover**. Here, the percentage of each quadrat covered by each species is recorded, either as a numerical value or using an abundance scale such as the ACFOR scale.

The ACFOR Abundance Scale

A = Abundant (30% +)
C = Common (20-29%)
F = Frequent (10-19%)
O = Occasional (5-9%)
R = Rare (1-4%)

The AFCOR scale could be used to assess the abundance of species in this wildflower meadow. Abundance scales are subjective, but it is not difficult to determine which abundance category each species falls into.

1. Describe one difference between the methods used to assess species abundance in plant and in animal communities:

2. Identify the main consideration when determining appropriate quadrat size:

3. Identify the main consideration when determining number of quadrats:

4. Explain two main disadvantages of using the ACFOR abundance scale to record information about a plant community:

(a)

(b)

Related activities: Sampling Populations, Quadrat Sampling
Web links: Ecological Sampling Methods

Sampling a Leaf Litter Population

The diagram on the following page represents an area of leaf litter from a forest floor with a resident population of organisms. The distribution of four animal species as well as the arrangement of leaf litter is illustrated. Leaf litter comprises leaves and debris that have dropped off trees to form a layer of detritus. This exercise is designed to practice the steps required in planning and carrying out a sampling of a natural population. It is desirable, but not essential, that students work in groups of 2–4.

1. **Decide on the sampling method**

   For the purpose of this exercise, it has been decided that the populations to be investigated are too large to be counted directly and a quadrat sampling method is to be used to estimate the average density of the four animal species as well as that of the leaf litter.

2. **Mark out a grid pattern**

   Use a ruler to mark out 3 cm intervals along each side of the sampling area (area of quadrat = 0.03 x 0.03 m). **Draw lines** between these marks to create a 6 x 6 grid pattern (total area = 0.18 x 0.18 m). This will provide a total of 36 quadrats that can be investigated.

3. **Number the axes of the grid**

   Only a small proportion of the possible quadrat positions are going to be sampled. It is necessary to select the quadrats in a random manner. It is not sufficient to simply guess or choose your own on a 'gut feeling'. The best way to choose the quadrats randomly is to create a numbering system for the grid pattern and then select the quadrats from a random number table. Starting at the *top left hand corner*, **number the columns** and **rows** from 1 to 6 on each axis.

4. **Choose quadrats randomly**

   To select the required number of quadrats randomly, use random numbers from a random number table. The random numbers are used as an index to the grid coordinates. Choose 6 quadrats from the total of 36 using table of random numbers provided for you at the bottom of the facing page. Make a note of which column of random numbers you choose. Each member of your group should choose a different set of random numbers (i.e. different column: A–D) so that you can compare the effectiveness of the sampling method.

   Column of random numbers chosen: ______

   NOTE: Highlight the boundary of each selected quadrat with coloured pen/highlighter.

5. **Decide on the counting criteria**

   Before the counting of the individuals for each species is carried out, the criteria for counting need to be established.

   There may be some problems here. You must decide before sampling begins as to what to do about individuals that are only partly inside the quadrat. Possible answers include:

   (a) Only counting individuals if they are completely inside the quadrat.

   (b) Only counting individuals that have a clearly defined part of their body inside the quadrat (such as the head).

   (c) Allowing for 'half individuals' in the data (e.g. 3.5 snails).

   (d) Counting an individual that is inside the quadrat by half or more as one complete individual.

   Discuss the merits and problems of the suggestions above with other members of the class (or group). You may even have counting criteria of your own. Think about other factors that could cause problems with your counting.

6. **Carry out the sampling**

   Carefully examine each selected quadrat and **count the number of individuals** of each species present. Record your data in the spaces provided on the facing page.

7. **Calculate the population density**

   Use the combined data TOTALS for the sampled quadrats to estimate the average density for each species by using the formula:

   Density =

   $$\frac{\textbf{Total number in all quadrats sampled}}{\textbf{Number of quadrats sampled} \times \textbf{area of a quadrat}}$$

   Remember that a total of 6 quadrats are sampled and each has an area of 0.0009 m^2. The density should be expressed as the number of individuals *per square metre (no. m^{-2})*.

   Woodlouse: ☐ False scorpion: ☐

   Centipede: ☐ Leaf: ☐

   Springtail: ☐

8. (a) In this example the animals are not moving. Describe the problems associated with sampling moving organisms. Explain how you would cope with sampling these same animals if they were really alive and very active:

   ____

   ____

   ____

   (b) Carry out a direct count of all 4 animal species and the leaf litter for the whole sample area (all 36 quadrats). Apply the data from your direct count to the equation given in (7) above to calculate the actual population density (remember that the number of quadrats in this case = 36):

   Woodlouse: ☐ Centipede: ☐ False scorpion: ☐ Springtail: ☐ Leaf: ☐

   Compare your estimated population density to the actual population density for each species:

   ____

   ____

   ____

Related activities: Quadrat Sampling, Quadrat-Based Estimates

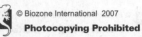

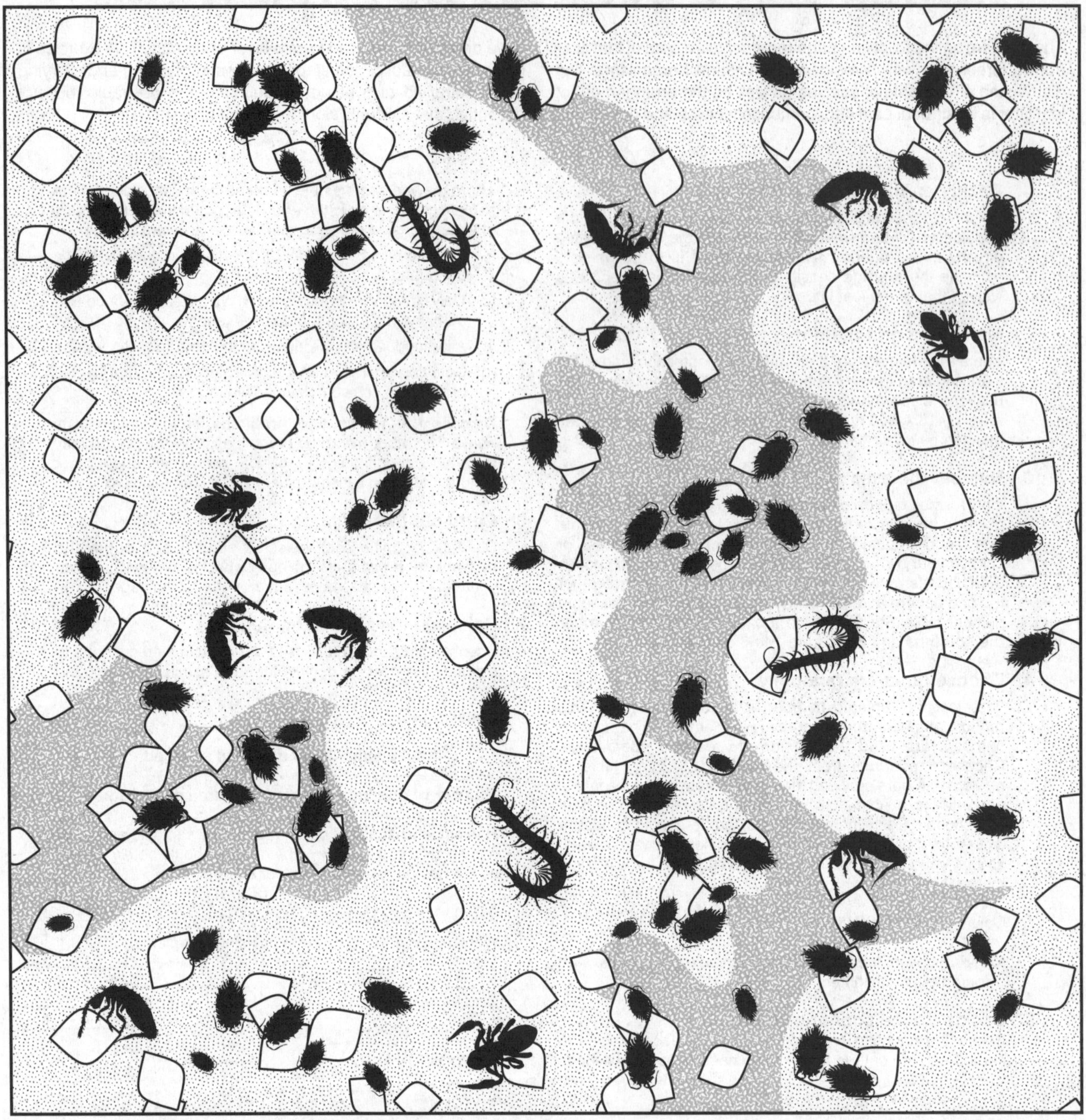

Coordinates for each quadrat	Woodlouse	Centipede	False scorpion	Springtail	Leaf
1:					
2:					
3:					
4:					
5:					
6:					
TOTAL					

Table of random numbers

A	B	C	D
2 2	3 1	6 2	2 2
3 2	1 5	6 3	4 3
3 1	5 6	3 6	6 4
4 6	3 6	1 3	4 5
4 3	4 2	4 5	3 5
5 6	1 4	3 1	1 4

The table above has been adapted from a table of random numbers from a statistics book. Use this table to select quadrats randomly from the grid above. Choose one of the columns (A to D) and use the numbers in that column as an index to the grid. The first digit refers to the row number and the second digit refers to the column number. To locate each of the 6 quadrats, find where the row and column intersect, as shown below:

Example: **5 2** refers to the 5th row and the 2nd column

Transect Sampling

A **transect** is a line placed across a community of organisms. Transects are usually carried out to provide information on the **distribution** of species in the community. This is of particular value in situations where environmental factors change over the sampled distance. This change is called an **environmental gradient** (e.g. up a mountain or across a seashore). The usual practice for small transects is to stretch a string between two markers. The string is marked off in measured distance intervals, and the species at each marked point are noted. The sampling points along the transect may also be used for the siting of quadrats, so that changes in density and community composition can be recorded. Belt transects are essentially a form of continuous quadrat sampling. They provide more information on community composition but can be difficult to carry out. Some transects provide information on the vertical, as well as horizontal, distribution of species (e.g. tree canopies in a forest).

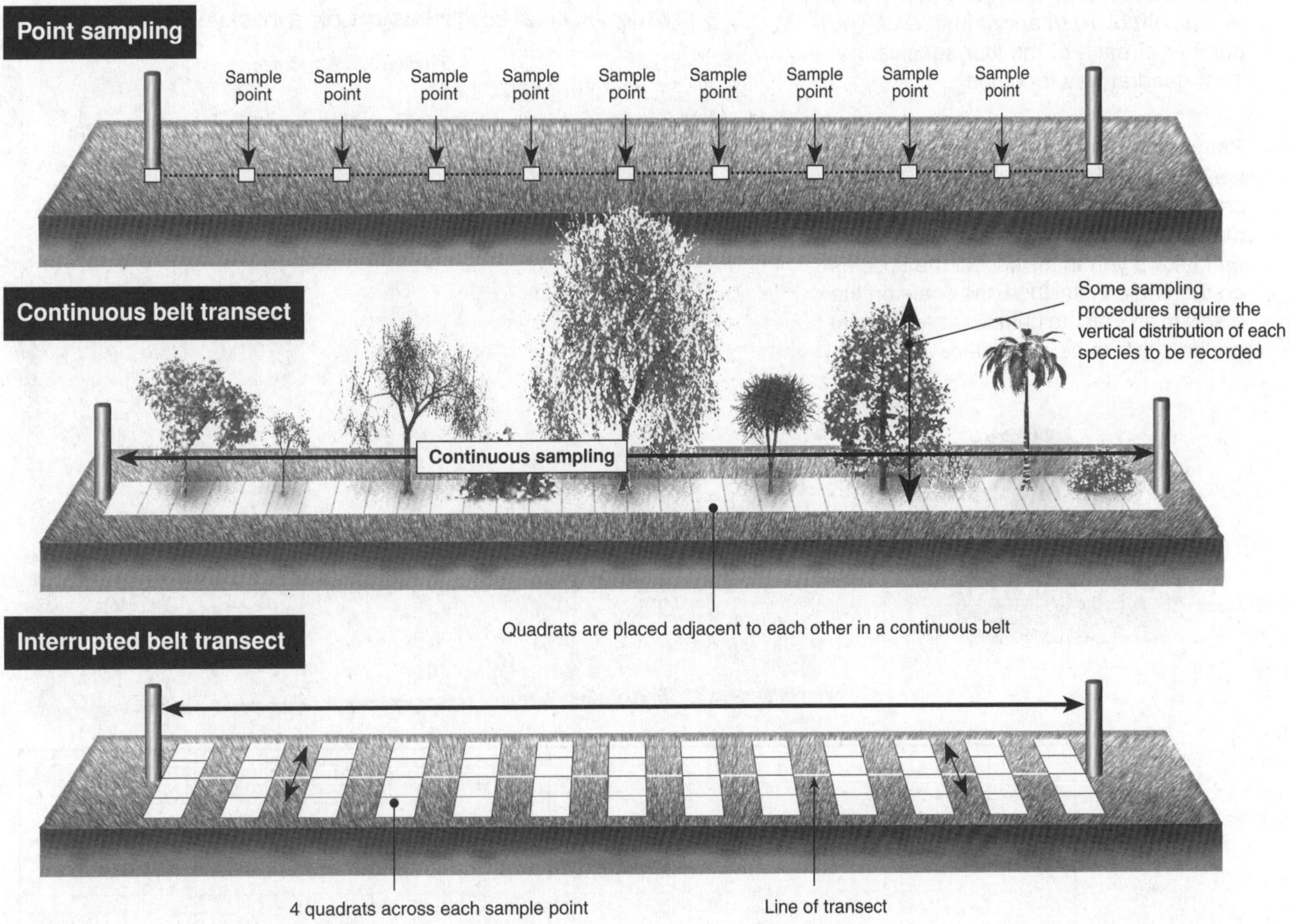

1. Belt transect sampling uses quadrats placed along a line at marked intervals. In contrast, point sampling transects record only the species that are touched or covered by the line at the marked points.

   (a) Describe one disadvantage of belt transects: ______

   (b) Explain why line transects may give an unrealistic sample of the community in question: ______

   (c) Explain how belt transects overcome this problem: ______

   (d) Describe a situation where the use of transects to sample the community would be inappropriate: ______

2. Explain how you could test whether or not a transect sampling interval was sufficient to accurately sample a community:

______

Kite graphs are an ideal way in which to present distributional data from a belt transect (e.g. abundance or percentage cover along an environmental gradient. Usually, they involve plots for more than one species. This makes them good for highlighting probable differences in habitat preference between species. Kite graphs may also be used to show changes in distribution with time (e.g. with daily or seasonal cycles).

3. The data on the right were collected from a rocky shore field trip. Periwinkles from four common species of the genus *Littorina* were sampled in a continuous belt transect from the low water mark, to a height of 10 m above that level. The number of each of the four species in a 1 m^2 quadrat was recorded.

   Plot a **kite graph** of the data for all four species on the grid below. Be sure to choose a scale that takes account of the maximum number found at any one point and allows you to include all the species on the one plot. Include the scale on the diagram so that the number at each point on the kite can be calculated.

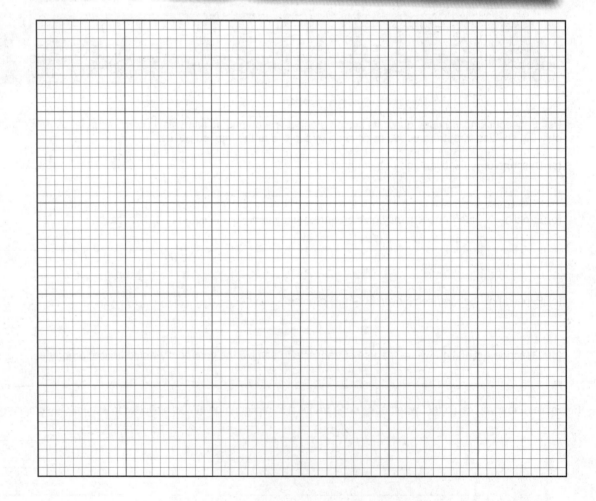

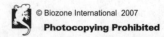

Field data notebook

Numbers of periwinkles (4 common species) showing vertical distribution on a rocky shore

Height above low water (m)	Periwinkle species: *L. littorea*	*L. saxatalis*	*L. neritoides*	*L. littoralis*
0-1	0	0	0	0
1-2	1	0	0	3
2-3	3	0	0	17
3-4	9	3	0	12
4-5	15	12	0	1
5-6	5	24	0	0
6-7	2	9	2	0
7-8	0	2	11	0
8-9	0	0	47	0
9-10	0	0	59	0

Mark and Recapture Sampling

The mark and recapture method of estimating population size is used in the study of animal populations where individuals are highly mobile. It is of no value where animals do not move or move very little. The number of animals caught in each sample must be large enough to be valid. The technique is outlined in the diagram below.

First capture | **Release back into the natural population** | **Second capture**

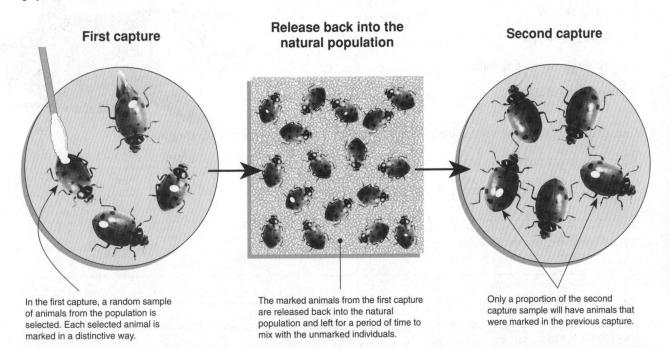

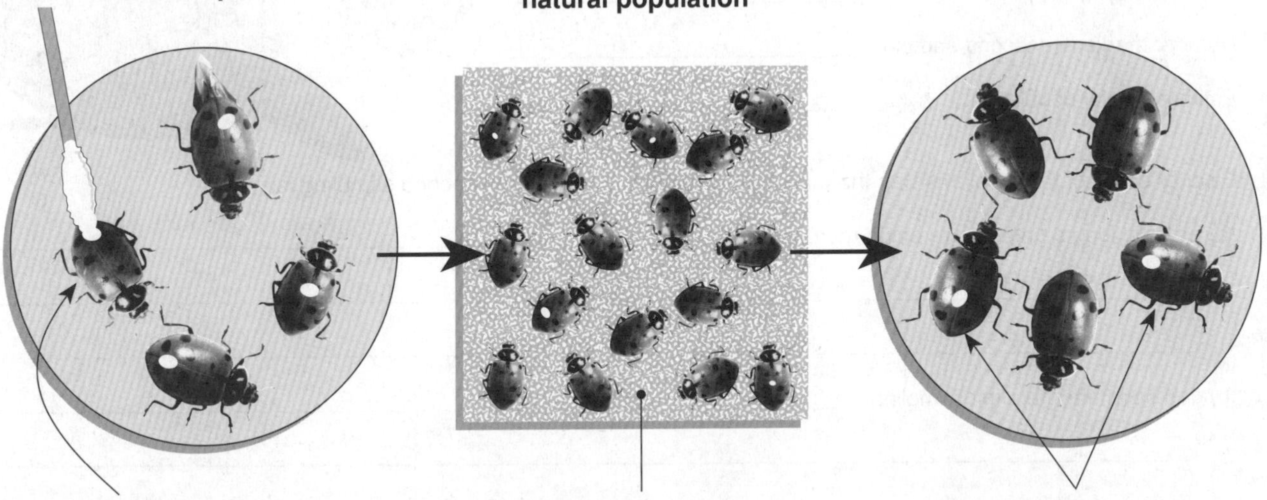

In the first capture, a random sample of animals from the population is selected. Each selected animal is marked in a distinctive way.

The marked animals from the first capture are released back into the natural population and left for a period of time to mix with the unmarked individuals.

Only a proportion of the second capture sample will have animals that were marked in the previous capture.

The Lincoln Index

$$\text{Total population} = \frac{\text{No. of animals in 1st sample (all marked)} \times \text{Total no. of animals in 2nd sample}}{\text{Number of marked animals in the second sample (recaptured)}}$$

The mark and recapture technique comprises a number of simple steps:

1. The population is sampled by capturing as many of the individuals as possible and practical.
2. Each animal is marked in a way to distinguish it from unmarked animals (unique mark for each individual not required).
3. Return the animals to their habitat and leave them for a long enough period for complete mixing with the rest of the population to take place.
4. Take another sample of the population (this does not need to be the same sample size as the first sample, but it does have to be large enough to be valid).
5. Determine the numbers of marked to unmarked animals in this second sample. Use the equation above to estimate the size of the overall population.

1. For this exercise you will need several boxes of matches and a pen. Work in a group of 2-3 students to 'sample' the population of matches in the full box by using the mark and recapture method. Each match will represent one animal.

   (a) Take out 10 matches from the box and mark them on 4 sides with a pen so that you will be able to recognise them from the other unmarked matches later.
   (b) Return the marked matches to the box and shake the box to mix the matches.
   (c) Take a sample of 20 matches from the same box and record the number of marked matches and unmarked matches.
   (d) Determine the total population size by using the equation above.
   (e) Repeat the sampling 4 more times (steps b–d above) and record your results:

	Sample 1	Sample 2	Sample 3	Sample 4	Sample 5
Estimated population					

   (f) Count the actual number of matches in the matchbox : ____________________

   (g) Compare the actual number to your estimates. By how much does it differ: ____________________

2. In 1919 a researcher by the name of Dahl wanted to estimate the number of trout in a Norwegian lake. The trout were subject to fishing so it was important to know how big the population was in order to manage the fish stock. He captured and marked 109 trout in his first sample. A few days later, he caught 177 trout in his second sample, of which 57 were marked. Use the Lincoln index (on the previous page) to estimate the total population size:

Size of first sample: ______________________

Size of second sample: ______________________

Number marked in second sample: ______________________

Estimated total population: ______________________

3. Discuss some of the problems with the mark and recapture method if the second sampling is:

(a) Left too long a time before being repeated: ______________________

(b) Too soon after the first sampling: ______________________

4. Describe two important assumptions being made in this method of sampling, which would cause the method to fail if they were not true:

(a) ______________________

(b) ______________________

5. Some types of animal would be unsuitable for this method of population estimation (i.e. would not work).

(a) Name an animal for which this method of sampling would not be effective: ______________________

(b) Explain your answer above: ______________________

6. Describe three methods for marking animals for mark and recapture sampling. Take into account the possibility of animals shedding their skin, or being difficult to get close to again:

(a) ______________________

(b) ______________________

(c) ______________________

7. At various times since the 1950s, scientists in the UK and Canada have been involved in computerised tagging programmes for Northern cod (a species once abundant in Northern Hemisphere waters but now severely depleted). Describe the type of information that could be obtained through such tagging programmes:

____

Sampling Animal Populations

Unlike plants, most animals are highly mobile and present special challenges in terms of sampling them **quantitatively** to estimate their distribution and abundance. The equipment available for sampling animals ranges from various types of nets and traps (below), to more complex electronic devices, such as those used for radio-tracking large mobile species.

Plankton net

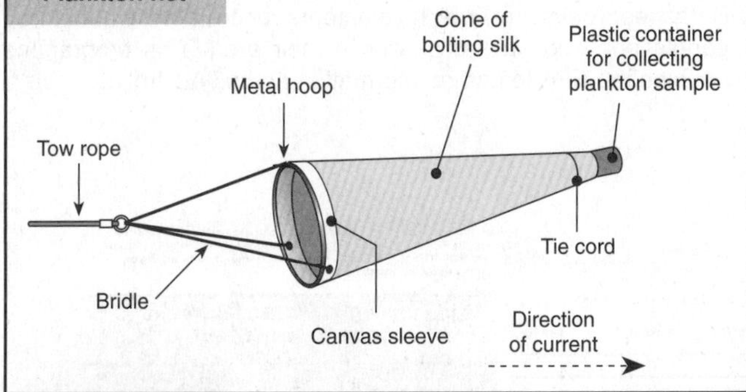

Beating tray

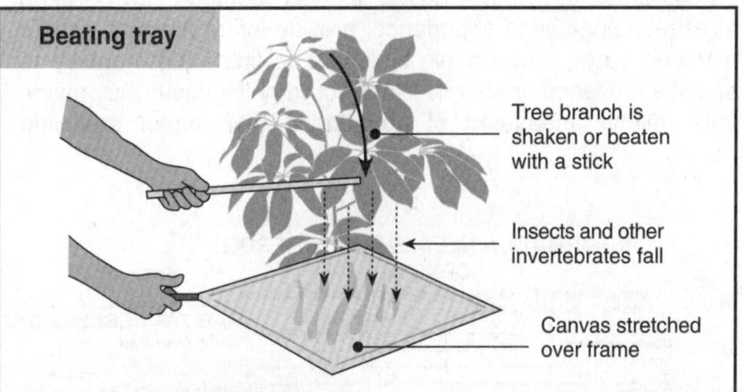

Kick sampling

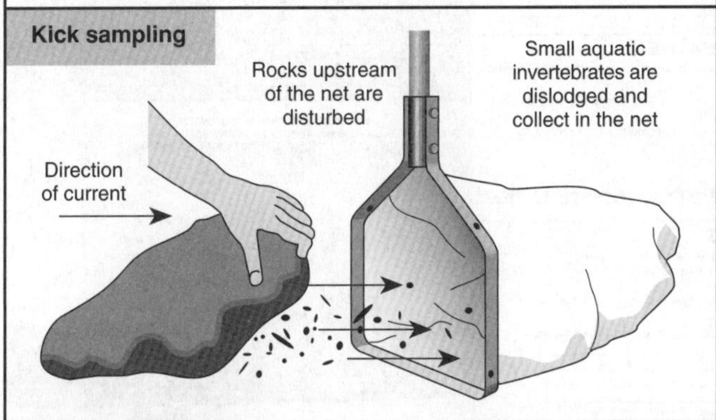

Tullgren funnel

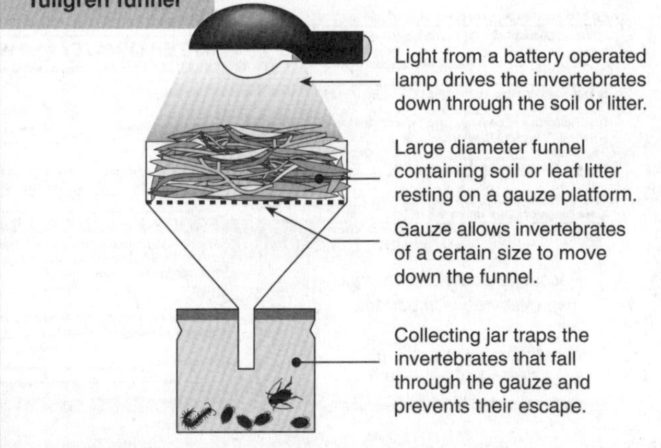

Pooter (aspirator)

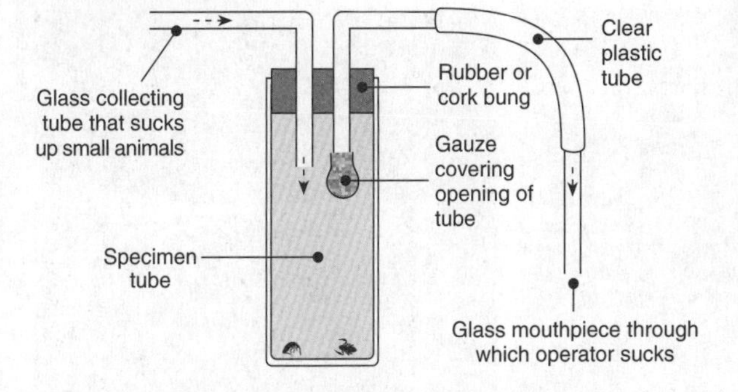

Pitfall trap

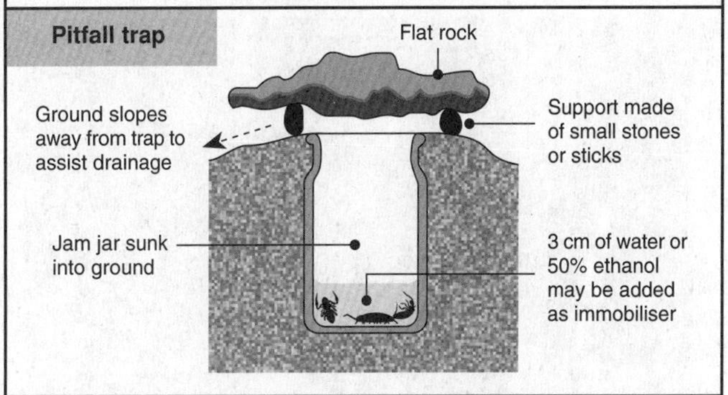

1. Describe which of the sampling techniques pictured above provides the best **quantitative** method for sampling invertebrates in vegetation. Explain your answer:

2. Explain why pitfall traps are not recommended for estimates of population density:

3. (a) Explain how mesh size could influence the sampling efficiency of a plankton net:

   (b) Explain how this would affect your choice of mesh size when sampling animals in a pond:

Related activities: Mark & Recapture Sampling, Sampling a Leaf Litter Population

Indirect Sampling

If populations are small and easily recognised they may be monitored directly quite easily. However, direct measurement of elusive, easily disturbed, or widely dispersed populations is not always feasible. In these cases, indirect methods can be used to assess population abundance, provide information on habitat use and range, and enable biologists to link habitat quality to species presence or absence. Indirect sampling methods provide less reliable measures of abundance than direct sampling methods, such as mark and recapture, but are widely used nevertheless. They rely on recording the signs of a species, e.g. scat, calls, tracks, and rubbings or markings on vegetation, and using these to assess population abundance. In Australia, the Environmental Protection Agency (EPA) provides a Frog Census Datasheet (below) on which volunteers record details about frog populations and habitat quality in their area. This programme enables the EPA to gather information across Australia.

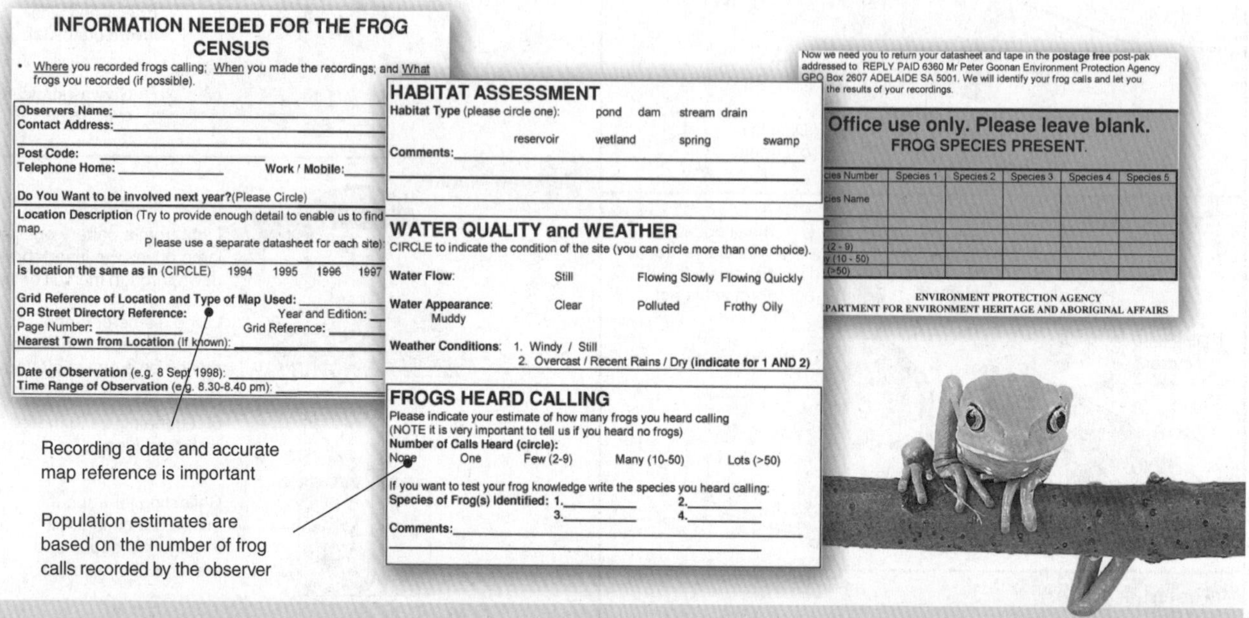
INFORMATION NEEDED FOR THE FROG CENSUS
• Where you recorded frogs calling; When you made the recordings; and What frogs you recorded (if possible).
Observers Name:
Contact Address:
Post Code:
Telephone Home:
Work / Mobile:
Do You Want to be involved next year?(Please Circle)
Location Description (Try to provide enough detail to enable us to find
map.
Please use a separate datasheet for each site):
is location the same as in (CIRCLE) 1994 1995 1996 1997
Grid Reference of Location and Type of Map Used:
OR Street Directory Reference:
Year and Edition:
Page Number:
Grid Reference:
Nearest Town from Location (If known):
Date of Observation (e.g. 8 Sept 1998):
Time Range of Observation (e.g. 8.30-8.40 pm):
HABITAT ASSESSMENT
Habitat Type (please circle one): pond dam stream drain reservoir wetland spring swamp
Comments:
WATER QUALITY and WEATHER
CIRCLE to indicate the condition of the site (you can circle more than one choice).
Water Flow: Still Flowing Slowly Flowing Quickly
Water Appearance: Clear Polluted Frothy Oily Muddy
Weather Conditions: 1. Windy / Still
2. Overcast / Recent Rains / Dry (indicate for 1 AND 2)
FROGS HEARD CALLING
Please indicate your estimate of how many frogs you heard calling
(NOTE it is very important to tell us if you heard no frogs)
Number of Calls Heard (circle):
None One Few (2-9) Many (10-50) Lots (>50)
If you want to test your frog knowledge write the species you heard calling:
Species of Frog(s) Identified: 1. 2. 3. 4.
Comments:
Now we need you to return your datasheet and tape in the postage free post-pak addressed to REPLY PAID 6360 Mr Peter Goonan Environment Protection Agency GPO Box 2607 ADELAIDE SA 5001. We will identify your frog calls and let you
the results of your recordings.
Office use only. Please leave blank.
FROG SPECIES PRESENT.
cies Number Species 1 Species 2 Species 3 Species 4 Species 5
cies Name
(2 - 9)
y (10 - 50)
(>50)
ENVIRONMENT PROTECTION AGENCY
PARTMENT FOR ENVIRONMENT HERITAGE AND ABORIGINAL AFFAIRS
Recording a date and accurate map reference is important
Population estimates are based on the number of frog calls recorded by the observer

COD

VOLUME
ON/OFF
FREQUENCY
kHz
Stag Electronics
BATBOX III

To sample nocturnal, highly mobile species, e.g. bats, electronic devices, such as the bat detector above, can be used to estimate population density. In this case, the detector is tuned to the particular frequency of the hunting clicks emitted by specific bat species. The number of calls recorded per unit time can be used to estimate numbers per area.

The analysis of animal tracks allows wildlife biologists to identify habitats in which animals live and to conduct population surveys. Interpreting tracks accurately requires considerable skill as tracks may vary in appearance even when from the same individual. Tracks are particularly useful as a way to determine habitat use and preference.

Sam Banks

Wombat scat

All animals leave scats (faeces) which are species specific and readily identifiable. Scats can be a valuable tool by which to gather data from elusive, nocturnal, easily disturbed, or highly mobile species. Faecal analyses can provide information on diet, movements, population density, sex ratios, age structure, and even genetic diversity.

1. (a) Describe the kind of information that could be gathered from the Frog Census Datasheet:

(b) Identify the benefits of linking a measure of abundance to habitat assessment:

2. Describe one other indirect method of population sampling and outline its advantages and drawbacks:

Web links: New Zealand Frog Survey

Monitoring Change in an Ecosystem

Much of the importance we place on ecosystem change stems ultimately from what we want from that ecosystem. Ecosystems are monitored for changes in their status so that their usefulness can be maintained, whether that use is for agriculture, industry, recreation, or conservation. Never is this so apparent as in the monitoring of aquatic ecosystems. Aquatic environments of all types provide aesthetic pleasure, food, habitat for wildlife, water for industry and irrigation, and potable water. The different uses of aquatic environments demand different standards of **water quality**. For any water body, this is defined in terms of various chemical, physical, and biological characteristics. Together, these factors define the 'health' of the aquatic ecosystem and its suitability for various desirable uses. Water quality is determined by measurement or analysis on-site or in the laboratory. Other methods, involving the use of **indicator species**, can also be used to biologically assess the health of a water body.

Techniques for Monitoring Water Quality

Some aspects of water quality, such as black disk clarity measurements (above), must be made in the field.

The collection of water samples allows many quality measurements to be carried out in the laboratory.

Telemetry stations transmit continuous measurements of the water level of a lake or river to a central control office.

Temperature and dissolved oxygen measurements must be carried out directly in the flowing water.

All Photos EW

Water Quality Standards in Aquatic Ecosystems

Water quality variable	Why measured	Standards applied:	
Dissolved oxygen	• A requirement for most aquatic life • Indicator of organic pollution • Indicator of photosynthesis (plant growth)	More than 80% saturation More than 5 gm^{-3}	**(F, FS, SG)** **(WS)**
Temperature	• Organisms have specific temperature needs • Indicator of mixing processes • Computer modelling examining the uptake and release of nutrients	Less than 25°C Less than 3°C change along a stretch of river	**(F)** **(AE, F, FS, SG)**
Conductivity	• Indicator of total salts dissolved in water • Indicator for geothermal input		
pH (acidity)	• Aquatic life protection • Indicator of industrial discharges, mining	Between pH 6 - 9	**(WS)**
Clarity - turbidity - black disk	• Aesthetic appearance • Aquatic life protection • Indicator of catchment condition, land use	Turbidity: 2 NTU Black disk: more than 1.6 m	**(AE, CR, A)**
Colour - light absorption	• Aesthetic appearance • Light availability for excessive plant growth • Indicator of presence of organic matter		
Nutrients (Nitrogen and phosphorus)	• Enrichment, excessive plant growth • Limiting factor for plant and algal growth	DIN: less than 0.100 gm^{-3} DRP: less than 0.030 gm^{-3} NO_3^-: less than 10 gm^{-3}	**(AE, A)** **(WS)**
Major ions (Mg^{2+}, Ca^{2+}, Na^+, K^+, Cl^-, HCO_3^-, SO_4^{2-})	• Baseline water quality characteristics • Indicator for catchment soil types, geology • Water hardness (magnesium/calcium) • Buffering capacity for pH change (HCO_3^-)		
Organic carbon	• Indicator of organic pollution • Catchment characteristics	BOD: less than 5 gm^{-3}	**(AE, CR, A)**
Faecal bacteria	• Indicator of pollution with faecal matter • Disease risk for swimming etc.	ENT: less than 33 cm^{-3} FC: less than 200 cm^{-3}	**(CR)**

Fly fishing is a pursuit which demands high water quality.

Spawning salmon require high oxygen levels for egg survival.

Standards refer to specified water uses: **AE** = aquatic ecosystem protection, **A** = aesthetic, **CR** = contact recreation, **SG** = shellfish gathering, **WS** = water supply, **F** = fishery, **FS** = fish spawning, **SW** = stock watering.

Key to abbreviations: NTU = a unit of measurement for turbidity, DIN = dissolved inorganic nitrogen, DRP = dissolved reactive phosphorus, BOD = biochemical oxygen demand, ENT = enterococci, FC = faecal coliform.

1. Explain why dissolved oxygen, temperature, and clarity measurements are made in the field rather than in the laboratory:

Calculation and Use of Diversity Indices

One of the best ways to determine the health of an ecosystem is to measure the variety (rather than the absolute number) of organisms living in it. Certain species, called **indicator species**, are typical of ecosystems in a particular state (e.g. polluted or pristine). An objective evaluation of an ecosystem's biodiversity can provide valuable insight into its status, particularly if the species assemblages have changed as a result of disturbance.

Diversity can be quantified using a **diversity index (DI)**. Diversity indices attempt to quantify the degree of diversity and identify indicators for environmental stress or degradation. Most indices of diversity are easy to use and they are widely used in ecological work, particularly for monitoring ecosystem change or pollution. One example, which is a derivation of **Simpson's index**, is described below. Other indices produce values ranging between 0 and almost 1. These are more easily interpreted because of the more limited range of values, but no single index offers the "best" measure of diversity: they are chosen on their suitability to different situations.

Simpson's Index for finite populations

This diversity index (DI) is a commonly used inversion of Simpson's index, suitable for finite populations.

$$DI = \frac{N(N - 1)}{\Sigma n(n - 1)}$$

After Smith and Smith as per IOB.

Where:

DI = Diversity index
N = Total number of individuals (of all species) in the sample
n = Number of individuals of each species in the sample

This index ranges between 1 (low diversity) and infinity. The higher the value, the greater the variety of living organisms. It can be difficult to evaluate objectively without reference to some standard ecosystem measure because the values calculated can, in theory, go to infinity.

Example of species diversity in a stream

The example describes the results from a survey of stream invertebrates. The species have been identified, but this is not necessary in order to calculate diversity as long as the different species can be distinguished. Calculation of the DI using Simpson's index for finite populations is:

Species	No. of individuals
A (Common backswimmer)	12
B (Stonefly larva)	7
C (Silver water beetle)	2
D (Caddis fly larva)	6
E (Water spider)	5
Total number of individuals = 32	

$$DI = \frac{32 \times 31}{(12\times11) + (7\times6) + (2\times1) + (6\times5) + (5\times4)} = \frac{992}{226} = 4.39$$

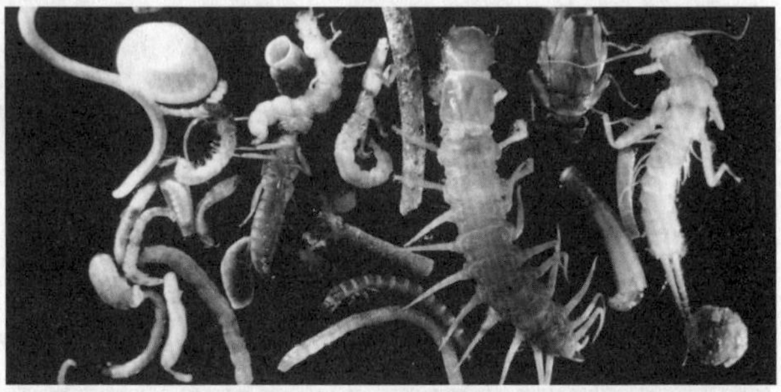

A stream community with a high macroinvertebrate diversity (above) in contrast to a low diversity stream community (below).

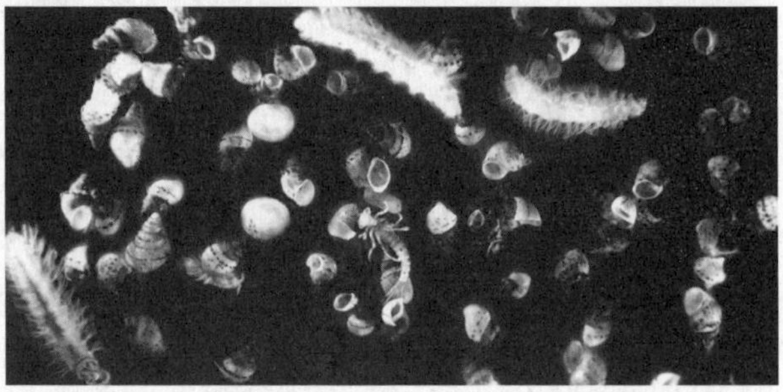

Photos: Stephen Moore

2. Discuss the link between water quality and land use: ______

3. Describe a situation where a species diversity index may provide useful information: ______

4. An area of forest floor was sampled and six invertebrate species were recorded, with counts of 7, 10, 11, 2, 4, and 3 individuals. Using Simpson's index for finite populations, calculate DI for this community:

   (a) DI= ______ DI = ______

   (b) Comment on the diversity of this community: ______

5. Explain how you could use indicator species to detect pollution in a stream: ______

Land, Water, and Energy

Global land and water use. Energy resources and consumption.

Global human nutrition, agriculture and the green revolution, fishing, mining, and the consumption and conservation of energy.

Learning Objectives

☐ 1. Compile your own glossary from the **KEY WORDS** displayed in **bold type** in the learning objectives below.

Plants and Agricultural Systems

Plants and human nutrition *(pages 99-102)*

☐ 2. Outline the importance of plants to human populations, with specific reference to each of the following: food, fuel, clothing, building materials, and aesthetic values. Provide examples to illustrate each of these uses.

☐ 3. Discuss the factors contributing to global starvation. Describe the key **macronutrients** required to achieve a balanced diet, and the implications of nutrient deficiencies. Explain why **malnutrition** is a disease of both developing and developed countries.

☐ 4. Explain what is meant by the **green revolution**, and distinguish its two phases. Discuss the differences between the initial green revolution, which modified agricultural practices to improve crop yields, and the second green revolution (or gene revolution), which has utilised advances in gene technology to improve crops.

☐ 5. Discuss some ways the **gene revolution** has been used to enhance major cereal crops. Include in your discussion how it has been used to:
- Improve yields and enhance nutritional value
- Improve resistance and environmental tolerance
- Deliver edible vaccines

☐ 6. Identify some of the most important cereal crops (in a global sense) and appreciate their importance in the human diet. Identify the regions where these crops are grown and their relative importance in those regions.

☐ 7. Describe the **cultivation** of a cereal plant of economic importance, e.g. wheat, maize, rice, or sorghum. Describe features that make the crop plant well suited to its region and method of cultivation.

Agricultural systems *(pages 103-104, 109-110)*

☐ 8. Using appropriate examples, describe the characteristic features of the major types of agricultural practice, with particular reference to **subsistence agriculture** and **intensive industrialised agriculture**.

☐ 9. Compare and contrast industrialised agriculture with agricultural systems based on sustainability and/or organic farming practices. Consider the following:
- Use of organic vs inorganic fertilisers
- Crop and land management practices
- Use of pesticides, growth regulators, feed additives
- Energy inputs required to maintain yield
- Amount of land used
- Pollution through fertiliser and pesticide run off
- Soil degradation and loss
- Long term sustainability

Pest management *(pages 105-106)*

☐ 10. Appreciate the crop losses attributed to **pests** in terms of quantifiable harvest. Understand the means by which pests and **weeds** reduce crop yields:
- Weeds: Compete with crops for water and nutrients.
- Insect pests: Damage and consume plant tissues, reduce photosynthesis, act as vectors for disease.

☐ 11. Compare and contrast **biological** and **chemical pest control**. Consider the efficacy of the control, the environmental risks, economic costs, long term sustainability, and availability of alternatives. With respect to pesticide use, explain the terms: **persistence**, **toxicity**, biomagnification and **specificity**.

☐ 12. Describe the characteristics of **integrated pest management** and discuss its environmental and economic advantages over other pest control methods.

Soil degradation *(pages 107-108, 134)*

☐ 13. Explain what constitutes a 'healthy soil', and discuss how human activities can deplete or contaminate soil. In your discussion, include reference to the effects of agriculture, landfills, flood control, and industry on soil quality. Describe potential options for soil conservation and restoration and comment on their feasibility.

☐ 14. Explain what is meant by **desertification**, and discuss its causes and its environmental and economic impacts.

Energy use and Conservation *(pages 111-114)*

☐ 15. Recall basic energy concepts; forms of energy, energy conversions, and the laws of thermodynamics. Distinguish between **renewable** and **non-renewable** energy resources and give examples of each.

☐ 16. Summarise the history of human energy consumption, with particular reference to the **Industrial Revolution** and our increasing dependence on fossil fuels, such as coal and oil. Examine current global energy use and projections for future energy needs, and comment on the sustainability of current energy usage patterns.

☐ 17. Discuss each of the following energy resource options, including reference to their sustainability and environmental impact:

(a) **Fossil fuels**, including formation of **coal**, **oil**, and **natural gas**, world reserves, and global demand.

(b) **Nuclear power**, including the nuclear fission process, reactor types, and safety and environmental issues.

(c) **Hydroelectric power**, including capacity for electricity generation and environmental impact.

(d) **Renewable energy sources**, including solar energy, hydrogen fuel cells, ocean waves and tidal energy, wind energy, geothermal energy (arguably renewable), and energy from fast-growing biomass and waste (**gasohol** and **biogas**).

- ☐ 18. Explain the concepts of **energy conservation** and **energy efficiency** and discuss their importance in meeting future energy requirements. Compare the energy efficiency and environmental impact of different methods of energy production.
- ☐ 19. Discuss the measures on both a local and national scale to improve energy efficiency. Consider:
  - Reducing and using waste heat
  - Saving energy in industry and transportation
  - Reducing energy usage in buildings
  - Saving energy in industry and transportation, including reference to the **CAFE** (Corporate Average Fuel Economy) regulations, the use of **hybrid vehicles**, and the use of public transport.

Fisheries Management *(pages 115-117)*

- ☐ 20. Describe the relevant laws and treaties pertaining to commercial fisheries in national and international waters. Explain what is meant by **maximum sustainable yield** and **by-catch**, and explain their significance to fisheries management and to the ecology of marine ecosystems.
- ☐ 21. Describe the ecological impacts of **commercial fishing** practices. Include in your discussion reference to its effect on the ocean environment, fish stocks, and the direct and indirect impacts on by-catch species. Fishing techniques that could be considered include:
  - (a) Netting, e.g. set netting and seining
  - (b) Trolling, using baited hooks or lures
  - (c) Lining, e.g. hand-lines and long-lines
  - (d) Trawling (mid-water and bottom trawling)
- ☐ 22. Explain how **stock indicator species** are used to monitor the health and sustainability of a fishery. Describe some of the measures which can be taken to aid recovery of fish levels.
- ☐ 23. Discuss the viability and environmental impact of **aquaculture** as an alternative to the fishing of natural populations (e.g. for shellfish, salmon, or crustaceans). In your discussion, comment on the situations in which aquaculture may or may not be ecological sound.

Supplementary Texts

See page 7 for additional details of these texts:

■ Miller, G.T, 2007. **Living in the Environment: Principles, Connections and Solutions.**

■ Raven *et. al.*, 2008. **Environment**.

■ Reiss, M. & J. Chapman, 2000. **Environmental Biology** (Cambridge University Press).

■ Smith, R. L. & T.M. Smith, 2001. **Ecology and Field Biology**.

Periodicals

See page 7 for details of publishers of periodicals:

STUDENT'S REFERENCE

■ **State of the Planet** National Geographic, 202(3), Sept. 2002, pp. 102-115. *Sustainable management of global resources after Rio.*

■ **Time to Rethink Everything** New Scientist 27 April-18 May 2002 (4 issues). *Globalisation, the impact of humans, & the sustainability of our future.*

■ **Ironing out Malnutrition** Biol. Sci. Rev., 13(2) Nov. 2000, pp. 19-22. *Improving iron uptake in plants for improved human nutrition.*

■ **Insecticides and the Conservation of Hedgerows** Biol. Sci. Rev., 16(4) April 2004, pp. 28-31. *Well managed hedgerows can reduce the need for insecticides in adjacent crops by encouraging natural pest control agents.*

■ **Food, Crop Protection. and the Environment** Biol. Sci. Rev., 19(2) Nov. 2006, pp. 21-25. *Intensive agriculture is required to meet the tripling of global food demands. Problems arising as a result of monocultural systems can be potentially overcome using integrated pest management programmes.*

■ **Birds, Bees and Superweeds** Biol. Sci. Rev., 17(2) Nov. 2004, pp. 24-27. *The history of using genetic modification in modern agriculture. Discusses the advantages & concerns of GE crops.*

■ **Genetic Manipulation of Plants** Biol. Sci. Rev., 15(1) Sept. 2002, pp. 10-13. *The genetic modification of crop plants to improve tolerance to herbicides, pests & environmental stress.*

■ **The Adaptations of Cereals** Biol. Sci. Rev., 13(3) Jan. 2001, pp. 30-33. *A look at the world's major cereal crops: production and adaptations.*

■ **Green Dreams** National Geographic, 212(4), October 2007, pp. 38-59. *An account of the merits and potential environmental effects of making ethanol biofuel from fast growing biomass.*

■ **Biodiesel: Tomorrow's Liquid Gold** Biol. Sci. Rev., 45(1) Feb. 1998, pp. 17-21. *A discussion on the production, advantages and use of biodiesel.*

■ **The High Cost of Coal** National Geographic, 209(3) March 2006, pp. 96-103. *Coal as a resource: our dependence on it as a fuel and the environmental effects of burning and mining it.*

■ **Nuclear Power Risking a Comeback** National Geographic, 209(4), April 2006, pp. 54-63. *Despite its hazards, nuclear energy is a promising alternative to fossil fuels in terms of reducing greenhouse emissions and still being cost-effective.*

■ **Oceans of Nothing** Time Magazine, 13 Nov. 2006, pp. 42-43. *A study of the impact of human activity on fish populations over 50 years.*

■ **Everybody Loves Atlantic Salmon-Here's the Catch** National Geographic, 204(1), July 2003, pp. 100-123. *As wild stocks decline, salmon farming is increasing, but there are many problems associated with this lucrative business.*

TEACHER'S REFERENCE

■ **Future Farming - A Return to Roots** Scientific American, August 2007, pp. 66-73. *Modern agriculture's intensive land use quashes natural ecosystems and biodiversity. An increasing human population will require more land to be cultivated to supply needs. A shift to growing perennials could conserve soil and allow cultivation of wheat and grain crops in areas currently considered marginal.*

■ **Cassava Comeback** New Scientist, 21 April. 2007, pp. 38-39. *Cassava is a food staple for many but is vulnerable to viral disease. Cross breeding strains offers a chance of resistance to disease.*

■ **Thought for Food** New Scientist, 7 Aug. 2004, p. 17. *A short account of the ways in which agricultural science and technology are being used to reduce world hunger and poverty.*

■ **Manna or Millstone** New Scientist, 18 Sept. 2004, pp. 29-31. *The evolution of humans from hunter/gathers to farmers.*

■ **Quick and Dirty** New Scientist, 11 Aug. 2007, pp. 33-35. *An account of the effect of agriculture on global soil erosion. Methods of soil conservation and generating new soil are discussed.*

■ **Kicking the Habit** New Scientist, 25 Nov. 2000, pp. 34-42. *The pressing need for alternative fuels.*

■ **An Efficient Solution - Energy Efficiency** Scientific American, Sept. 2006, pp. 40-43. *The quickest, easiest way to reduce carbon emissions is to avoid losses during energy conversions. Improving the energy efficiency of buildings and industries offers impressive savings.*

■ **Who Needs Oil?** New Scientist, 7 Jul. 2007, pp. 28-31. *An account of our global dependence on oil, both as a fuel and as a precursor of plastics. The potential of some of the suggested alternative energy sources is also discussed and their viability for long term power examined.*

■ **Can Biofuels Rescue American Prairies?** New Scientist, 18 Aug. 2007, pp. 6-7. *Alternative fuel sources and their impact on the environment.*

■ **High Hopes for Hydrogen-A Hydrogen Economy** Scientific American, Sept. 2006, pp. 70-77. *Hydrogen-fuel-cell cars could become commercially feasible if automakers succeed in developing safe, inexpensive, durable models that can travel long distances before refuelling.*

■ **Counting the Last Fish** Scientific American, July 2003, pp. 34-39. *Overfishing has reduced the world's fish stocks to an all time low. There is an urgent need for effective, cooperative management.*

Internet

See pages 4-5 for details of how to access **Bio Links** from our web site: **www.thebiozone.com**. From Bio Links, access sites under the topics:

RESOURCE MANAGEMENT & AGRICULTURE > **General sites**: • World Resources Institute • European Environmental Agency ... *and others* > **Land Management**: • CSIRO land and water • Landcare Research **Fisheries and Aquaculture**: • NOAA fisheries • EEA fisheries • EUROPA The Common Fisheries Policy ... *and others* > **Agriculture**: • Agriculture 21 • Food and Agriculture Organisation of the UN • Sustainable agriculture education page ...*and others* > **Human Population Growth**: • Dimensions of need > **Crop Production**: • Crop and Grassland Service > **Pests and Pest Control**: • Biological control • Integrated pest management • IPM in and around the home • The pesticide management education program ...*and others*

HEALTH AND DISEASE > **Human Health Issues**: • Institute of Crop and Food Research

ECOLOGY > **Energy flows and nutrient cycles**: • Bioaccumulation • Human alteration of the global nitrogen cycle

CONSERVATION > **Habitat loss**: • Soil erosion • Soil erosion and water infiltration

HUMAN IMPACT > • Human alteration of the global nitrogen cycle • Human impact • Human impact on the natural environment

The Importance of Plants

Via the process of photosynthesis, plants provide oxygen and are also the ultimate source of food and metabolic energy for nearly all animals. Besides foods (e.g. grains, fruits, and vegetables), plants also provide people with shelter, clothing, medicines, fuels, and the raw materials from which innumerable other products are made.

Plant tissues provide the energy for almost all heterotrophic life. Many plants produce delicious fruits in order to spread their seeds.

Plant tissues can be utilised to provide shelter in the form of framing, cladding, and roofing on both temporary and permanent structures.

Many plants provide fibres for a range of materials including cotton (above), linen (from flax), and coir (from coconut husks).

Plant extracts, including latex from rubber trees (above) and resins from gymnosperms, can be utilised in many ways as an important material in manufacturing or artwork.

Coal, petroleum, and natural gas are fossil fuels which were formed from the dead remains of plants and other organisms. Together with wood, they provide important sources of fuel.

All photos: AT

Plants produce many beneficial and not so beneficial substances. Over 25% of all modern medicines are derived from plant extracts. Many recreational drugs are also of plant origin (above).

1. Using examples, describe how plant species are used by people for each of the following:

   (a) Food: ______

   (b) Fuel: ______

   (c) Clothing: ______

   (d) Building materials: ______

   (e) Aesthetic value: ______

   (f) "Recreational" drugs: ______

   (g) Therapeutic drugs (medicines): ______

2. Outline three reasons why the destruction of native forests is of concern:

   (a) ______

   (b) ______

   (c) ______

Related activities: Global Human Nutrition

Global Human Nutrition

Globally, 854 million people are undernourished and, despite improvements in agricultural methods and technologies, the number of hungry people in the world continues to rise. The majority of these people live in developing nations, but 9 million live in industrialised countries. Over 6 million people die annually from starvation, while millions of others suffer debilitating diseases as a result of malnutrition. Protein deficiencies (such as kwashiorkor), are common amongst the world's malnourished, because the world's poorest nations consume only a fraction of the world's protein resources, surviving primarily on cereal crops. Political and environmental factors contribute significantly to the world's hunger problem. In some countries, food production is sufficient to meet needs, but inadequate distribution methods cause food shortages in some regions. Advances in agricultural practices (fertiliser and pesticide application, genetically enhanced stock) improve agricultural productivity and food supply, but can have detrimental effects through loss of biodiversity, soil and water pollution, and increased levels of greenhouse gases.

Human Nutritional Requirements

A **balanced diet,** taken from the components below, is essential for human growth, development, metabolism, and good health. In many developing countries, deficiency diseases and starvation are prevalent either because of an absolute scarcity of food or because of inadequate nutrition. In many developed Western nations, an oversupply of cheap, nutritionally poor and highly processed food is contributing to an increase of diet-related diseases such as obesity, diabetes, and heart disease. **Malnutrition** (a lack of specific nutrients), once commonly associated with undernutrition, is now rising in developed nations over consuming on poor quality processed foods.

Proteins (supplied by beans and pulses, and animal products such as meat and fish) are essential to growth and repair of muscle and other tissues. Unlike animal protein, plant protein is incomplete and sources must be chosen to complement one another nutritionally. Deficiencies result in kwashiorkor or marasmus.

Carbohydrates (right) are supplied in breads, starchy vegetables, cereals, and grains. They form the staple of most diets and provide the main energy source for the body.

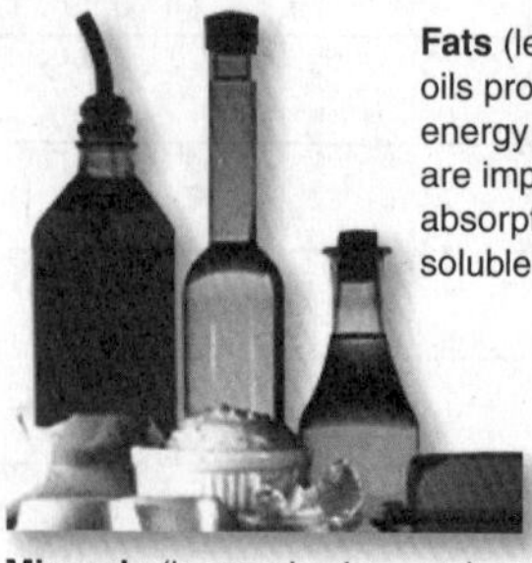

Fats (left) and oils provide an energy source and are important for absorption of fat soluble vitamins.

Minerals (inorganic elements) and **vitamins** (essential organic compounds) are both required for numerous normal body functions. They are abundant in fruit and vegetables (right)

Agricultural Practices

Essentially, there are two broadly different types of agriculture: industrialised, or high-input, agriculture and traditional or subsistence agricultural systems. These categories exclude the hunter-gatherer societies, such as the Inuit, which collect food directly from the wild using methods of foraging and hunting with little or no domestication of target foodstuffs.

Subsistence agriculture is low technology, low-input farming where only enough food is grown to supply the family unit. It has minimal environmental impact, but can be unsustainable in densely populated areas. Subsistence agriculture occurs mostly in Africa, Asia and South America, and parts of the Pacific (e.g. Niue, left).

Industrialised, intensive agriculture produces high yields per unit of land at cheaper prices to the consumer, but has a large environmental impact because of high inputs of energy, fertilisers, and pesticides. **Wheat production** (left) and animal "factory farming" are examples. Rice production is also an example of intensive agriculture, but remains largely traditional (not mechanised) in many parts of the world.

Plantation agriculture is a form of industrialised agriculture practised mainly in tropical countries solely for the production of a high value cash crop for sale in developed countries. Typical crops include bananas (left), cotton, coffee, sugarcane, tobacco, and cocoa.

1. Discuss the differences between subsistence and industrialised agriculture with respect to relative inputs of land, labour, financial capital, and fossil fuel energy:

___

___

___

___

___

2. *One of the likely effects of a global fuel crisis would be food shortage.* Explain this statement: ___

___

___

___

The Green Revolution

Since the 1950s, most increases in global food production have come from increased yields per unit area of cropland rather than farming more land. The initial **green revolution** increased the intensity and frequency of cropping, and applied liberal amounts of fertilisers, pesticides, and water to improve yields. The **second green revolution** began in the 1960s and improved production by developing high yielding crop varieties. The countries whose crop yields per unit of land area have increased during the two green revolutions are illustrated below. Several agricultural research centres and **seed** or **gene banks** also play a key role in developing high yielding crop varieties. Most of the world's gene banks store the seeds of the hundred or so plant species that collectively provide approximately 90% of the food consumed by humans. However, some banks are also storing the seeds of species threatened with extinction or a loss of genetic diversity. Producing more food on less land is an important way of protecting biodiversity by saving large areas of natural habitat from being used to grow food.

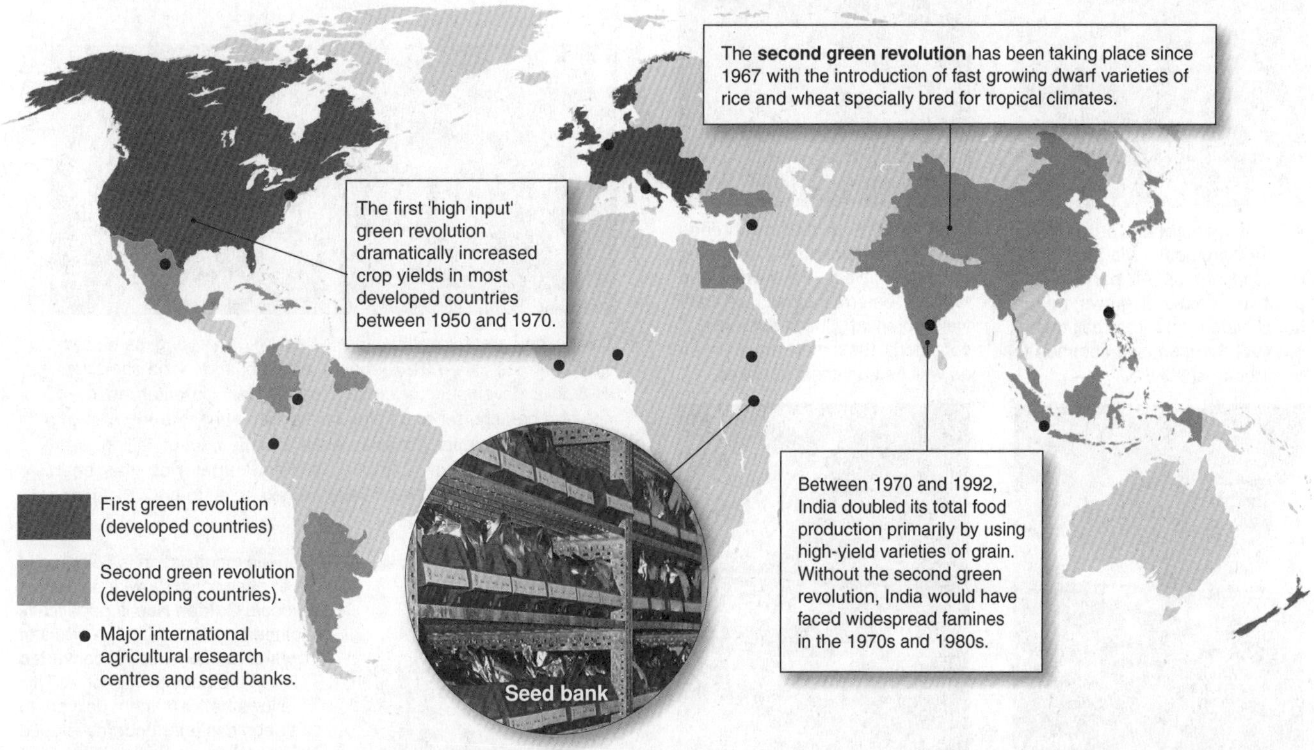
The **second green revolution** has been taking place since 1967 with the introduction of fast growing dwarf varieties of rice and wheat specially bred for tropical climates.
The first 'high input' green revolution dramatically increased crop yields in most developed countries between 1950 and 1970.
Seed bank
Between 1970 and 1992, India doubled its total food production primarily by using high-yield varieties of grain. Without the second green revolution, India would have faced widespread famines in the 1970s and 1980s.
First green revolution (developed countries)
Second green revolution (developing countries).
Major international agricultural research centres and seed banks.

High-input, intensive agriculture uses large amounts of fossil fuel energy, water, commercial inorganic fertilisers, and pesticides to produce large quantities of single crops (monocultures) from relatively small areas of land. At some point though, outputs diminish or even decline.

There are approximately 30 000 plant species with parts suitable for human consumption, but just three grain crops (wheat, rice, and corn) provide more than half the calories the world's population consumes. These crops have been the focus of the second green revolution.

Increased yields from industrialised agriculture depend on the extensive use of fossil fuels to run machinery, produce and apply fertilisers and pesticides, and pump water for irrigation. Since 1950, the use of fossil fuels in agriculture has increased four-fold.

1. Describe how the technologies of the first and second green revolutions differ: ____________________

____

____

____

____

____

____

Related activities: Global Human Nutrition, Cereal Crop Production, The Impact of Farming **Web links**: Green Revolution: Curse or Blessing?

The second green revolution (also called the **gene revolution**) is based on developments in **selective breeding** and **genetic engineering**. It has grown rapidly in scope and importance since it began in 1967. Initially, it involved the development of fast growing, high yielding varieties of rice, corn, and wheat, which were specially bred for tropical and subtropical climates to meet global food demand (below left). More recently, genetically modified seeds have been used to create plants with higher yields and specific tolerances (e.g. pest resistance, herbicide tolerance, or drought tolerance). GM seed is also used to improve the nutritional quality of crops (e.g. by increasing protein or vitamin levels), or to produce plants for edible vaccine delivery.

Recent Crop Developments

A new potential crop plant is the tropical winged bean (*Psophocarpus*). All parts of the plant are edible, it grows well in hot climates, and it is resistant to many of the diseases common to other bean species.

Most green revolution breeds are "high-responders", requiring optimum levels of water and fertiliser before they realise their yield potential. Under sub-optimal conditions they may not perform as well as traditional varieties.

Improvements in crop production have come from the modification of a few, well known species. Future research aims to maintain genetic diversity in high-yielding, disease resistant varieties.

A century ago, yields of maize (corn) in the USA were around 25 bushels per acre. In 1999, yields from hybrid maize were five to ten times this, depending on the growing conditions.

Improving Rice Crops

Rice is the world's second most important cereal crop, providing both a food and an income source to millions of people worldwide. Traditional rice strains lack many of the essential vitamins and minerals required by humans for good health and are susceptible to crop failure and low yields if not tended carefully. Advances in plant breeding, biotechnology, and genetic engineering (below) have helped to overcome these problems.

The second green revolution produced a high-yielding, semi-dwarf variety of rice called **IR-8** (above) in response to food shortages. IR-8 was developed by cross breeding two parental strains of rice and has shorter and stiffer stalks than either parent, allowing the plant to support heavier grain heads without falling over. More recently, a new improved variety, '**super rice**' has been developed to replace IR-8. Yields are expected to be 20% higher.

Genetic modification is being used to alter rice for a wide variety of purposes. **Golden rice** is genetically engineered to contain high levels of beta-carotene, which is converted in the body to vitamin A. This allows better nutrient delivery to people in poor underdeveloped countries where rice is the food staple. Other companies are focusing on improving the resistance of rice crops to insects, bacteria, and herbicides, improving yields, or delivering edible hepatitis and cholera vaccines in the rice.

2. Using examples, explain how the technologies of the second green revolution are being used to:

   (a) Improve crop yields: ______________________________

   ____

   ____

   (b) Improve the nutritional quality of crops: ______________________________

   ____

   ____

3. (a) Explain how countries currently suffering from food shortages might benefit from recent crop developments:

   ____

   ____

   ____

   (b) Describe the constraints that might exist on developing countries taking advantage of these potential benefits:

   ____

   ____

Cereal Crop Production

Agricultural ecosystems may be industrialised (high-input) or traditional. Industrialised agriculture uses large amounts of fossil fuel energy, water, fertilisers, and pesticides to increase net production (crop yield). Despite the high diversity of edible plants, the world's population depends on just 30 crops for 95% of its food. Four crops: wheat, rice, maize, and potato, account for a bigger share than all other crops combined. Since 1950, most of the increase in global food production has resulted from increased yields per unit of farmed land. This increase was termed the **green revolution**. More recently, a second green revolution has being taking place, with the use of fast growing, high yielding varieties of rice, corn, and wheat, specially bred for tropical and subtropical climates. Producing more food from less land increases the per capita food production while at the same time protecting large areas of potentially valuable agricultural land from development. Although food production has nearly tripled since the 1950s, the rate of this increase has started to slow, and soil loss and degradation are taking a toll on formerly productive land. Sustainable farming practices provide one way in which to reduce this loss of productivity.

Wheat (*Triticum* spp.)

Wheat is the most important world cereal crop and is extensively grown in temperate regions. Bread (common) wheat is a soft wheat with a high gluten (protein) content. It is cultivated for the grain, which is used both whole or ground. Durum wheat is a hard (low gluten) wheat used primarily for the manufacture of pasta. Key areas for wheat production are the prairies of Canada and the USA, Europe, and Russia (the former Soviet wheat belt). The economic stability of many nations is affected by the trade in wheat and related commodities. ***New developments***: Wheat cultivars are selected for particular nutritional qualities or high yield in local conditions. Research focuses on breeding hardy, disease resistant, and high yielding varieties.

Maize (corn, *Zea mays*)

Maize is a widely cultivated tropical and subtropical C_4 cereal crop, second only to wheat in international importance as a food grain. The USA corn belt produces nearly half the world's maize. Some is exported, but now 85% is used within the USA as animal feed (as grain and silage). Maize is also a major cereal crop in Africa but is second to rice in importance in Asian countries. Nutritionally, maize is poor in the essential amino acids tryptophan and lysine. Recent breeding efforts have been aimed at addressing these deficiencies. ***New developments***: Plant breeding has produced high lysine hybrid varieties with better disease resistance and higher yields. Most countries have cultivars suited to local conditions.

Rice (*Oryza sativa*)

Rice is the basic food crop of monsoon Asia. It is highly nutritious and requires relatively little post-harvest processing. The most common paddy (*japonica*) varieties are aquatic and are often grown under irrigation. Its cultivation is labour intensive. Upland (*indica*) varieties have similar requirements to other cereal crops. Most rice is grown in China, mainly for internal consumption. Other major producers include India, Pakistan, Japan, Thailand and Vietnam. ***New developments***: Much effort has gone into producing fast growing, disease resistant, high yielding cultivars which will crop up to 3 times a season. Genetic engineering to increase tolerance to high salinity is extending the range for cultivation in the upland varieties of rice.

Sorghum (*Sorghum bicolor*)

Sorghum is a frost-sensitive, tropical C_4 plant, well adapted to arid conditions. It has low soil nutrition and water requirements, reflecting its origin in the sub-Saharan Sudan region of Africa. Sorghum is now widely cultivated in Africa, the middle East to India and Myanmar, and parts of Australia, the Americas and Southern Europe. It is nutritious and is used as a human foodstuff in Asia and Africa. In other regions, it is used mainly as animal feed and as an industrial raw material (for oil, starch, and fiber). ***New developments***: New hybrids are high yielding, low-growing, and ripen uniformly. Further breeding aims to improve grain quality, and combine high yield properties with the disease resistance of the African wild stocks.

World Production of Major Food Crops

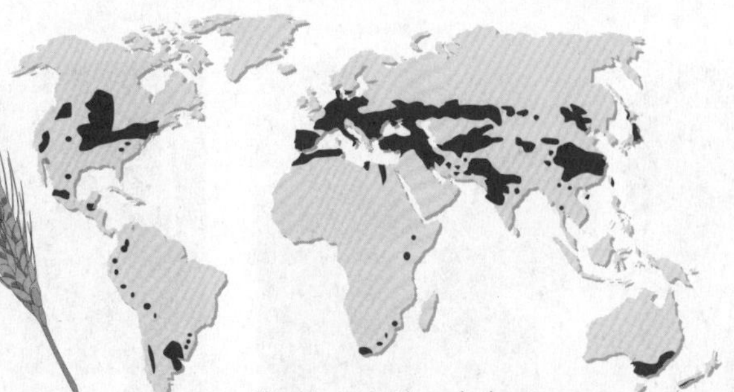
World production of wheat

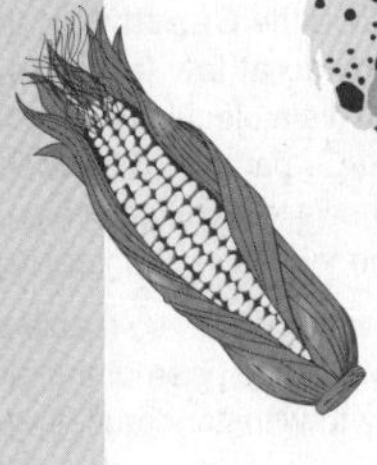

World production of maize

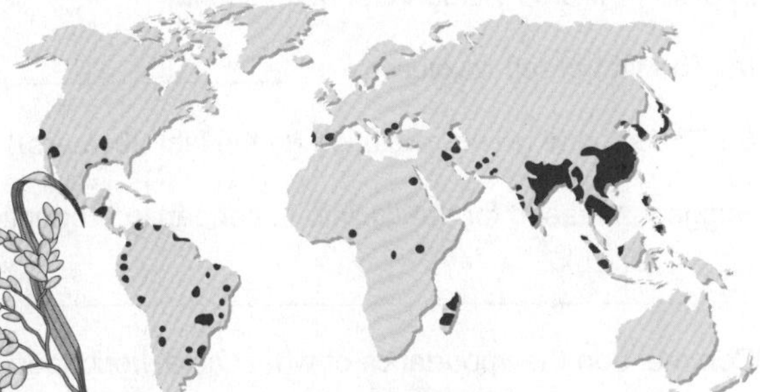
World production of rice

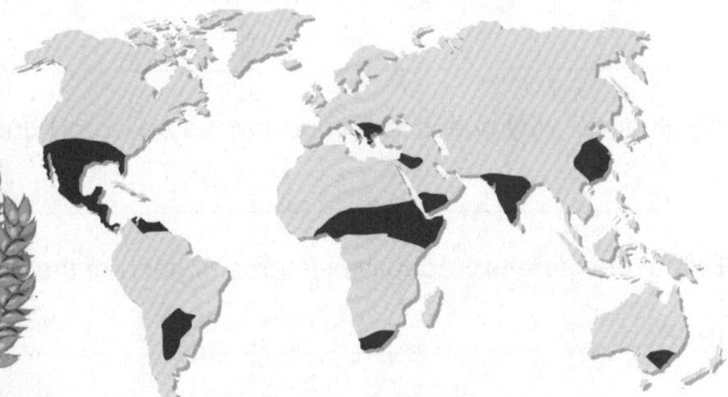
World production of sorghum

Related activities: Global Human Nutrition, The Green Revolution
Web links: FAO: Crop and Grassland Service

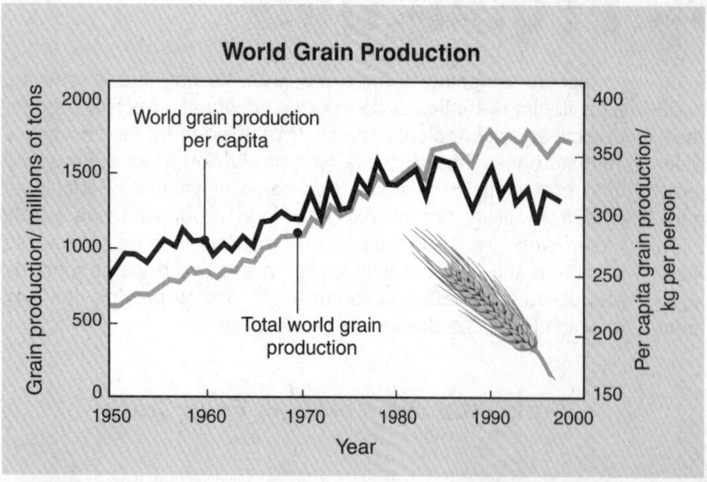

Cropping properties of major world crop plants

Crop plant	Yield/ kg grain ha^{-1}	Specific requirements for growth
Maize	1000 - 4000	Warm, frost free climate, fertile soil, drought intolerant
Wheat	1000 - 14 500	Adapted to a wide range of temperate climates and soils
Rice	4500 (paddy) 1500 (indica)	Tropical, paddy varieties are aquatic, drought intolerant
Sorghum	300 - 2000. As high as 6500 for irrigated hybrids	Wide range of soils. Drought tolerant. Grown in regions too dry for maize.

Sorghum is able to grow well in the very hot, dry regions of tropical Africa and central India. Adaptations include:

- A **dense root system** that is very efficient at extracting water from the soil.
- A thick **waxy cuticle** that prevents evaporative water loss through the leaf surface.
- The presence of special cells (called **motor cells**) on the underside of the leaf that cause the leaf to roll inwards in dry conditions. This traps moist air in the rolled leaf and reduces water loss.
- Reduced number of sunken stomata on leaves.

Maize grows well where temperature and light intensity are high. Adaptations include:

- A slightly different biochemical pathway for photosynthesis than that in most cooler climate plants. Called the **C_4 pathway**, the plant can fix carbon dioxide at low levels as a four-carbon molecule. This molecule is used to boost CO_2 in the regular C_3 pathway. This mechanism allows photosynthesis to continue at high rates (primarily through the inhibition of photorespiration).
- The roots are shallow, so maize often has small **aerial roots** at the base of the stem to increase their ability to withstand buffeting by wind.

Most of the **rice** in southeast Asia is grown partly submerged in paddy fields. Adaptations include:

- The stem of a rice plant has **large air spaces** (hollow aerenchyma) running the length of the stem. This allows oxygen to penetrate through to the roots which are submerged in water.
- The roots are also very shallow, allowing access to oxygen that diffuses into the surface layer of the waterlogged soil.
- When oxygen levels fall too low, the root cells respire anaerobically, producing ethanol. Ethanol is normally toxic to cells, but the root cells of rice have an unusually high tolerance to it.

1. Explain how crop yields were increased in:

   (a) The first green revolution: ______

   (b) The second green revolution (in the last 30 years): ______

2. Suggest a reason for the decline in per capita production of grain in the last decade: ______

3. Comment on the importance of wheat as a world food crop: ______

4. (a) Suggest when sorghum is a preferable crop to maize: ______

   (b) Suggest why rice is less important as an export crop than wheat or maize: ______

5. Briefly describe two adaptive features of each of the cereal crops below:

   (a) Rice: ______

   (b) Maize: ______

Pest Control

Pest control refers to the regulation or management of a species defined as a pest because of perceived detrimental effects on other species, the environment, or the economy. Pests can be managed through **biological controls**, which exploit natural existing ecological relationships, and **chemical controls** (pesticides). Opponents of pesticide use believe that the harmful effects of pesticides outweigh the benefits, especially given an increasing resistance to pesticides by target organisms. When pesticide resistance develops, more frequent applications and larger doses are often recommended. This leads to a **pesticide treadmill**, where farmers pay more and more for a pest control programme that becomes less and less effective. Newer, **integrated pest management** (IPM) programmes attempt to circumvent the pesticide treadmill by evaluating each crop and its pests as part of an ecological system and then developing a control programme that includes a sequence of crop management, and biological and chemical controls. The aim is not pest eradication but a reduction in crop damage to an economically tolerable level. Well managed IPM programmes have outstanding success and are recognised as being economically and ecologically sound.

Chemical Control

Pesticides, radioactive isotopes, heavy metals, and industrial chemicals such as PCBs can be taken up by organisms via their food or be absorbed from the surrounding medium. The **toxicity** of a chemical is a measure of how poisonous it is to both target and non-target organisms. Its **specificity** describes how selective it is in targeting a pest, while its **persistence** describes how long it stays in the environment. Many highly persistent pesticides show progressive concentration in food chain; an undesirable feature of their use called **biomagnification**.

Pesticide type	Examples	Environmental persistence	Biomagnification
Insecticides			
Organochlorines	*DDT*, dieldrin*	2-15 yrs	Yes
Organophosphates	*Malathion*	1-2 weeks/years	No
Carbamates	*Carbaryl*	Days to weeks	No
Botanicals	*Pyrethrum, camphor*	Days to weeks	No
Microbials	*Microorganisms*	Days to weeks	No
Fungicides			
Various chemicals	*Methyl bromide*	Days	No
Herbicides			
Contact§ chemicals	*Paraquat*	Days to weeks	No
Systemic¶ chemicals	*2,4-D, 2,4,5-T, glyphosphate*	Days to weeks	No
Soil sterilants	*Butylate*	Days	No
Fumigants			
Various chemicals	*Methyl bromide*	Years	Yes

* Now banned in most developed countries
¶ Systemic chemicals: Effective when absorbed into general circulation
§ Contact chemicals: Effective after contact with surface tissue

Source of data: Miller (2000) Living in the Environment, Brooks/Cole

Biomagnification of DDT in an aquatic ecosystem

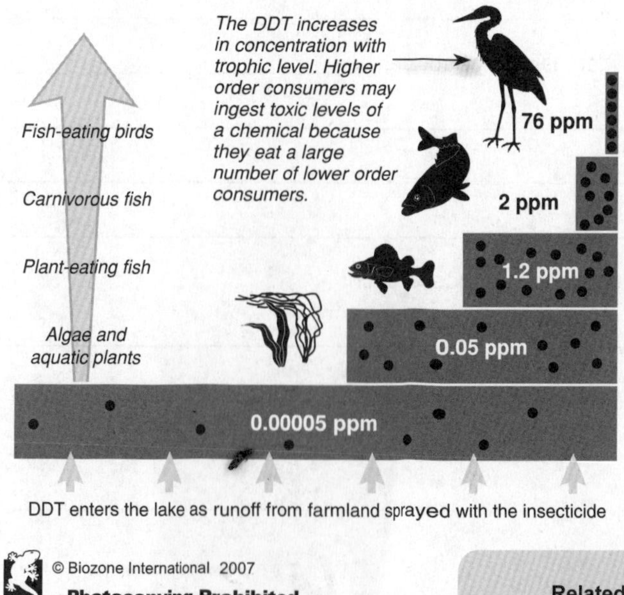
The DDT increases in concentration with trophic level. Higher order consumers may ingest toxic levels of a chemical because they eat a large number of lower order consumers.
Fish-eating birds
Carnivorous fish
Plant-eating fish
Algae and aquatic plants
76 ppm
2 ppm
1.2 ppm
0.05 ppm
0.00005 ppm

DDT enters the lake as runoff from farmland sprayed with the insecticide

Biological Control

Biological control (biocontrol) is a management tool for controlling pests using parasites, predators, pathogens, and weed feeders. Some control agents with a botanical or microbial origin are called **biopesticides**. Others, such as pheromone traps, are also sometimes included as biocontrols. Biological control serves as an important alternative to conventional pesticide use and it is an important component of **integrated pest management**. Thorough investigation of a biocontrol agent is required in order to predict its behaviour in a new environment. Some biological control agents may even become pests themselves. Some biocontrol programs attempt total elimination of a pest species, while others aim only to maintain pest numbers at acceptably low numbers.

Biological Control of Greenhouse Whitefly (*Trialeurodes vaporariorum*)

Adult whiteflies resemble tiny moths and are about 3 mm long. Their young appear as scales on the undersides of many glasshouse plants where they feed by sucking the sap. Whitefly can over-winter in a glasshouse on crops or weeds and the scales (the immobile nymph and pupal stages) can withstand the occasional frost. The young excrete a sticky "honeydew" on which sooty moulds develop. The mould reduces the amount of light reaching the leaves, thereby reducing photosynthetic rate and also crop yield. Two biocontrol agents are in common use for whitefly. The ladybird *Delphastus* feeds voraciously on the eggs and larvae of whitefly, consuming up to 150 whitefly eggs a day. The tiny parasitic wasp *Encarsia formosa* parasitises the whitefly scale and also feeds on them directly, further helping to reduce the whitefly numbers.

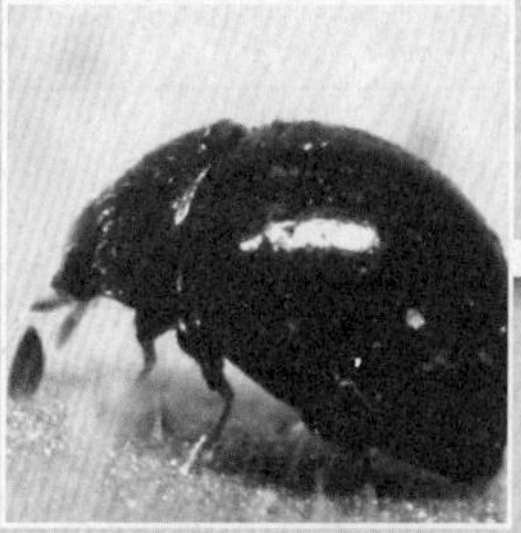

***Delphastus** can consume up to 150 whitefly eggs in a day.*

Adult whitefly *produce 30-500 eggs in a 1-2 month life span.*

Photos: Dr. John Dale, Defenders Ltd

Encarsia can parasitise up to 300 whitefly scales in 30 days.

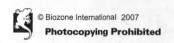

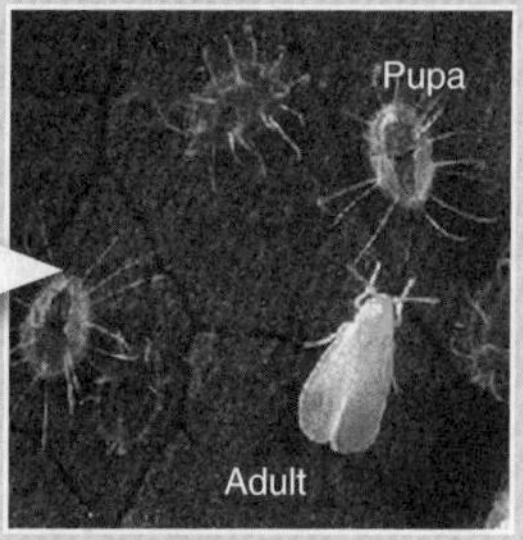
Pupa
Adult

Once the whitefly nymphs settle they become immobile.

Land, Water, and Energy

Features of Integrated Pest Management

Crop monitoring (ongoing)

Careful crop management and monitoring of pest levels. When damage is unacceptable, farmers implement control measures (below).

IPM may be slower to take effect than conventional controls and, for a programme to work, expert knowledge of crop and pest ecology is required. However, well designed programmes can reduce pest control costs and pesticide use by 50 - 90%. IPM can also reduce crop losses and the need for fertilisers, and slow the development of pesticide resistance. Where chemical pesticide use is required, IPM attempts to use ecologically sensitive pesticides. These pesticides tend to be specific to the target pest but harmless to non-target species, less likely to cause resistance, non-persistent, and low-cost.

An international success story

- In 1986, the Indonesian government banned the use of 57 of 66 pesticides used on rice and phased out pesticide subsidies over 2 years.
- The money saved from subsidy reduction was used to launch a nationwide programme of IPM.
- Between 1987 and 1992, pesticide use fell by 65%, rice production rose by 15%, and 250 000 farmers were trained in IPM.
- By 1993 the programme had saved the Indonesian government over $1.2 billion in pesticide costs; enough to fund IPM.

Stage 1: Cultivation controls

Cultivation controls, e.g., vacuuming up pests, and hand, hot water, or flame weeding, start the pest and weed control programme.

Stage 2: Biological controls

Sex attractants and biological controls, e.g. natural predators, are used to reduce pest populations. The Colorado potato beetle pest (above) is controlled using a fungal pathogen combined with release and conservation of the beetle's natural predators.

Stage 3: Targeted pesticide use

Small amounts of narrow spectrum pesticides are applied as a last resort if other methods do not achieve adequate control. A variety of chemicals may be used at different times to slow the development of pest resistance and preserve natural predator populations.

1. Explain why top consumers are most at risk of the toxic effects of pesticide **bioaccumulation**:

2. (a) Explain the general principle underlying the **biological control** of pests:

   (b) Explain what precautions should be taken when implementing a biocontrol programme:

3. Explain why persistence is an important property to consider when using a pesticide:

4. Describe the basic aim of **integrated pest management**:

5. Describe the main features of an IPM programme, briefly stating the importance of each:

   (a)

   (b)

   (c)

   (d)

Soil Degradation

Soil is a rather fragile resource and can be easily damaged by inappropriate farming practices. Some soils, such as those under the Amazon rainforest, are very vulnerable to human interference. Attempts to clear the forest and bring it into agricultural production are proving a disaster. These soils are very thin and nutrient poor; after only a few years farming they may be abandoned due to poor production. Overgrazing and deforestation may cause **desertification**. Chemically intensive agricultural practices, which call for ever-increasing doses of herbicides, insecticides, fungicides, and fertilisers, often result in high crop yields, but soils that are compacted and lacking structure. Healthy soils are 'alive' with a diverse community of organisms, including bacteria, fungi, and invertebrates. These organisms improve soil structure and help to create humus. Repeated chemical applications kill soil organisms and eventually result in a soil that is hard, lacking in organic material, and unproductive.

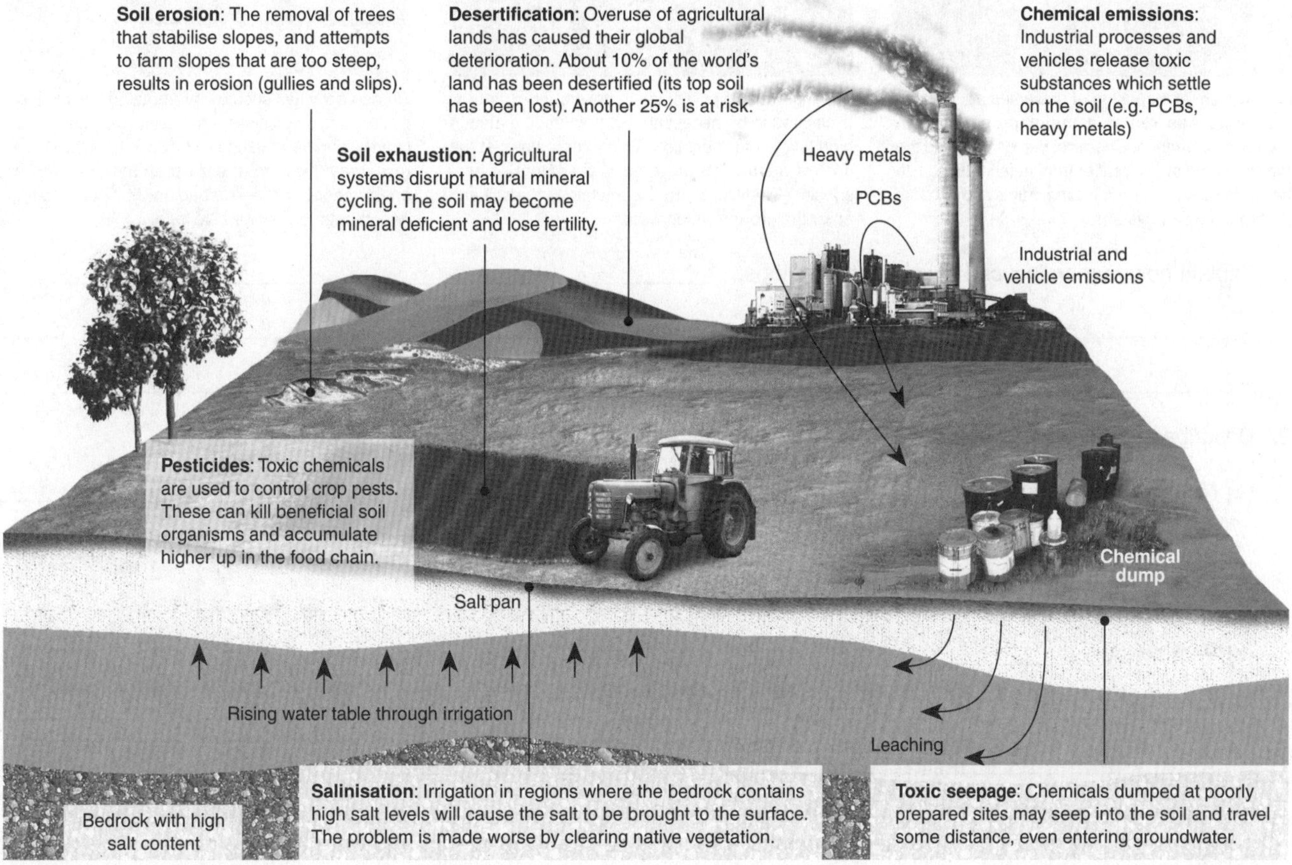
Soil erosion: The removal of trees that stabilise slopes, and attempts to farm slopes that are too steep, results in erosion (gullies and slips).
Desertification: Overuse of agricultural lands has caused their global deterioration. About 10% of the world's land has been desertified (its top soil has been lost). Another 25% is at risk.
Chemical emissions: Industrial processes and vehicles release toxic substances which settle on the soil (e.g. PCBs, heavy metals)
Soil exhaustion: Agricultural systems disrupt natural mineral cycling. The soil may become mineral deficient and lose fertility.
Heavy metals
PCBs
Industrial and vehicle emissions
Pesticides: Toxic chemicals are used to control crop pests. These can kill beneficial soil organisms and accumulate higher up in the food chain.
Chemical dump
Salt pan
Rising water table through irrigation
Leaching
Bedrock with high salt content
Salinisation: Irrigation in regions where the bedrock contains high salt levels will cause the salt to be brought to the surface. The problem is made worse by clearing native vegetation.
Toxic seepage: Chemicals dumped at poorly prepared sites may seep into the soil and travel some distance, even entering groundwater.

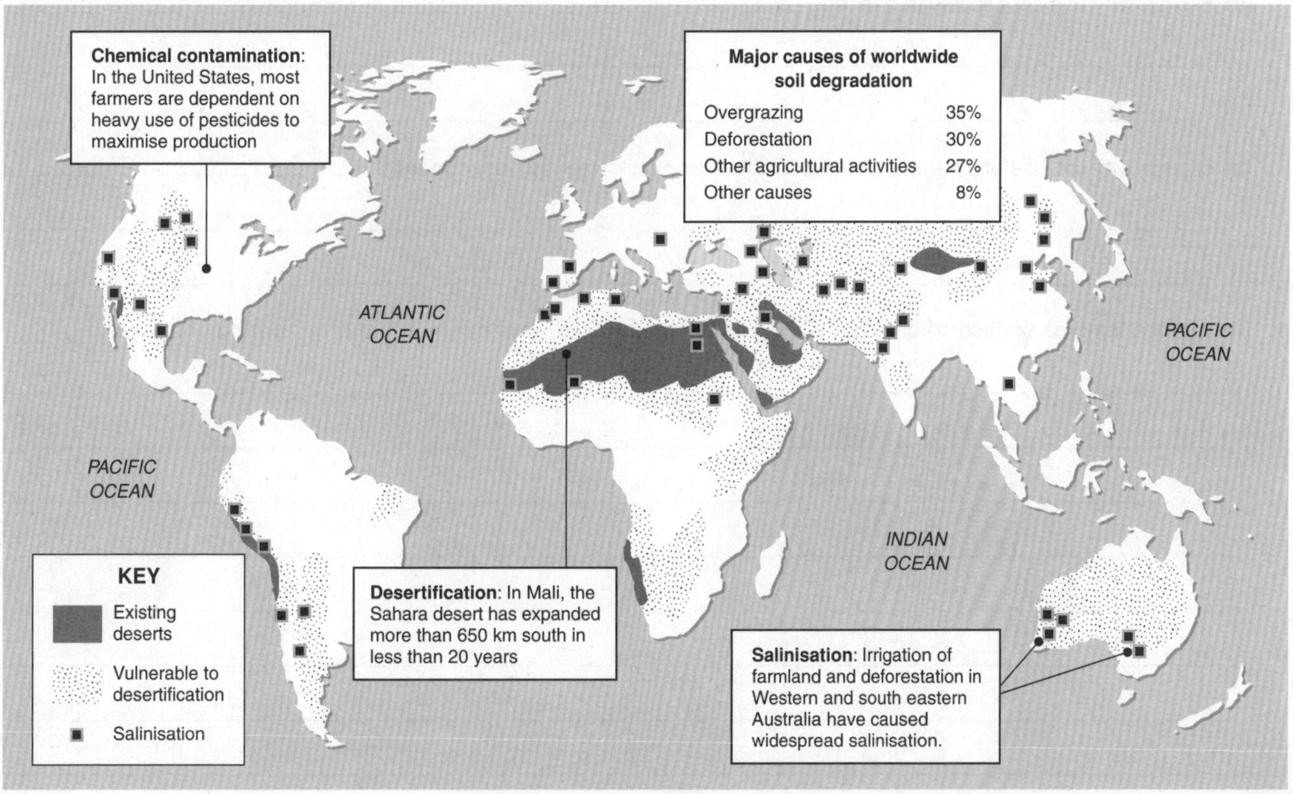
Chemical contamination: In the United States, most farmers are dependent on heavy use of pesticides to maximise production
Major causes of worldwide soil degradation
Overgrazing 35%
Deforestation 30%
Other agricultural activities 27%
Other causes 8%
ATLANTIC OCEAN
PACIFIC OCEAN
PACIFIC OCEAN
INDIAN OCEAN
KEY
Existing deserts
Vulnerable to desertification
Salinisation
Desertification: In Mali, the Sahara desert has expanded more than 650 km south in less than 20 years
Salinisation: Irrigation of farmland and deforestation in Western and south eastern Australia have caused widespread salinisation.

Related activities: Pest Control, The Impact of Farming, Sustainable Land Use
Web links: Soil Erosion and Conservation, Soil Degradation Maps

The problem of disposing of unwanted agricultural chemicals has reached major proportions in developed countries. Chemical dumps (such as the one illustrated above) suffer from deterioration, with the contents spilling from rusting drums and entering the ground water system.

The high use of pesticides in developed countries is claimed to be necessary by growers to maintain high levels of production. This often comes at the cost of destroying the natural predators of pest species. Pesticides can accumulate in the soil and enter the ground water system.

Human activities such as overgrazing livestock on pasture and deforestation may cause regional climate changes and a marked reduction in rainfall. This in turn may lead to the formation of a desert environment or the encroachment of an existing desert onto formerly arable land.

1. Explain how human induced salinisation develops:

2. Describe alternative farming practices that do not use chemically intensive methods for:

   (a) Nutrient enrichment:

   (b) Pest control:

3. The synthetic insecticide DDT was in wide use after 1945 by farmers around the world. In the 1970s, the use of this chemical was banned by most Western countries. It remains as a residue in soils and waterways.

   (a) Explain why this chemical was banned:

   (b) Explain why the use of synthetic pesticides is becoming increasingly uneconomic:

4. Discuss the features of **desertification**, outlining its causes and the ways in which it may be averted or reversed:

The Impact of Farming

Global crop production depends on soil, yet soil loss via erosion and pollution threatens ecosystems throughout the world. As the world population grows and more land is cultivated for agriculture, sustainable farming practices will become essential to our continued survival. Current industrialised farming practices are considered unsustainable, even though they increase yields from relatively small amounts of land. Over-use of inorganic fertiliser, erosion of top soil, and high water demands are the leading causes of this unsustainability. Finding solutions is not easy; if farmers suddenly stopped using inorganic fertilisers, the world would experience widespread famine. The implications of two contrasting farming practices are outlined below.

Industrialised Intensive Agriculture

Intensive farming techniques flourished after World War II. Using **high-yielding hybrid cultivars** and large inputs of **inorganic fertilisers**, **chemical pesticides**, and **farm machinery**, crop yields increased to 3 or 4 times those produced using the more extensive (low-input) methods of 5 decades ago. Large areas planted in monocultures (single crops) are typical. Irrigation and fertiliser programmes are often extensive to allow for the planting of several crops per season. Given adequate irrigation and continued fertiliser inputs, yields from intensive farming are high. Over time, these yields decline as soils are eroded or cannot recover from repeated cropping. These problems can be alleviated with good crop management.

Intensive agriculture relies on the heavy use of irrigation, inorganic fertilisers (produced using fossil fuels), pesticides, and farm machinery. Such farms may specialise in a single crop for many years.

Impact on the Environment

- Pesticide use is escalating yet pesticide effectiveness against target pest species is decreasing. Non-target species are often adversely affected and species diversity declines.
- Mammals and birds may be affected by **bioaccumulation** of pesticides in the food chain and loss of food sources as invertebrate species diminish.
- Fertiliser use is increasing, resulting in a continued decline in soil and water quality.
- More fertiliser leaches from the soil and enters groundwater as a pollutant, relative to organic farming practices.
- Large fields lacking hedgerows or vegetated borders create an impoverished habitat and cause the isolation of remaining wooded areas.
- A monoculture regime leads to reduced biodiversity.

Sustainable Agricultural Practices

Organic farming is a sustainable form of agriculture based on the avoidance of chemicals and applied inorganic fertilisers. It relies on mixed (crop and livestock) farming and crop management, combined with the use of environmentally friendly pest controls (e.g. biological controls and flaming), and livestock and green manures. Organic farming uses **crop rotation** and **intercropping**, in which two or more crops are grown at the same time on the same plot, often maturing at different times. If well cultivated, these plots can provide food, fuel, and natural pest control and fertilisers on a sustainable basis. Yields are typically lower than on intensive farms, but the produce can fetch high prices, and pest control and fertiliser costs are reduced.

Some traditional farms use low-input agricultural practices similar to those used in modern organic farming. However, many small farming units find it difficult to remain economically viable.

Impact on the Environment

- Pesticides do not persist in the environment nor accumulate in the food chain.
- Produce is pesticide free and produced in a sustainable way.
- Alternative pest controls, such as using natural predators and pheromone traps, reduce the dependence on pesticides.
- The retention of hedgerows or vegetated borders increases habitat diversity and produces corridors for animal movement between forested areas.
- Crop rotation (alternation of various crops, including legumes) prevents pests and disease species building up to high levels.
- Conservation tillage (ploughing crop residues into the topsoil) as part of the crop rotation cycle improves soil structure.
- More land may need to be cultivated to achieve the same biomass, so less land can be left unfarmed.

1. Discuss how intensive farming practices could be modified to reduce their impact on the environment:

Related activities: Pest Control, Sustainable Land Use Practices

Sustainable Agriculture

Sustainable agriculture refers to the long-term ability of a farm to produce food without irreversibly damaging ecosystem health. The key to reducing world hunger and the harmful environmental effects of high input agriculture is to develop sustainable farming systems and phase them in over a manageable period of time. Two key issues in sustainable agriculture are **biophysical** and **socio-economic**. Biophysical issues centre on soil health and the biological processes essential to crop productivity. Socio-economic issues centre on the long-term ability of farmers to manage resources, such as labour, and obtain inputs, such as seed. Sustainable agriculture relies on good soil conservation practices, and the use of manure, compost, and other forms of organic matter over inorganic fertilisers. It also uses an integrated programme of pest management, emphasising biological and physical means of pest control and using chemical controls only as a last resort. Such regimes are increasingly proving more profitable than high-input systems because less money is spent on inputs of irrigation water, fertilisers, and pesticides.

Sustainable agricultural practices are economically viable and environmentally sound. Increasingly, farmers are investigating methods by which they can earn a reasonable living from the land, while remaining less reliant on government subsidies, and petroleum and chemical inputs. As in intensive farming, there are many approaches to sustainable agriculture. That pictured here just one.

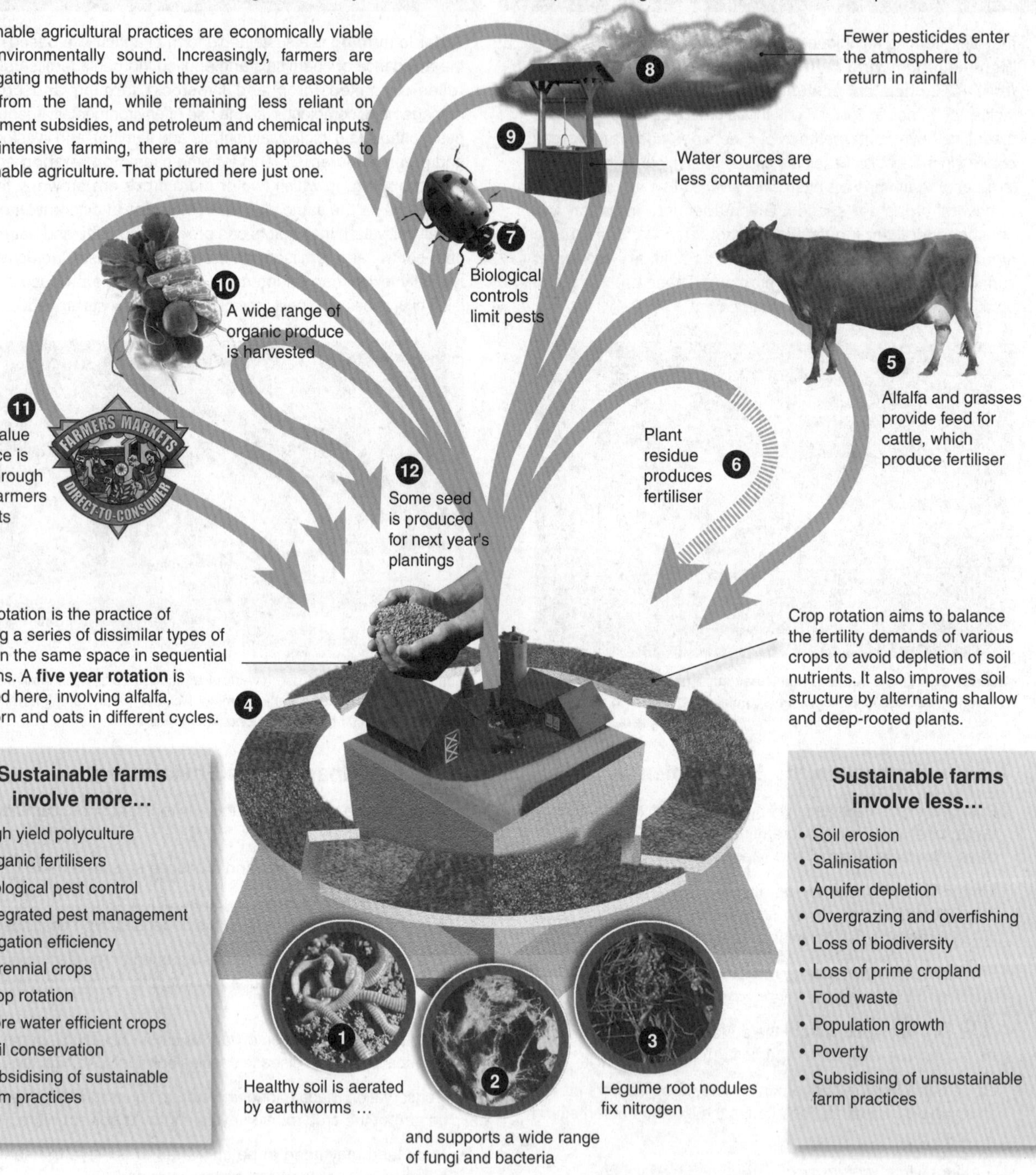
8
Fewer pesticides enter the atmosphere to return in rainfall
9
Water sources are less contaminated
7
Biological controls limit pests
10
A wide range of organic produce is harvested
5
Alfalfa and grasses provide feed for cattle, which produce fertiliser
11
High value produce is sold through local farmers markets
FARMERS MARKETS
DIRECT-TO-CONSUMER
12
Some seed is produced for next year's plantings
6
Plant residue produces fertiliser
Crop rotation is the practice of growing a series of dissimilar types of crops in the same space in sequential seasons. A **five year rotation** is pictured here, involving alfalfa, rye, corn and oats in different cycles.
4
Crop rotation aims to balance the fertility demands of various crops to avoid depletion of soil nutrients. It also improves soil structure by alternating shallow and deep-rooted plants.
1
Healthy soil is aerated by earthworms ...
2
and supports a wide range of fungi and bacteria
3
Legume root nodules fix nitrogen

Sustainable farms involve more...

- High yield polyculture
- Organic fertilisers
- Biological pest control
- Integrated pest management
- Irrigation efficiency
- Perennial crops
- Crop rotation
- More water efficient crops
- Soil conservation
- Subsidising of sustainable farm practices

Sustainable farms involve less...

- Soil erosion
- Salinisation
- Aquifer depletion
- Overgrazing and overfishing
- Loss of biodiversity
- Loss of prime cropland
- Food waste
- Population growth
- Poverty
- Subsidising of unsustainable farm practices

1. Describe the core principle of sustainable agriculture: ______________________

____

2. Explain why farmers practising sustainable agriculture can find it more profitable than intensive farming in the long term:

____

____

____

Related activities: The Impact of Farming

Energy Resources

The types and amounts of energy humans use are major factors in determining quality of life and harmful environmental effects. The sun provides most of the energy used to heat the Earth and also provides several forms of renewable energy. A tiny 1% is generated and sold to supplement this solar input. Most of this **commercial energy** comes from extracting and burning mineral resources obtained from the Earth's crust (primarily fossil fuels). Fossil fuels (petroleum, natural gas, and coal) provide about 85% of the all commercial energy in the world. **Biomass fuels** (wood, peat, charcoal, and manure) contribute about 6% of commercial energy, while renewable sources (e.g. hydroelectricity) and **nuclear power** make up 4-5% each. These contributions are not made equally across the globe. In many poorer countries, biomass fuels provide most of the energy used for heating and cooking, while in more developed countries nuclear power supplies up to 20% of all electric power. Per capita consumption is an important consideration too. The twenty wealthiest countries in the world constitute less than a fifth of the world's population yet use more than half the world's commercial energy supply. World-wide, affluence equates to high energy consumption.

Commercial Energy Resources from the Earth's Crust

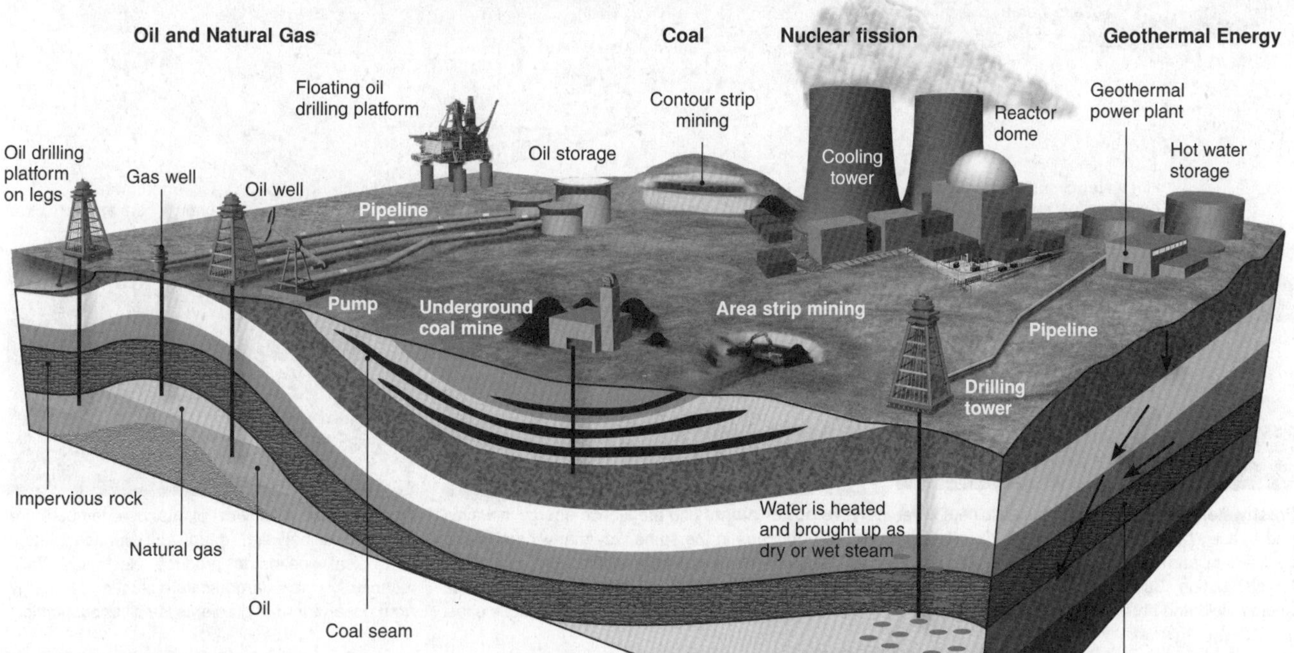

Important commercial energy resources from the Earth's crust include geothermal energy, coal, oil, and natural gas. Processed **uranium ore** is important as a fuel in nuclear reactors. With the possible exception of geothermal energy, essentially all these resources are non-renewable and finite.

Wikimedia Commons

In geothermal power plants, steam, heat, or hot water from **geothermal reservoirs** drives turbines to produce electricity. The used geothermal water is then returned down an injection well into the reservoir to be reheated, to maintain pressure, and to sustain the reservoir.

A nuclear power plant uses uranium-235 or plutonium-239 as fuel in a controlled nuclear fission reaction to release energy for propulsion, heat, and electricity generation. While nuclear power releases no CO_2, the safe storage and disposal of nuclear waste remains a challenge.

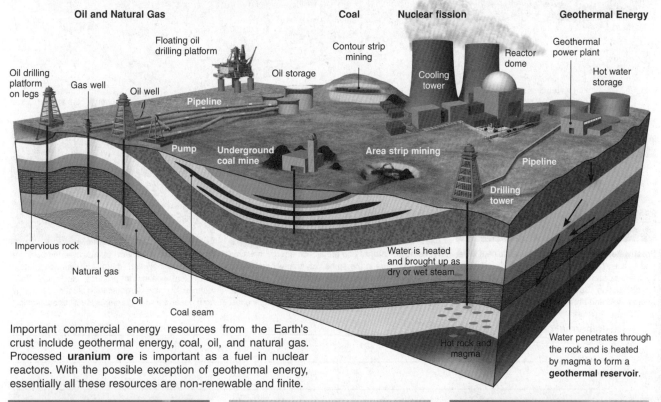

Fossil fuels are hydrocarbons ranging from very volatile materials, such as methane, to liquid petroleum and nonvolatile materials like coal. Fossil fuels are non-renewable because they take millions of years to form and are being depleted at a much faster rate than they are being formed.

1. Describe the biological basis and historical basis of each of the following sources of power generation:

(a) Coal: ______________________________

____

____

(b) Oil: ______________________________

____

____

Related activities: Biofuels, Global Warming, Energy Conservation

Renewable Energy Resources

Wave and tidal power: Regular tidal flows and the kinetic energy in waves can be used to spin turbines and produce electricity. Tidal power is more predictable than wind or solar power but the number of suitable sites is limited.

Solar thermal plants collect sunlight and focus it on oil-filled pipes running through the middle of curved solar collectors. These systems can generate temperatures high enough to run industrial processes or run turbines and produce electricity.

Wind power has been gaining popularity since the 1980s. Wind power is a virtually unlimited source of energy at favourable sites; its global potential is about five times the current world energy use and has significant environmental advantages over nuclear power.

Hydroelectric power generation uses the energy of flowing water generate electricity. A dam is built across a large river to create a reservoir and stored water is released at controlled rates to spin turbines and produce electricity.

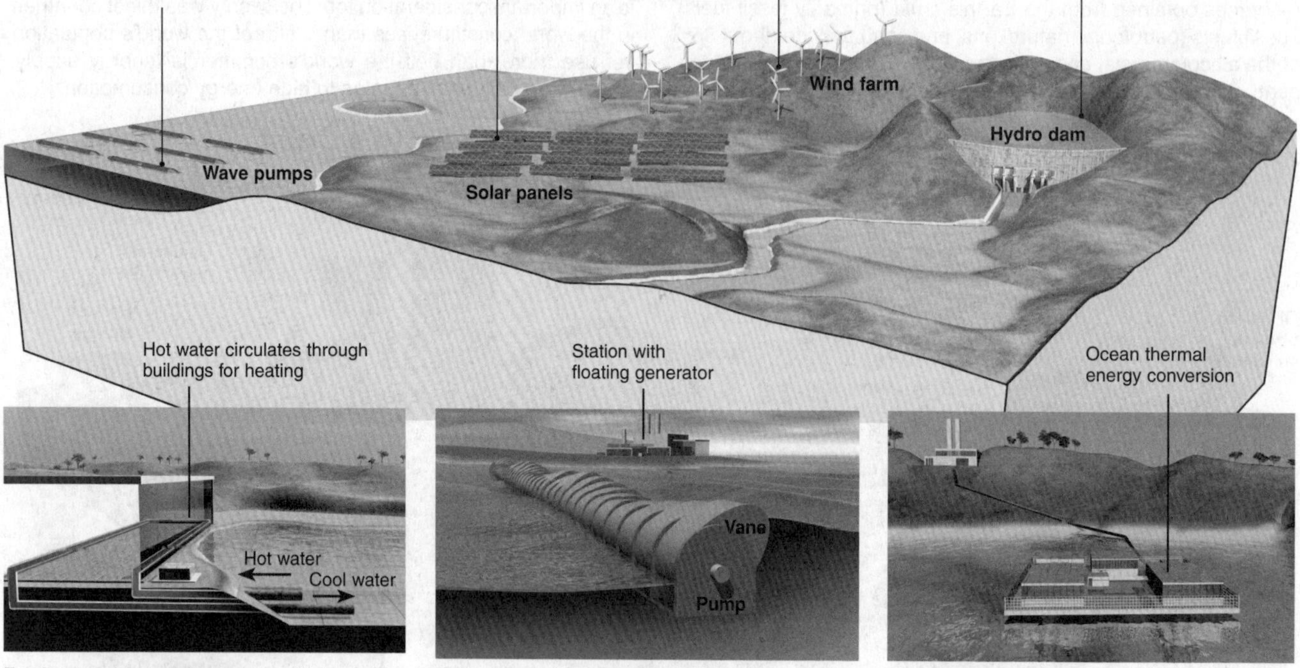

Freshwater solar ponds can be used to heat water and space. Water, in insulated ponds, is heated by the sun and then pumped to insulated tanks for distribution. Such systems have a moderate energy yield and are environmentally friendly.

Wave power plants use the kinetic energy in waves to drive turbines in the same way that windmills use air. Tidal and wave power probably won't contribute much to global electricity production because of the high construction costs and the lack of suitable sites.

Ocean thermal power plants use temperature differentials between deep and surface waters in tropical oceans to produce electricity. While technically viable, large scale extraction of energy from ocean thermal gradients is still uneconomic.

(c) Geothermal power: ______________________________

____

____

2. Choose one of the world's current major commercial energy sources (fossil fuels, biomass fuels, hydroelectric power, or nuclear power) and discuss its sustainablility, economic cost per unit energy yield, and environmental impact:

____

____

____

____

____

____

3. Choose one of the current primary sustainable sources of power (solar, tidal or wave, hydroelectric, or wind) and discuss its merits and disadvantages as a significant source of global energy in the future:

____

____

____

____

____

____

Biofuels

There is a global energy crisis looming. The combined effect of the dwindling reserves of fossil fuels, combined with their rather catastrophic long term effect on the world's climate, make the search for renewable energy sources imperative. Alternative energy sources such as solar power and the requisite high efficiency batteries have yet to become efficient enough and cheap enough to be serious replacements for fossil fuels.

Renewable biomass energy resources may provide a useful supplement to traditional fuels such as **coal**, **gas**, and **oil** (including its refined products of diesel, petrol, and kerosene). **Biofuels** include ethanol, **gasohol** (a blend of petrol and ethanol), methanol, and diesel made from a blend of plant oils and traditional diesel oil. **Biogas** (methane) is an important renewable gas fuel made by fermenting wastes in a digester.

Gasohol

Gasohol is a blend of finished motor gasoline containing alcohol (generally ethanol but sometimes methanol). In Brazil, gasohol consists of 24% ethanol mixed with petrol.

Advantages

- Cleaner fuel than petrol
- Renewable resource
- Creates many jobs in rural areas

Disadvantages

- Ethanol burns hotter than petrol so petrol engines tend to overheat and they need to be modified
- Fuel tank and pipes need coating to prevent corrosion by ethanol
- Fuel consumption 20% greater compared with petrol

Sources of biomass for ethanol production

- *Sugar cane* (ethanol is produced in this way in Brazil).
- *Corn starch* (in the USA).
- Grass, certain waste materials (paper, cardboard), and from wood. Fast-growing hardwood trees can be treated to release cellulose. Once released, it may be converted to simple glucose by hydrolytic enzymes and then fermented to produce ethanol.

Biogas

Methane gas is produced by anaerobic fermentation of organic wastes such as sewage sludge at sewage waste treatment stations, animal dung, agricultural wastes, or by the rotting contents of landfill sites.

Stages in methane production

Saprophytic bacteria (facultative anaerobes) break down fats, proteins, and polysaccharides.

↓

Acid-forming bacteria break down these monomers to short-chain organic acids.

↓

Methanogen bacteria (strict anaerobes) produce methane gas.

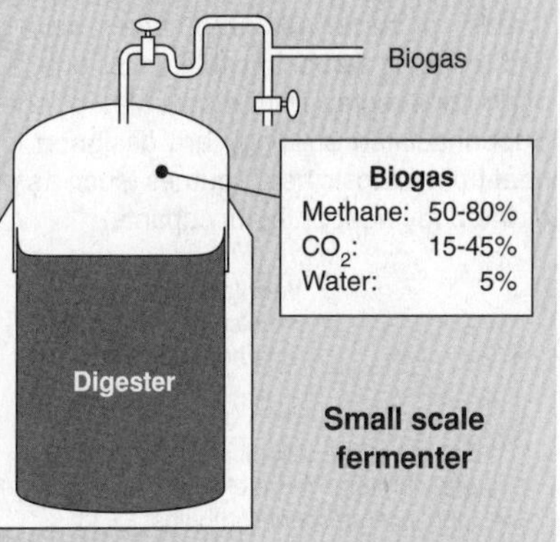

Sources: *Biological Sciences Review*, Sep 2000, pp.27-29; *Biologist*, Feb 1998, pp. 17-21; Microorganism & Biotechnology, 1997, Chenn, P. (John Murray Publishers).

Traditional sources of renewable energy include animal dung, which is collected and then dried in the sun and used as fuel.

Fuels such as petrol, diesel, LPG, and CNG are derived from oil and natural gas extracted from non-renewable geological deposits.

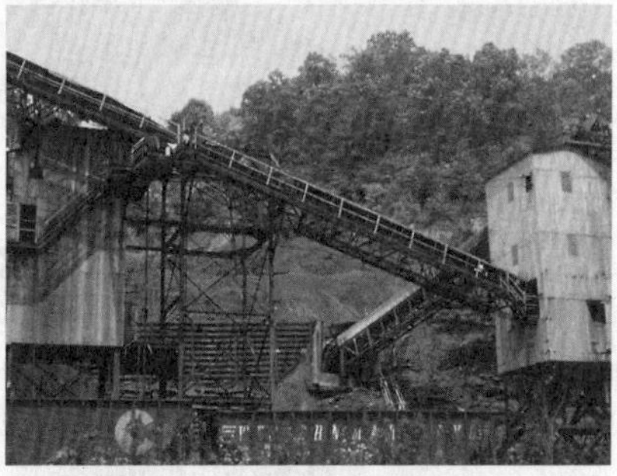

Coal provided the energy for the industrial revolution. It is now regarded as a dirty fuel with many health hazards associated with its use.

1. Explain the nature of the following renewable fuels:

   (a) Biogas: ______________________________

   ____

   (b) Gasohol: ______________________________

   ____

2. Describe two disadvantages of using pure ethanol as a motor fuel: ______________________________

   ____

3. Suggest how a small biogas fermenter could be used on a farm to reduce waste and provide a fuel source:

   ____

   ____

4. Comment on the potential conflicts associated with producing fuels from food crops (such as maize):

   ____

   ____

Energy Conservation

An **energy conservation** drive is required to make better use of the energy sources we have. Developed nations waste large amounts of energy. In the United States 84% of commercial energy is wasted (41% as unavoidable waste and 43% through the use of energy inefficient systems and poorly designed and insulated buildings). Traditionally, the solution for our global energy requirements has been to produce more energy, but a more **energy efficient** solution is now in demand. Energy efficiency involves improving products or systems so that they do more work and waste less energy, thus conserving energy overall. General improvements in efficiency can be achieved by reducing energy use, improving the energy efficiency of processes, appliances, and vehicles, and increased use of public transport. Energy experts also advise that producing and using the most economical energy sources first, before moving on to more expensive forms, conserves both energy and resources.

Energy Efficiency at Home

Most of the energy used in domestic or commercial buildings is for heating, air conditioning, and lighting. Most buildings are highly energy inefficient, leaking energy as heat. New technologies and products enable the construction of energy efficient buildings (below), or **superinsulated** homes, saving the home owner money and reducing carbon dioxide emissions. Superinsulated homes are often constructed from strawbales or sheltered (in part) by earth. Superinsulated buildings are designed to leak no heat, and gain heat from intrinsic heat sources (such as waste heat from appliance or the body heat of the occupants).

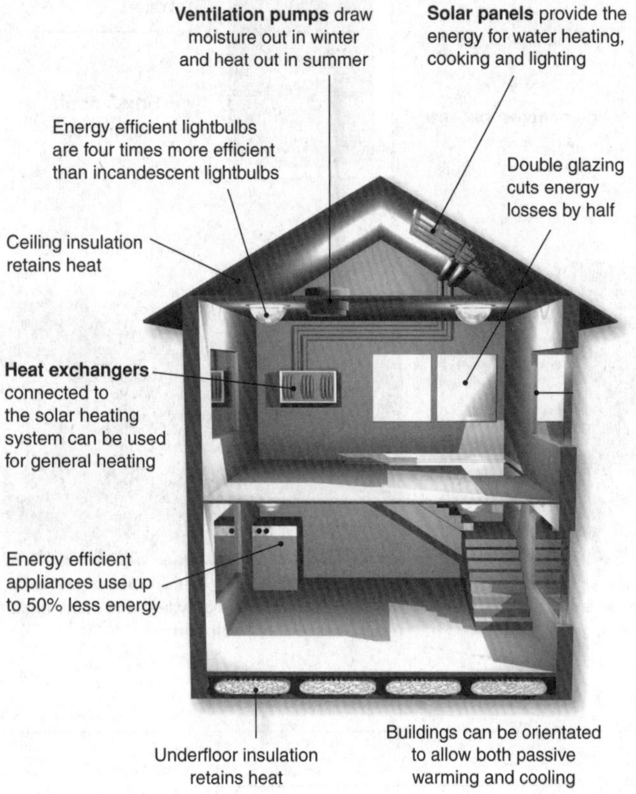

Ventilation pumps draw moisture out in winter and heat out in summer
Solar panels provide the energy for water heating, cooking and lighting
Energy efficient lightbulbs are four times more efficient than incandescent lightbulbs
Double glazing cuts energy losses by half
Ceiling insulation retains heat
Heat exchangers connected to the solar heating system can be used for general heating
Energy efficient appliances use up to 50% less energy
Underfloor insulation retains heat
Buildings can be orientated to allow both passive warming and cooling

Energy Efficiency in Transportation

20% of the world's global energy is used for transportation, 90% of which is wasted because it can not be utilised by internal combustion engines. The **Corporate Average Fuel Economy** (**CAFE**) regulation is designed to improve the fuel economy of cars and light trucks sold in the US. In 2002, fuel economy was 14% better than it would have been without the standards in place. The use of lighter, stronger materials in car manufacturing, coupled with improved aerodynamics, and the inclusion of heavier vehicles into the CAFE regulation will also aid fuel efficiency.

Hybrid vehicles (right) use two or more different power sources for propulsion, with the combination of combustion engine and electric batteries being the most common. Energy savings are gained by capturing the energy released during braking, storing energy in the batteries, using the electric engine during idling, and using both the petrol and electric motors for peak power needs (which reduces fuel consumption).

Yellow Cab

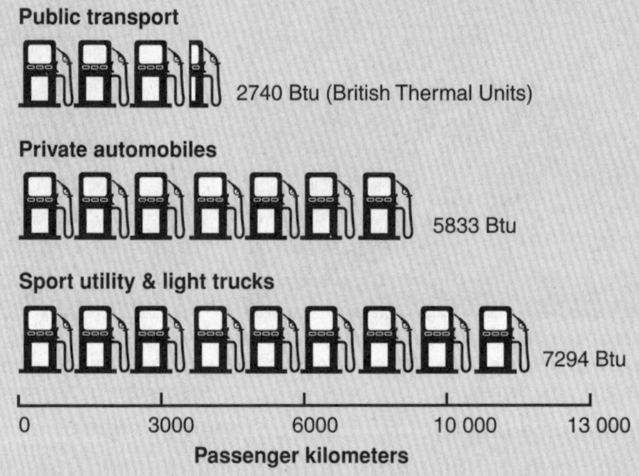

Public transport
2740 Btu (British Thermal Units)
Private automobiles
5833 Btu
Sport utility & light trucks
7294 Btu
0
3000
6000
10 000
13 000
Passenger kilometers

Despite being at least twice as fuel efficient as travel by a private vehicles, public transport remains unpopular in many developed nations. Increased use of public transport would reduce the costs of running private vehicles, building and maintaining roads, and maintaining air quality.

1. (a) Explain why reducing energy use is seen as important in this century: ______________________

   (b) Discuss the methods by which we can save energy in transportation: ______________________

2. Describe how an energy efficient building is designed to reduce energy waste: ______________________

Ecological Impacts of Fishing

Fishing is an ancient human tradition that not only satisfies our need for food, but is economically, socially, and culturally important. Today, however, tradition has been transformed in a worldwide resource extraction industry. Several decades of overfishing in all of the world's oceans has pushed commercially important species (such as cod) into steep decline. The United Nation's Food and Agriculture Organisation (FAO) reports that almost seven out of ten of the ocean's commercially targeted marine fish stocks are either fully or heavily exploited (44%), over-exploited (16%), depleted (6%), or very slowly recovering from previous overfishing (3%). The **maximum sustainable yield** has been exceeded by too many fishing vessels catching too many fish, often using wasteful and destructive methods.

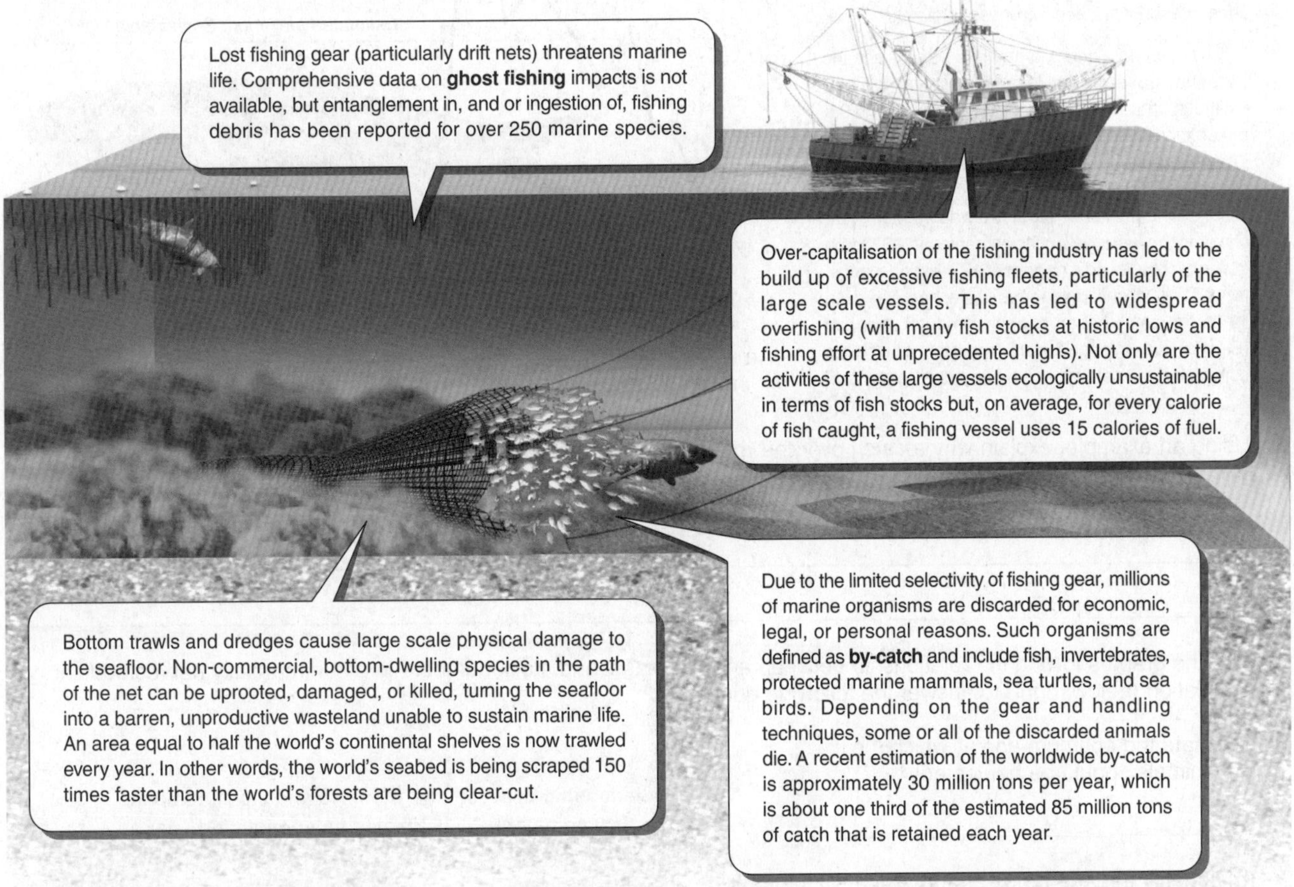
Lost fishing gear (particularly drift nets) threatens marine life. Comprehensive data on ghost fishing impacts is not available, but entanglement in, and or ingestion of, fishing debris has been reported for over 250 marine species.
Over-capitalisation of the fishing industry has led to the build up of excessive fishing fleets, particularly of the large scale vessels. This has led to widespread overfishing (with many fish stocks at historic lows and fishing effort at unprecedented highs). Not only are the activities of these large vessels ecologically unsustainable in terms of fish stocks but, on average, for every calorie of fish caught, a fishing vessel uses 15 calories of fuel.
Bottom trawls and dredges cause large scale physical damage to the seafloor. Non-commercial, bottom-dwelling species in the path of the net can be uprooted, damaged, or killed, turning the seafloor into a barren, unproductive wasteland unable to sustain marine life. An area equal to half the world's continental shelves is now trawled every year. In other words, the world's seabed is being scraped 150 times faster than the world's forests are being clear-cut.
Due to the limited selectivity of fishing gear, millions of marine organisms are discarded for economic, legal, or personal reasons. Such organisms are defined as by-catch and include fish, invertebrates, protected marine mammals, sea turtles, and sea birds. Depending on the gear and handling techniques, some or all of the discarded animals die. A recent estimation of the worldwide by-catch is approximately 30 million tons per year, which is about one third of the estimated 85 million tons of catch that is retained each year.

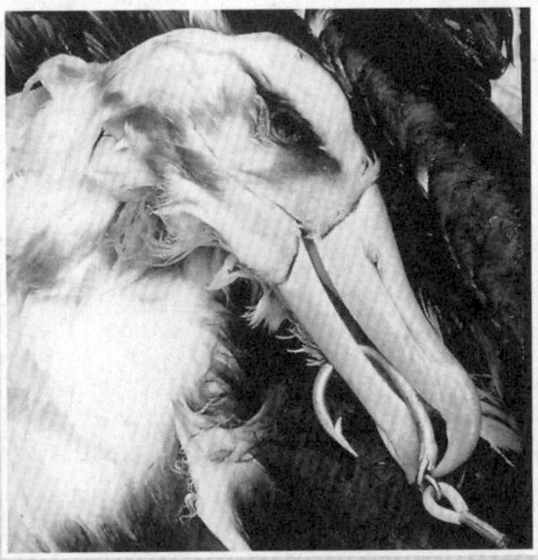

Photo: Jane Ussher

Longline fishing (mainly for tuna) results in the death of 100 000 albatrosses and petrels every year in the southern Pacific alone. Six of the world's twenty albatross species are in serious decline and longline fishing is implicated in each case.

NOAA

Over-harvesting of abundant species, or removal of too many reproductive individuals from a population, can have far reaching ecological effects. Modern boats, with their sophisticated fish-finding equipment, have the ability to catch entire schools of fish.

Fish farming, once thought to be the solution to the world's overfishing problems, actually accelerates the decline of wild fish stocks. Many farmed fish are fed meal made from wild fish, but it takes about one kilo of wild fish to grow 300 g of farmed fish. Some forms of fish farming destroy natural fish habitat and produce large scale effluent flows.

1. Explain the term **over-exploitation** in relation to commercial fisheries management: ______

2. Define the term **by-catch**: ______

The Peruvian Anchovy Fishery: An Example of Over-Exploitation

Before 1950, fish in Peru were harvested mainly for human consumption. The total annual catch was 86 000 tonnes. In 1953, the first fish meal plants were developed. Within nine years, Peru became the number one fishing nation in the world by volume; 1700 purse seiners exploited a seven month fishing season and Peru's economy was buoyant.

In 1970, fearing a crash, a group of scientists in the Peruvian government issued a warning. They estimated that the sustainable yield was around 9.5 million tonnes, a number that was being surpassed. The government decided to ignore this; due to the collapse of the Norwegian and Icelandic herring fisheries the previous year, Peru was the dominant player in the lucrative anchovy market. In 1970, the government allowed a harvest of 12.4 million tonnes. In 1971, 10.5 million tonnes were harvested. In 1972, the combination of environmental changes (El Niño) and prolonged overfishing led to a complete collapse of the fishery, which has never recovered.

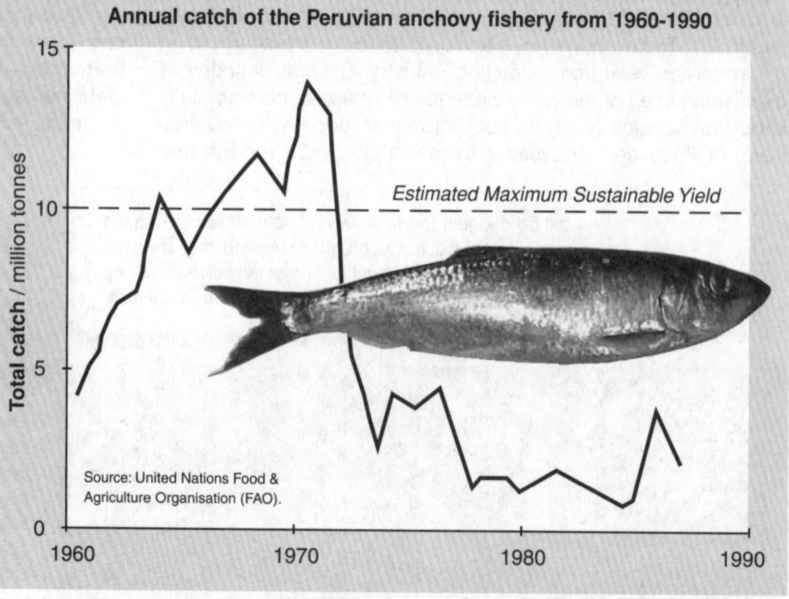

3. Using an example, explain why a catch over the **maximum sustainable yield** will result in the collapse of a fishery:

4. Use the graph showing the relationship between age, biomass, and stock numbers in a commercially harvested fish population (below, right) to answer the following questions:

(a) State the optimum age at which the animals should be harvested:

(b) Identify the age range during which the greatest increase in biomass occurs:

(c) Suggest what other life history data would be required by fisheries scientists when deciding on the management plan for this population:

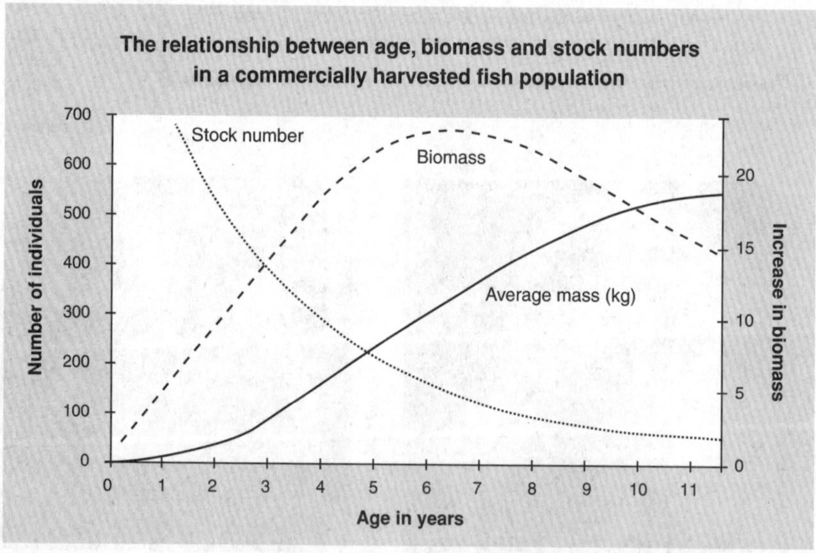

5. Discuss three methods by which fish populations can be conserved:

6. (a) Outline two advantages of marine fish farming:

(b) Outline two disadvantages of marine fish farming:

Fisheries Management

The stock of North Sea cod (*Gadus morhua*) is one of the world's six large populations of this economically important species. As one of the most intensively studied, monitored, and exploited fish stocks in the North Sea, it is considered a highly relevant indicator of how well sustainable fisheries policies are operating. Stocks of commercially fished species must be managed carefully to ensure that the catch (take) does not undermine the long term sustainability of the fishery. This requires close attention to **stock indicators**, such as catch per unit of fishing effort, stock recruitment rates, population age structure, and spawning biomass. Currently, the North Sea cod stock is below safe biological limits and stocks are also depleted in all waters adjacent to the North Sea, where the species is distributed. Recent emergency measures plan to arrest this decline.

Total international landings of North Sea cod

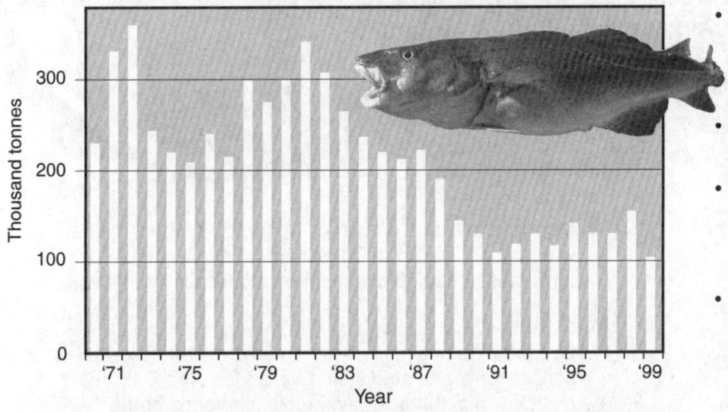

Recruitment and spawning stock biomass of North Sea cod

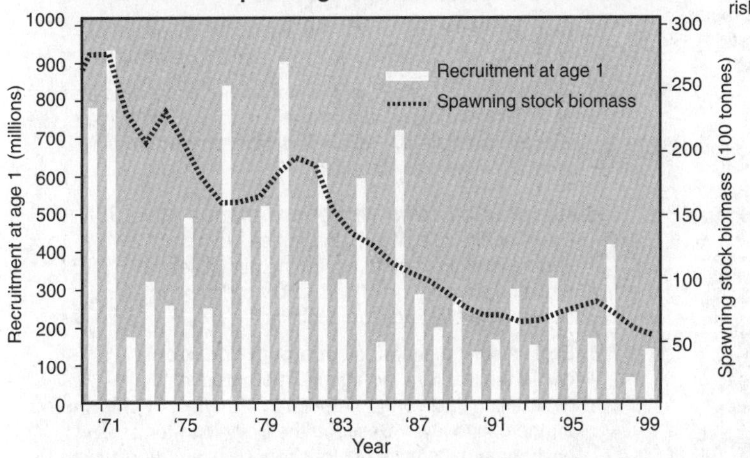

The state of the fishery

- Fishing mortality has increased gradually, and has been above the precautionary limit (that considered to be a safe take) since 1980 (left).
- Recruitment has been generally poor since 1987 (below, left).
- The number of spawning adults has fallen to levels below those required to recruit new individuals into the stock (below, left).
- ICES (the International Council for the Exploration of the Sea) advised that the spawning stock biomass (an indicator of the number of breeding adults) reached a new historic low in 2001, and that the risk of stock collapse is high.

What has been done?

- A large part of the North Sea was closed for cod fishing between February and April 2001, to protect juvenile cod .
- The TAC has been set at approximately half that set for the year 2000. Further regulations, such as increasing net mesh size and reducing the volume of fish discarded, are planned, and will further restrict the effort of fishing fleets until (if) the stock recovers.
- The ICES has recommended a recovery plan that will ensure recovery of the spawning stock to a level of more than 150 000 t. Reductions in TAC alone are insufficient to stop the declines.

Some important definitions

Stock: The part of the population from which catches are taken in a fishery

Stock recruitment: The entry of juvenile fish into the fish stock

Total Allowable Catch (TAC): The catch that can be legally taken from the stock

Stock collapse: Population level at which the fish stock cannot recover

Sources: European Environmental Agency (EEA), CEFAS (The Centre for Environment, Fisheries, and Aquaculture Science), and the ICES.

1. It has been known for more than a decade that the stock of cod in the North Sea has been declining drastically and that fishing takes were not sustainable. With reference to the data above, discuss the evidence to support this statement:

2. Using the information provided above for guidance, describe the state the North Sea cod fishery, summarizing the main points below. If required, develop these as a separate report. Identify:

   (a) The location of the fishery:

   (b) The current state of the fishery (including stock status, catch rates, TAC, and quota):

   (c) Features of the biology of cod that are important in the management of the fishery (list):

   (d) Methods used to assess sustainability (list):

   (e) Management options for the fishery (list):

Related activities: Ecological Impacts of Fishing
Web links: **Web links**: Starfish, Fisheries and the Environment

Pollution and Global Change

Investigating human Induced changes in ecosystems and conservation measures

Human impact on resources: water, air and land pollution, biodiversity and conservation.

Learning Objectives

☐ 1. Compile your own glossary from the **KEY WORDS** displayed in **bold type** in the learning objectives below.

Human Impact and Resources *(pages 75-76, 120)*

☐ 2. Describe the impacts of humans on the environment, including **pollution** and loss of biodiversity. Recognise that these impacts are the result of resource use, rapid population growth and **urbanisation**, large population size, and disproportionate distribution of resources.

Water Pollution *(pages 95-96, 121-123)*

☐ 3. Explain what is meant by **water pollution**. Describe different causes of water pollution and explain their effects. Distinguish between **point sources** and **diffuse** (or non-point) **sources** of water pollutants.

☐ 4. Describe the effects of **organic effluent** (e.g. sewage or milk) and **fertiliser run-off** (nitrates and phosphates) on aquatic ecosystems. Include reference to: water quality, **biochemical oxygen demand**, effects on biodiversity, **eutrophication** and algal blooms, spread of pathogens, and (toxic) nitrate load in the groundwater.

☐ 5. Recognise the link between intensive (industrialised) agricultural systems, soil degradation, fertiliser and pesticide misuse, and subsequent pollution of water sources and **eutrophication**.

☐ 6. Discuss measures for the prevention, reduction, or mitigation of water pollution, e.g. **sewage treatment**. If required, describe an example of recovery of a formerly polluted water body following removal of the pollutant.

☐ 7. Appreciate that country-specific legislation provides water quality standards for different water uses. Describe the role of **indicator organisms** in the assessment of water quality. Identify some common indicator organisms for freshwater systems and explain what their presence indicates.

Atmospheric Pollution *(pages 125-126)*

☐ 8. Explain what is meant by **atmospheric pollution** and recognise both natural and anthropogenic sources. Describe different air **pollutants** and, for each, identify its effect on the environment. Recognise **acid rain** and stratospheric **ozone depletion** as consequences of atmospheric pollution.

Global warming *(pages 117-118)*

☐ 9. Recognise the **greenhouse effect** as a natural phenomenon responsible for the Earth's equable temperature and distinguish this from the accelerated **global warming** occurring as a result of increased levels of **greenhouse gases**. Describe the causes of increasing greenhouse gas levels and document their predicted ecological and economic effects.

☐ 10. Discuss measures that could be taken to reduce global warming or its impact. Consider both short and long term measures and comment on the feasibility of these.

Acid rain *(pages 125, 131)*

☐ 11. Describe the origin, causes (formation), and biological consequences of **acid rain**. Explain the global distribution of the acid rain problem in terms of the original source of the pollution.

☐ 12. Discuss measures for the reduction and mitigation of acid rain, and comment on the feasibility of these.

☐ 13. Explain how air quality is monitored and appreciate that air quality standards regulate air emissions. Describe, with an example, the role of **indicator organisms** in the assessment of air quality.

Stratospheric ozone depletion *(pages 129-130)*

☐ 14. Describe the role of the atmospheric (strictly stratospheric) ozone in absorbing ultraviolet (UV) radiation. Outline the effects of UV radiation on living tissues and biological productivity.

☐ 15. Explain what is meant by **stratospheric ozone depletion**. Identify the agents implicated in its destruction, including reference to the effect of **chlorine** on the ozone layer. Describe the likely long term environmental effects of stratospheric ozone depletion.

☐ 16. Discuss measures to reduce the rate of ozone depletion. Emphasise methods for reducing the manufacture and release of ozone-depleting chemicals.

☐ 17. Distinguish between stratospheric ozone and localised ozone pollution in the lower atmosphere.

Land Management Issues *(pages 103-104, 134)*

☐ 18. Discuss the causes and effects of **deforestation**, emphasising its impact on the **biodiversity** and stability of forest ecosystems, and on carbon and nitrogen cycling. Identify regions (globally or locally) where deforestation is a major problem.

☐ 19. Discuss the conflict between the need to **conserve** forests and the rights of humans to obtain a living from the land. Describe methods by which forests may be conserved to ensure the **sustainable provision** of forestry resources. Comment on the feasibility of these.

☐ 20. Discuss the links between land clearance, agriculture, **soil degradation** and **desertification**, and loss of biodiversity. Describe the social and economic factors driving unsustainable land management practices.

Economic Costs of Pollution *(pages 132)*

☐ 21. Discuss the economic impact of pollution on societies. Consider the **direct costs**, such as the costs of environmental clean-up, as well as the **indirect costs**, such as the long term costs to human and agricultural health and environmental health and aesthetics.

☐ 22. Explain how decisions regarding environmental clean-up and restoration are made by weighing up the costs and benefits involved. Discuss the advantages and disadvantages in this approach.

☐ 23. Recognise that effective waste management systems carry a cost and that this is offset by the benefits gained by reducing pollution and waste. Discuss the view of solid waste as a potential resource (in part) and describe methods for **integrated waste management**, including reducing waste, and recycling organic waste and valuable commodities such as paper and glass.

Declining Biodiversity *(pages 133-139)*

☐ 24. Explain what is meant by **biodiversity** and discuss the importance of preserving and managing it as you would any resource. Identify and discuss factors important in the decline of the world's biodiversity, including:
(a) The impact of alien species on natural biota
(b) Hunting and collecting pressure
(c) Pollution and habitat loss

☐ 25. Identify some of the regions of naturally-occurring high **biodiversity** and describe the importance of these regions to global ecology. Using **rainforests** as an example, discuss the ethical, ecological, economic, and aesthetic reasons for the conservation of biodiversity.

☐ 26. Distinguish between **threatened** and **endangered species** and identify factors that cause species to become endangered. Identify some examples of endangered species and predict the short and long term consequences of species loss.

☐ 27. Using an example of a locally or globally endangered species (e.g. African elephant), describe its conservation status, potential for recovery, and management (see the strategies outlined in #28).

☐ 28. Using examples, discuss the advantages and application of the following conservation measures:
- *In-situ* conservation methods such as protection of terrestrial or aquatic nature reserves.
- Management programs for nature reserves, including control of alien species, **habitat restoration**, control of human exploitation, and **species recovery plans**.
- *Ex-situ* conservation methods such as **captive breeding** (and release) of animals, **botanic gardens**, and **seed** and sperm (gene) **banks**.
- The actions of international agencies (CITES, WWF).

☐ 29. Describe the role of **National Parks**, reserve lands, marine parks, and wildlife refuges in conserving biodiversity in your country. Understand the significance of legislative measures to protect species and the habitats they rely on.

Supplementary Texts

See page 7 for additional details of these texts:

■ Christopherson, R.W, 2007. **Elemental Geosystems**, chpt. 7, 16-17.

■ Miller, G.T, 2007. **Living in the Environment: Principles, Connections and Solutions**, chpt. 11, 14-17, 19, 21.

■ Raven *et. al.*, 2002. **Environment**, chpt. 11-14, 16-22.

■ Reiss, M. & J. Chapman, 2000. **Environmental Biology** (Cambridge University Press), pp. 5-7, 9-12, 24-25, chpt. 3.

■ Smith, R. L. & T.M. Smith, 2001. **Ecology and Field Biology**, reading as required.

Periodicals

STUDENT'S REFERENCE

See page 7 for details of publishers of periodicals:

■ **Global Warming** Time, special issue, 2007. *A special issue on global warming: the causes, perils, solutions, and actions. Comprehensive and well illustrated, this account provide up-to-date information at a readable level.*

■ **The Big Thaw** National Geographic, Sept. 2004, pp. 12-75. *Part of a special issue providing an up-to-date, readable account of the state of global warming and climate change.*

■ **Unlocking the Climate Puzzle** National Geographic, 193(5) May 1998, pp. 38-71. *Earth's climate, including global warming & desertification.*

■ **Water Pressure** National Geographic, Sept. 2002, pp. 2-33. *The demand for freshwater for human consumption and hygiene and the problems associated with increased pressure on supplies.*

■ **Biodiversity and Ecosystems** Biol. Sci. Rev., 11(4) March 1999, pp. 18-21. *The importance of biodiversity to ecosystem stability and sustainability.*

■ **Hot Spots** New Scientist, 4 April 1998, pp. 32-36. *An examination of the reasons for the very high biodiversity observed in the tropics.*

■ **Biodiversity: Taking Stock of Life** National Geographic, 195(2) Feb. 1999 (entire issue). *A special issue exploring the Earth's biodiversity and what we can do to preserve it.*

■ **In Search of Solutions** National Geographic, Feb. 1999, pp. 72-87. *The impact of deforestation and measures possible to restore the damage.*

■ **Last of the Amazon** National Geographic, 211(1) Jan. 2007, pp. 40-71. *The current state of the Amazon forest, one of the world's most biologically diverse regions.*

■ **Tropical Rainforest Regeneration** Biol. Sci. Rev., 17(2) Nov. 2004, pp. 34-37. *Tropical rainforests: causes of and reasons for their destruction, and the role of the many complex biotic interactions in forest regeneration.*

■ **China's Growing Pains** National Geographic, March 2004, pp. 68-95. *China's growing population is putting increasing pressure on the environment.*

■ **The Long Shadow of Chernobyl** National Geographic, April 2006, pp. 32-53. *20 years on from the disaster at Chernobyl, the environmental and health effects are still lingering.*

TEACHER'S REFERENCE

■ **Can Sustainable Management Save Tropical Rainforests?** Scientific American, April 1997, pp. 34-39. *The difficulties of sustainable management of rainforests and the implications for conservation.*

■ **Wading in Waste** Scientific American, June 2006, pp. 42-49. *The impervious nature of urban environments reduces water penetration and results in high volumes of runoff into waterways.*

■ **Conservation for the People** Scientific American, October 2007, pp. 26-33. *Preserving biodiversity in ecological hotspots is not working as a conservation strategy. We need to protect ecosystems like wetlands that are vital to ecosystem and human health.*

■ **How Did Humans First Alter Global Climate?** Scientific American, March 2005, pp. 34-41. *A bold new hypothesis suggests that humans began altering the global climate thousands of years before their more recent use of fossil fuels.*

■ **Time to Rethink Everything** New Scientist, 27 April-18 May 2002 (4 issues). *Globalisation, the impact of humans, & the sustainability of our future.*

■ **Abrupt Climate Change** Scientific American, Nov. 2004, pp. 40-47. *Slow, steady changes to ambient conditions may push major climate drivers, such as ocean currents, to a critical point, triggering sudden and dramatic shifts in climate.*

■ **Fall of the Wild** Scientific American, May 2006, pp. 42-77. *Oil reservoirs and the effects they are having on the fragile ecology of Alaska's north slope. Community opinion is split between oil revenue and the traditional way of life. This account provides a good examination of the conflict between conservation and development.*

■ **The Last Menageries** New Scientist, 19 Jan. 2002, pp. 40-43. *The role of zoos today in conservation, education, and research.*

Internet

See pages 4-5 for details of how to access **Bio Links** from our web site: **www.thebiozone.com** From Bio Links, access sites under the topics:

BIODIVERSITY > **Biodiversity**: • Convention on biological diversity • Ecology and biodiversity • World atlas of biodiversity ... *and others*

CONSERVATION: > **Habitat Loss**: • Rainforest Information Centre > **Conservation Issues**: • Environment Australia online ... *and others*

HUMAN IMPACT: > **Endangered Species**: • African elephant database homepage • Endangered species ... *and others* > **Habitat loss**: • Causes of habitat loss and species endangerment • Habitat loss • Rainforest destruction • Soil erosion • We need our forests ... *and others* > **Conservation Issues**: • CITES • Conservation International • WWF ... *and others* > **Pollution**: • Chernobyl Assessment Project • US EPA • Pollution online... *and others* > **Global Warming**: • The greenhouse effect (CSIRO) • Carbon Dioxide Information Analysis Center • The EPA global warming site ... *and others* > **Ozone Depletion**: • Ozone depletion resource center • Stratospheric ozone depletion ... *and others* > **Recycling and Waste Management**: • How does a sewage treatment plant work? • Aluminium recycling • A-Z of recycling... *and many others*

Presentation MEDIA to support this topic:
ECOLOGY
• Biodiversity & Conservation
• Human Impact

Types of Pollution

Any addition to the air, water, soil, or food that threatens the survival, health, or activities of organisms is called **pollution**. **Pollutants** can enter the environment naturally (e.g. from volcanic eruptions) or through human activities. Most pollution from human activity occurs in or around urban and industrial areas and regions of industrialised agriculture. Pollutants may come from single identifiable **point sources**, such as power plants, or they may enter the environment from non-point or **diffuse sources**, such as through land runoff. While pollutants often contaminate the areas where they are produced, they can also be carried by wind or water to other areas. Commonly recognised forms of pollution include air and water pollution, but other less obvious forms of pollution, including light and noise pollution, are also the result of concentrations of human activity. Some global phenomena, such as **global warming** and **ozone depletion** are the result of pollution of the Earth's stratosphere.

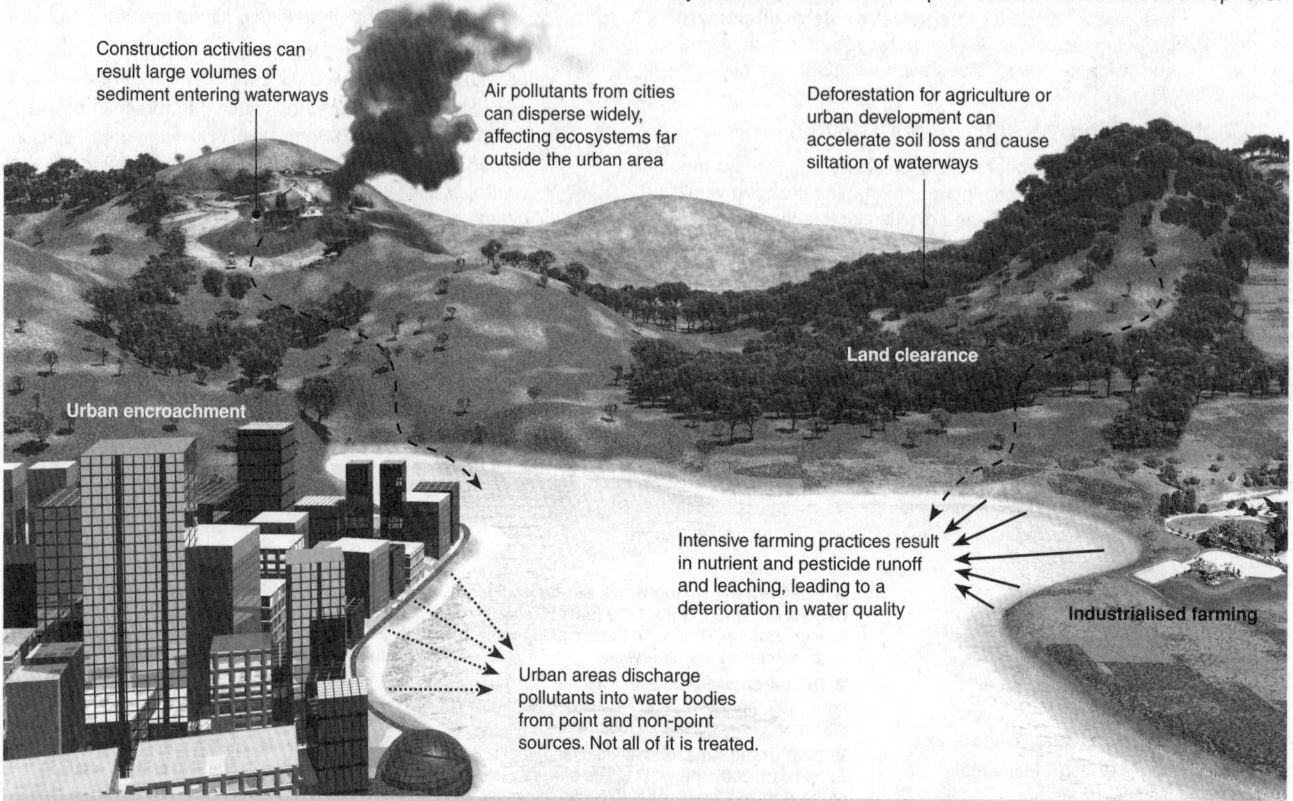
Construction activities can result large volumes of sediment entering waterways
Air pollutants from cities can disperse widely, affecting ecosystems far outside the urban area
Deforestation for agriculture or urban development can accelerate soil loss and cause siltation of waterways
Land clearance
Urban encroachment
Intensive farming practices result in nutrient and pesticide runoff and leaching, leading to a deterioration in water quality
Industrialised farming
Urban areas discharge pollutants into water bodies from point and non-point sources. Not all of it is treated.

Fertilisers, herbicides, and pesticides are major contaminants of soil and water in areas where agriculture is industrialised. Fertiliser runoff and leaching adds excess nitrogen and phosphorus to ground and surface water and leads to accelerated **eutrophication**.

Soil contamination occurs via chemical spills, leaching, or leakage from underground storage. The runoff from mining and metal processing operations can carry radioactive waste and heavy metals such as mercury, cadmium, and arsenic.

Together with vehicle exhausts, power plants and industrial emissions are a major source of air pollution. SO_2 and NO_2 from these primary sources mix with water vapour in the atmophere to form acids which may be deposited as rain, snow, or dry acid.

1. Identify the main sources of each of the following pollutants:

   (a) Pesticides and herbicides in waterways: ______________________

   (b) Sewage: ______________________

   (c) Oxides of sulfur and nitrogen: ______________________

   (d) Sedimentation and siltation of waterways: ______________________

2. Explain the impact of urbanisation on the pollution load in a given region: ______________________

____

____

Water Pollution

Water pollution can occur as a result of contamination from many sources, from urban and industrial to agricultural. Pollutants may first enter the groundwater where they are difficult to detect and manage. Some enter surface waterways directly through runoff from the land, but most are deliberately discharged at single (point) sources. Some pollutants alter the physical state of a water body (its temperature, pH, or turbidity). Others involve the addition of potentially harmful substances. Even substances that are beneficial at a low concentration may cause problems when their concentration increases. One such form of pollution involves excessive nutrient loading of waterways by organic effluent. This causes accelerated **eutrophication** (enrichment) or water bodies and results in excessive weed and algal growth. It also increases the uptake of dissolved oxygen by microorganisms that decompose the organic matter in the effluent. This reduces the amount of dissolved oxygen available to other aquatic organisms and may cause the death of many. An indicator of the polluting capacity of an effluent is known as the **biochemical** (or biological) **oxygen demand** or **BOD**. This is measured as the weight (mg) of oxygen used by one litre of sample effluent stored in darkness at 20°C for five days. Developing global and national initiatives to control water pollution is important because many forms of water pollution cross legislative boundaries. The US is the world's largest user of water but loses about 50% of the water it withdraws. Water conservation is required to enable more effective use of water, reduce the burden on wastewater systems, decrease pollution of surface and groundwater, and slow the depletion of aquifers.

Sources of Water Pollution

Sediment pollution: Soil erosion causes soil particles to be carried into waterways. The increased sediment load may cause choking of waterways, buildup behind dams, and the destruction of aquatic habitats.

Sewage: Water containing human wastes, soaps and detergents from toilets, washing machines, and showers are discharged into waterways such as rivers, lakes and the sea. Most communities apply some treatment.

Disease-causing agents: Disease-causing microbes from infected animals and humans can be discharged into waterways. This is particularly a problem during floods when human waste may mix with drinking water.

Inorganic plant nutrients: Fertiliser runoff from farmland adds large quantities of nitrogen and phosphorus to waterways. This nutrient enrichment accelerates the natural process of **eutrophication**, causing algal blooms and prolific aquatic weed growth.

Organic compounds: Synthetic, often toxic, compounds, may be released into waterways from oil spills (see above), the application of agrichemicals, and as the waste products of manufacturing processes (e.g. dioxin, PCBs, dieldrin, phenols, and DDT).

RA

Thermal pollution: Many industrial processes, including thermal power generation (above), release heated water into river systems. The increase in water temperature reduces oxygen levels and may harm the survival of river species.

Kurchatov Inst.

Radioactive substances: Mining and refinement of radioactive metals may discharge radioactive materials. Accidental spillages from atomic power stations, such as the Chornobyl nuclear accident of 1986, (pictured above) may contaminate waterways.

Inorganic chemicals: Acid drainage from mines and acid rain can severely alter the pH of waterways. The runoff from open-caste mining operations can be loaded with poisonous heavy metals such as mercury, cadmium, and arsenic.

Detecting Pollution

Water pollution can be monitored in several ways. The nutrient loading can be assessed by measuring the **BOD**. **Electronic probes** and **chemical tests** can identify the absolute levels of various inorganic pollutants (e.g. nitrates, phosphates, and heavy metals). The presence of **indicator species** can give an indication of the pollution status for a waterway. This method relies on an understanding of the *tolerance levels* to pollution of different species that should be living in the waterway (e.g. worms, insect larvae, snails, and crustaceans).

Related activities: Monitoring Change in an Ecosystem, Water Use, Soil Degradation Sewage Treatment **Web links**: Caring For Our Water

1. Explain the term **accelerated eutrophication** and its primary cause:

2. Describe three uses of water for each of the following areas of human activity:

   (a) Domestic use:

   (b) Industrial use:

   (c) Agricultural use:

3. (a) Explain what is meant by the term **biochemical oxygen demand** (BOD) as it is related to water pollution:

   (b) Describe how a very high BOD. in a body of water such as a lake or river may be created by human activity:

   (c) Describe the likely effect of a very high BOD on the invertebrates and fish living in a small lake:

   (d) Explain why, when measuring BOD, that the sample is kept in darkness:

4. Sewage effluent may be sprayed onto agricultural land to irrigate crops and plantations of trees:

   (a) Describe an advantage of utilising sewage in this way:

   (b) Describe a major drawback of using sewage effluent in this way:

   (c) Suggest an alternative treatment or use of the effluent:

5. When studying aquatic ecosystems, the species composition of the community (its biodiversity) in different regions of a water body or over time is often recorded. In general terms, suggest how a change in species composition of an aquatic community could be used to indicate water pollution:

Sewage Treatment

Once water has been used by household or industry, it becomes sewage. Sewage includes toilet wastes and all household water, but excludes storm water, which is usually diverted directly into waterways. In some cities, the sewerage and stormwater systems may be partly combined, and sewage can overflow into surface water during high rainfall. When sewage reaches a treatment plant, it can undergo up to three levels of processing (purification). Primary treatment is little more than a mechanical screening process, followed by settling of the solids into a sludge. Secondary sewage treatment is primarily a biological process in which aerobic and anaerobic microorganisms are used to remove the organic wastes. Advanced secondary treatment targets specific pollutants, particularly nitrates, phosphates, and heavy metals. Before water is discharged after treatment, it is always disinfected (usually by chlorination) to kill bacteria and other potential pathogens.

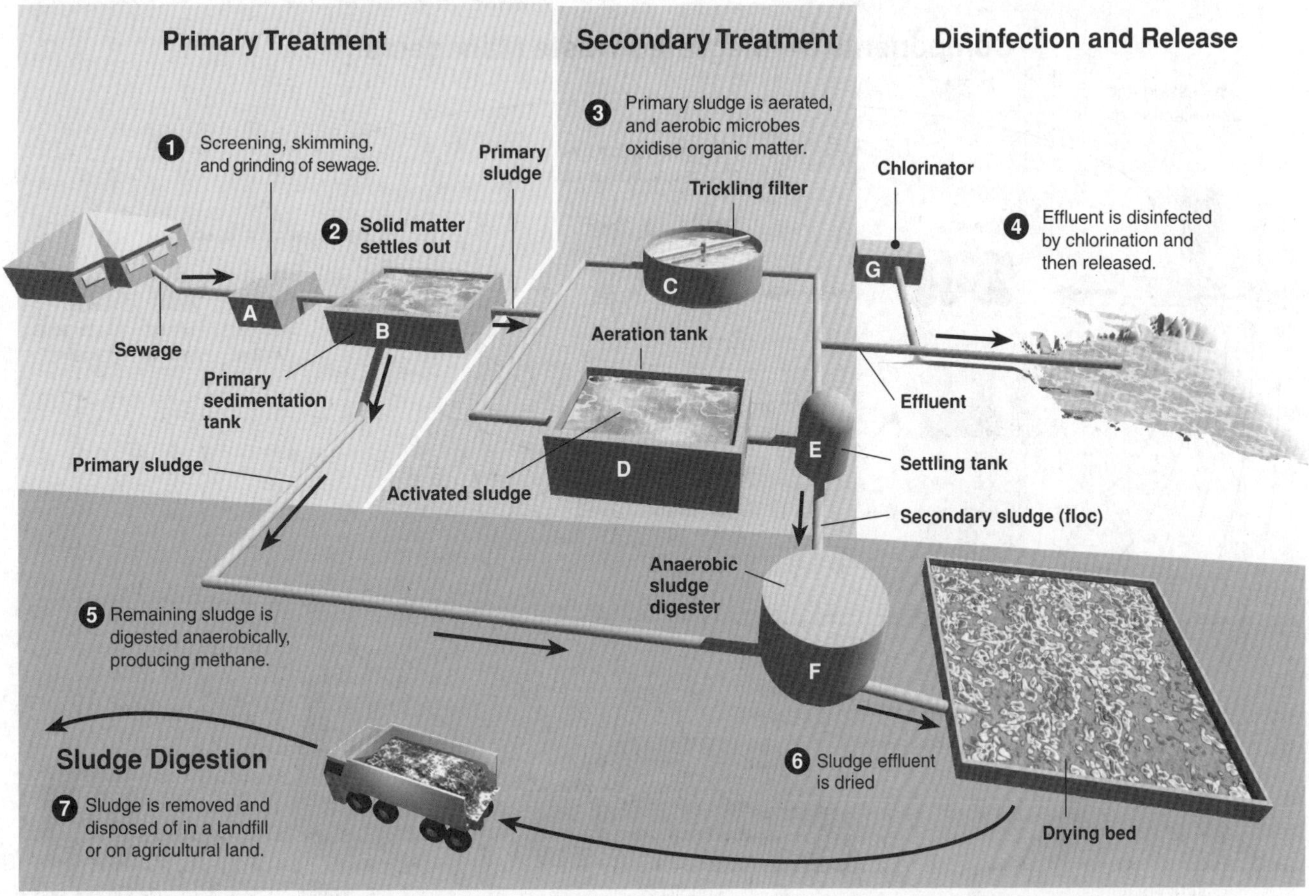

Primary Treatment
Secondary Treatment
Disinfection and Release
1 Screening, skimming, and grinding of sewage.
2 Solid matter settles out
Primary sludge
3 Primary sludge is aerated, and aerobic microbes oxidise organic matter.
Trickling filter
Chlorinator
4 Effluent is disinfected by chlorination and then released.
A
B
C
G
Sewage
Primary sedimentation tank
Aeration tank
Effluent
Primary sludge
Activated sludge
D
E
Settling tank
Secondary sludge (floc)
Anaerobic sludge digester
5 Remaining sludge is digested anaerobically, producing methane.
F
Sludge Digestion
7 Sludge is removed and disposed of in a landfill or on agricultural land.
6 Sludge effluent is dried
Drying bed

1. Using the information provided in the diagram and text above, classify each of the processes indicated A-G as either mechanical, biological, or chemical. If you wish, colour code these on the diagram for easy reference:

A: ______________________ D: ______________________ G: ______________________

B: ______________________ E: ______________________

C: ______________________ F: ______________________

2. Using the diagram above for reference, investigate the sewage treatment process in your own town or city, identifying the specific techniques and problems of waste water management in your area. Make a note of the main points to cover in the space provided below, and develop your discussion as a separate report. Identify:

(a) Your urban area and treatment station: ______________________

(b) The volume of sewage processed: ______________________

(c) The degree of purification: ______________________

(d) The treatment processes used (list): ______________________

(e) The discharge point(s): ______________________

(f) Problems of waste water management: ______________________

____

(g) Future options or plans: ______________________

____

Pollution and Global Change

Related activities: Types of Pollution, Water Pollution

Waste Management

The disposal of solid and hazardous waste is one of the most urgent problems of today's industrialised societies. Traditionally, solid waste has been disposed of in open dumps and, more recently, in sanitary, scientifically designed landfills. Even with modern designs and better waste processing, landfills still have the potential to contaminate soil and groundwater. In addition, they occupy valuable land, and their siting is often a matter of local controversy. More and more today, city councils and local authorities support the initiatives for the reduction, reuse, and recycling of solid wastes. At the same time, they must develop strategies for the safe disposal of hazardous wastes, which pose an immediate or potential threat to environmental and human health. A programme of integrated waste management (below) combines features of traditional waste management with new techniques to reduce and incinerate wastes. Such schemes will form the basis of effective waste management in the future.

Components of Integrated Waste Management

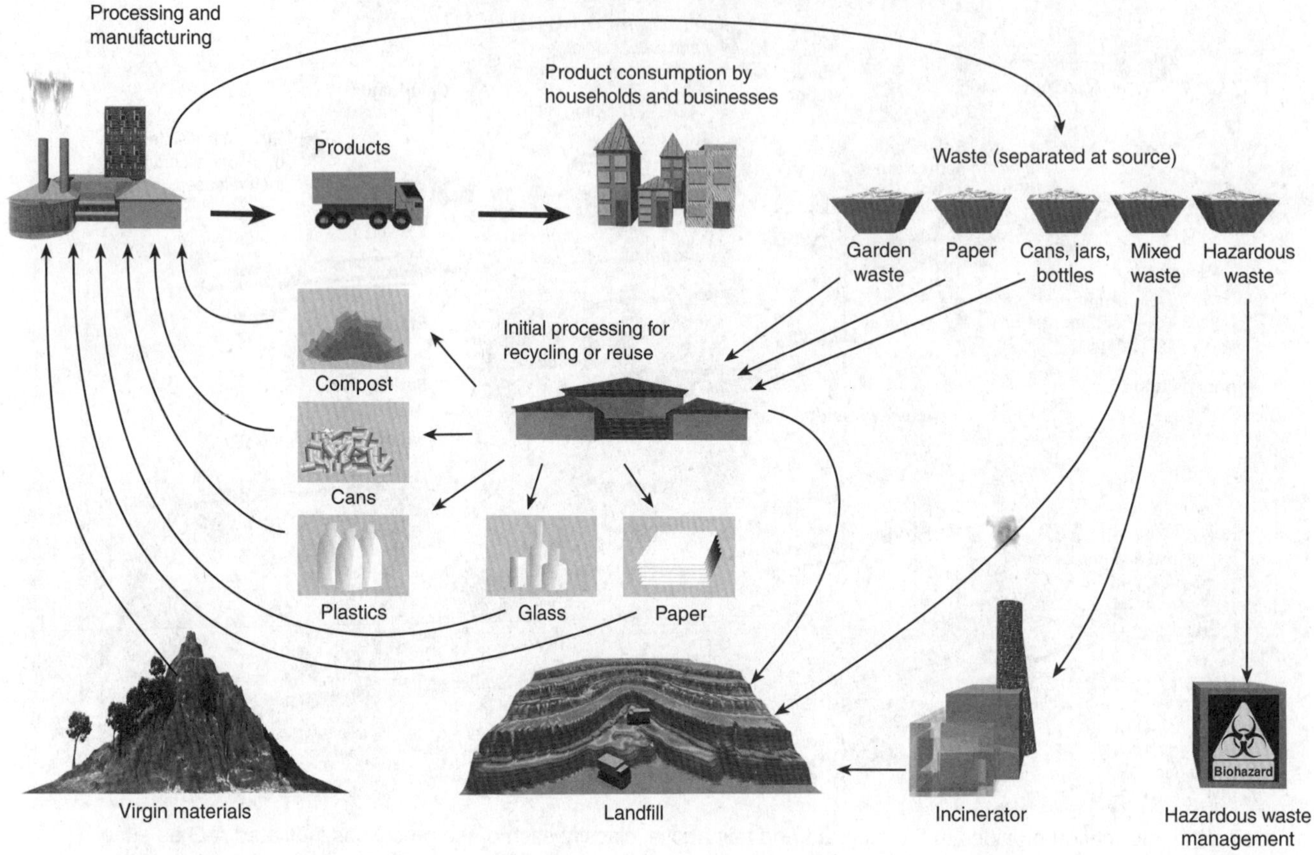

1. The diagram provides an overview of an **idealised management system** for waste materials from households and industries. It provides a starting point for comparing how different waste products could be disposed of or processed. Using the information provided for guidance, investigate the disposal, recycling, and post-waste processing options for each of the waste products listed below. List the important points in the spaces provided, including reference to disposal methods and particular problems associated with these, processing or recycling (if relevant), and useful end-products (if relevant). If required, develop this list, or part of it, as a separate report:

(a) Glass, plastic and paper waste: ___

(b) Metals and their alloys, e.g. aluminum, tin and steel: ___

(c) Organic waste: ___

(d) Hazardous waste (including medical): ___

2. Identify a waste product that is not part of an integrated waste management program: ___

Atmospheric Pollution

Air pollution consists of gases, liquids, or solids present in the atmosphere at levels high enough to harm living things (or cause damage to materials). Human activities make a major contribution to global air pollution, although natural processes can also be responsible. Lightning causes forest fires, oxidises nitrogen and creates ozone, while erupting volcanoes give off toxic and corrosive gases. Air pollution tends to be concentrated around areas of high population density, particularly in Western industrial and post-industrial societies. In the last few decades there has been a massive increase in air pollution in parts of the world that previously had little, such as Mexico city and some of the large Asian cities. Air pollution does not just exist outdoors. The air enclosed in spaces such as cars, homes, schools, and offices may have significantly higher levels of harmful air pollutants than the air outdoors.

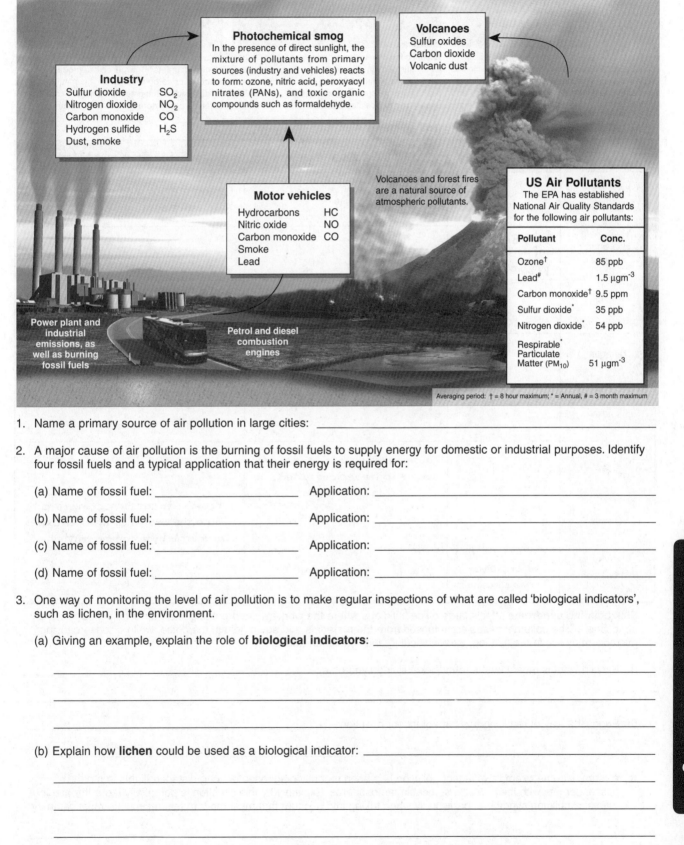

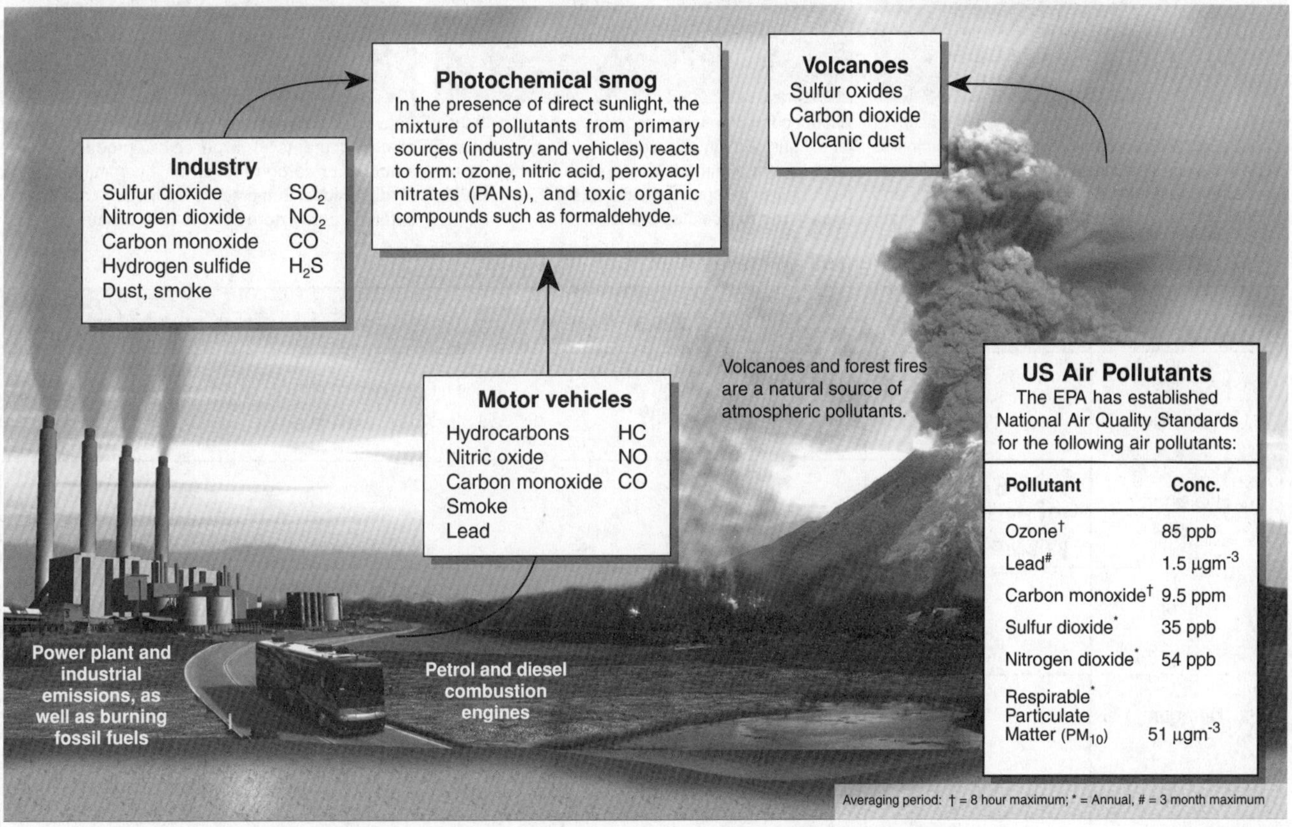
Industry
Sulfur dioxide SO_2
Nitrogen dioxide NO_2
Carbon monoxide CO
Hydrogen sulfide H_2S
Dust, smoke
Photochemical smog
In the presence of direct sunlight, the mixture of pollutants from primary sources (industry and vehicles) reacts to form: ozone, nitric acid, peroxyacyl nitrates (PANs), and toxic organic compounds such as formaldehyde.
Volcanoes
Sulfur oxides
Carbon dioxide
Volcanic dust
Motor vehicles
Hydrocarbons HC
Nitric oxide NO
Carbon monoxide CO
Smoke
Lead
Volcanoes and forest fires are a natural source of atmospheric pollutants.
Power plant and industrial emissions, as well as burning fossil fuels
Petrol and diesel combustion engines
US Air Pollutants
The EPA has established National Air Quality Standards for the following air pollutants:

Pollutant	Conc.
Ozone†	85 ppb
Lead#	1.5 μgm^{-3}
Carbon monoxide†	9.5 ppm
Sulfur dioxide*	35 ppb
Nitrogen dioxide*	54 ppb
Respirable* Particulate Matter (PM_{10})	51 μgm^{-3}

Averaging period: † = 8 hour maximum; * = Annual, # = 3 month maximum

1. Name a primary source of air pollution in large cities: ______

2. A major cause of air pollution is the burning of fossil fuels to supply energy for domestic or industrial purposes. Identify four fossil fuels and a typical application that their energy is required for:

   (a) Name of fossil fuel: ______ Application: ______

   (b) Name of fossil fuel: ______ Application: ______

   (c) Name of fossil fuel: ______ Application: ______

   (d) Name of fossil fuel: ______ Application: ______

3. One way of monitoring the level of air pollution is to make regular inspections of what are called 'biological indicators', such as lichen, in the environment.

   (a) Giving an example, explain the role of **biological indicators**: ______

   (b) Explain how **lichen** could be used as a biological indicator: ______

Related activities: Monitoring Change in an Ecosystem, Types of Pollution, Global Warming, Stratospheric Ozone Depletion, Acid Rain

Health officials are paying increasing attention to the *sick building syndrome*. This air pollution inside office buildings can cause eye irritations, nausea, headaches, respiratory infections, depression and fatigue. Gases, ozone and microbes are implicated.

Aircraft contribute to atmospheric pollution with their jet exhaust at high altitude (at 10 000 m). The cabin environment of aircraft is also often polluted. Some passengers may spread infections (e.g. TB and SARS) through the recirculated cabin air.

Automobiles are the single most important contributor of air pollutants in large cities, producing large amounts of carbon monoxide, hydrocarbons, and nitrous oxides. Some countries require cars to have **catalytic converters** fitted to their exhausts.

4. Complete the table below that summarises the main types of air pollutants:

Pollutant	Major sources	Harmful effects	Prevention or control
Carbon monoxide			Fit cars with catalytic converters and keep well tuned.
Hydrogen sulfides	Burning fuels, oil refineries, wood pulp processing.		
Sulfur oxides			Use alternative, sulfur free fuels such as natural gas and LPG (liquid petroleum gas).
Nitrogen oxides		Forms photochemical smog which irritates the eyes and nose. Retards plant growth.	
Smoke			
Lead		Causes convulsions, coma and damage to the nervous system.	
Ozone			Fit cars with catalytic converters to reduce the amount of nitrogen oxides and volatile hydrocarbons emitted.
Hydrocarbons	Incomplete combustion		

5. **Sick building syndrome** affects large office buildings where the workers are breathing in an air conditioned atmosphere. The pollutant gases are released from the materials and equipment in the office, while disease-causing microbes may live in the heating, air conditioning and ventilation ducts.

(a) Explain what is meant by the term 'sick building syndrome': ______________________

____

(b) Suggest a way of reducing this form of indoor pollution: ______________________

____

(c) A more extreme example of indoor pollution has been recently diagnosed on long distance flights in modern passenger jets with their pressurised cabin atmospheres. Explain why this situation is potentially more threatening than sick building syndrome, particularly when taking into account that most modern jets recirculate most cabin air:

____

____

Global Warming

The Earth's atmosphere comprises a mixture of gases including nitrogen, oxygen, and water vapour. Also present are small quantities of carbon dioxide (CO_2), methane, and a number of other "trace" gases. In the past, our climate has shifted between periods of stable warm conditions to cycles of glacials and interglacials. The current period of warming climate is partly explained by the recovery after the most recent ice age that finished 10 000 years ago. Eight of the ten warmest years on record (records kept since the mid-1800s) were in the 1980s and 1990s. Global surface temperatures in 1998 set a new record by a wide margin, exceeding those of the previous record year, 1995. Many researchers believe the current warming trend has been compounded by human activity, in particular, the release of certain gases into the atmosphere. The term '**greenhouse effect**' describes a process of global climate warming caused by the release of 'greenhouse gases', which act as a thermal blanket in the atmosphere, letting in sunlight, but trapping the heat that would normally radiate back into space. About three-quarters of the natural greenhouse effect is due to water vapour. The next most significant agent is CO_2. Since the industrial revolution and expansion of agriculture about 200 years ago, additional CO_2 has been pumped into the atmosphere. The effect of global warming on agriculture, other human activities, and the biosphere in general, is likely to be considerable.

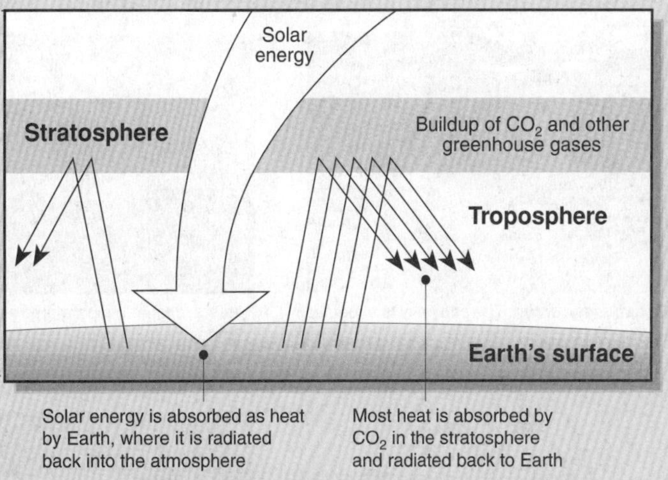

Sources of 'Greenhouse Gases'

Carbon dioxide
- Exhaust from cars
- Combustion of coal, wood, oil
- Burning rainforests

Methane
- Plant debris and growing vegetation
- Belching and flatus of cows

Chloro-fluoro-carbons (CFCs)
- Leaking coolant from refrigerators
- Leaking coolant from air conditioners

Nitrous oxide
- Car exhaust

Tropospheric ozone*
- Triggered by car exhaust (smog)

*Tropospheric ozone is found in the lower atmosphere (not to be confused with ozone in the stratosphere)

Greenhouse gas	Tropospheric conc. Pre-industrial 1860	Tropospheric conc. Present day (2004*)	Global warming potential *(compared to CO_2)¶*	Atmospheric lifetime *(years)§*
Carbon dioxide	288 ppm	377 ppm	1	120
Methane	848 ppb	1789 ppb	21	12
Nitrous oxide	285 ppb	318 ppb	310	120
CFCs	0 ppb	0.88 ppb	4000+	50-100
Tropospheric ozone	25 ppb	34 ppb	17	hours

ppm = parts per million; **ppb** = parts per billion; * Data from 2004 and current up to Jul 2006 ¶ Figures contrast the radiative effect of different greenhouse gases relative to CO_2, e.g. methane is 21 times more potent as a greenhouse gas than CO_2 § How long the gas persists in the atmosphere *Source: Carbon Dioxide Information Analysis Centre, Oak Ridge National Laboratory, USA.*

The graph on the right shows how the mean temperature for each year from 1860 until 2003 (grey bars) compared with the average temperature between 1961 and 1990. The thick black line represents the mathematically fitted curve and shows the general trend indicated by the annual data. Most anomalies since 1977 have been above normal; warmer than the long term mean, indicating that global temperatures are tracking upwards. In 1998 the global temperature exceeded that of the previous record year, 1995, by about 0.2°C.

Source: Hadley Centre for Prediction and Research

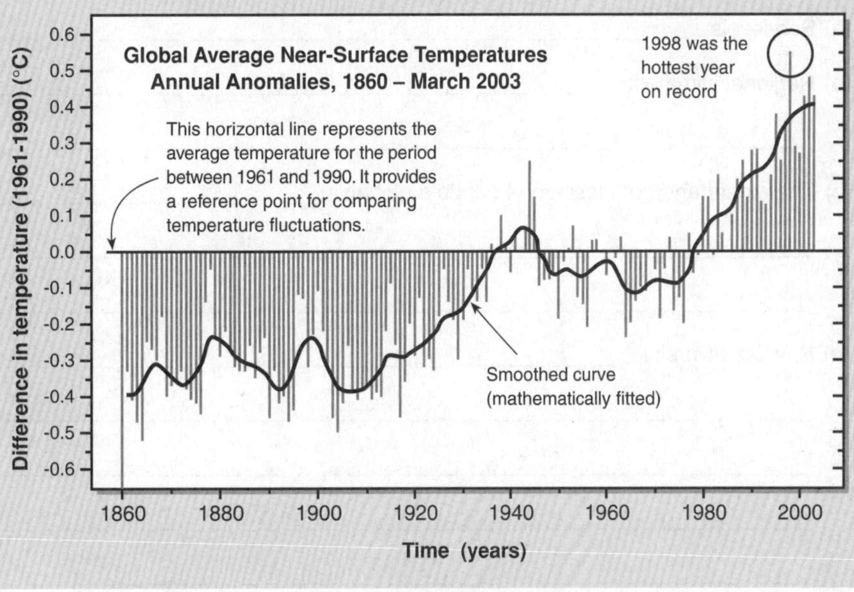

Related activities: The Atmosphere and Climate, Types of Pollution
Web links: The Greenhouse Effect

Changes in atmospheric CO_2 since 1000 AD

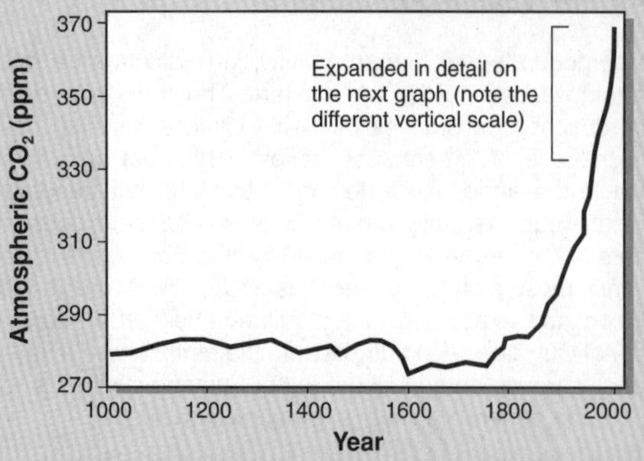

Changes in atmospheric CO_2 since 1955

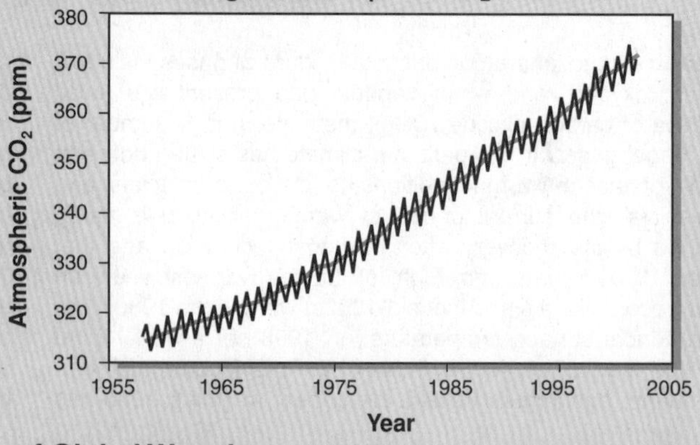

Sources: NASA Goddard Space Flight Center; NOAA / CMDL

Potential Effects of Global Warming

Coastal resources: The sea level is expected to rise by 50 cm by the year 2100. This is the result of ocean water expanding as it warms and by some melting of non-polar ice.

Forests: Higher temperatures and precipitation changes could increase forest susceptibility to fire, disease, and insect damage. A richer CO_2 atmosphere will reduce transpiration in plants.

Weather patterns: Global warming may cause regional changes in weather patterns such as El Niño and La Nina, as well as affecting the intensity and frequency of storms (hurricanes).

Water resources: Changes in precipitation and increased evaporation will affect water availability for irrigation, industrial use, drinking, and electricity generation.

Agriculture: Impacts of climate change in developing countries could be significant. Changes in climate may affect the viability of staple crops in some regions.

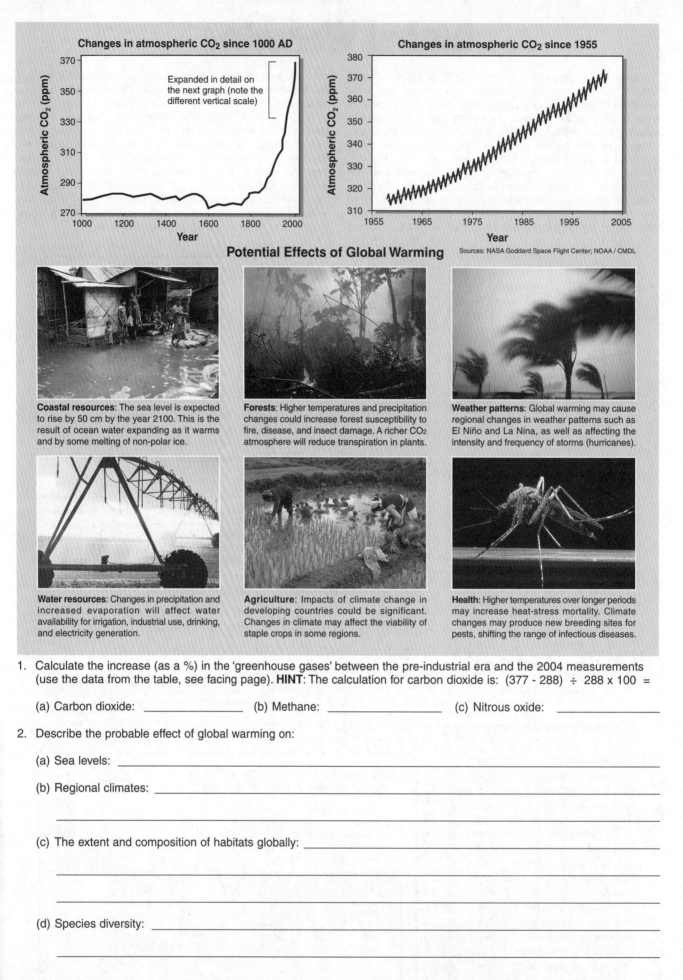

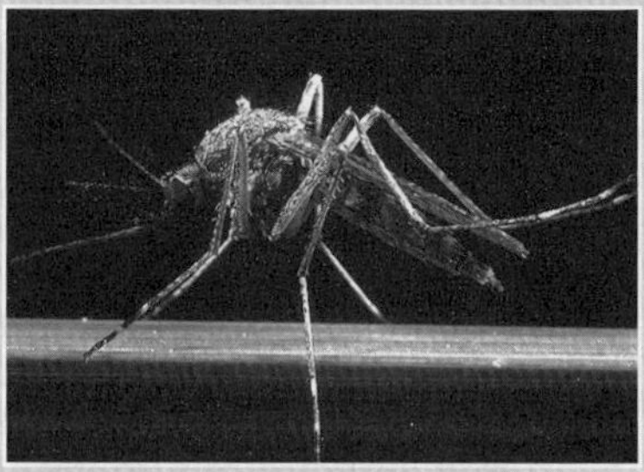

Health: Higher temperatures over longer periods may increase heat-stress mortality. Climate changes may produce new breeding sites for pests, shifting the range of infectious diseases.

1. Calculate the increase (as a %) in the 'greenhouse gases' between the pre-industrial era and the 2004 measurements (use the data from the table, see facing page). **HINT**: The calculation for carbon dioxide is: $(377 - 288) \div 288 \times 100 =$

   (a) Carbon dioxide: ______________ (b) Methane: ______________ (c) Nitrous oxide: ______________

2. Describe the probable effect of global warming on:

   (a) Sea levels: ______________________________

   (b) Regional climates: ______________________________

   ____

   (c) The extent and composition of habitats globally: ______________________________

   ____

   ____

   (d) Species diversity: ______________________________

   ____

   ____

   ____

   ____

Stratospheric Ozone Depletion

In a band of the upper stratosphere, 17-26 km above the Earth's surface, exists a thin veil of renewable **ozone** (O_3). This ozone absorbs about 99% of the harmful incoming UV radiation from the sun and prevents it from reaching the Earth's surface. Apart from health problems, such as increasingly severe sunburns, increase in skin cancers, and more cataracts of the eye (in both humans and other animals), an increase in UV-B radiation is likely to cause immune system suppression in animals, lower crop yields, a decline in the productivity of forests and surface dwelling plankton, more smog, and changes in the global climate. Ozone is being depleted by a handful of human-produced chemicals (ozone depleting compounds or ODCs). The problem of **ozone depletion** was first detected in 1984. Researchers discovered that ozone in the upper stratosphere over Antarctica is destroyed during the Antarctic spring and early summer (September–December). Rather than a "hole", it is more a thinning, where ozone levels typically decrease by 50% to 100%. In 2000, the extent of the hole above Antarctica was the largest ever, but depletion levels were slightly less than 1999. Severe ozone loss has also been observed over the Arctic. During the winter of 1999-2000, Arctic ozone levels were depleted by 60% at an altitude of 18 km, up from around 45% in the previous winter. The primary cause for ozone depletion appears to be the increased use of chemicals such as chloro-fluoro-carbons (**CFCs**). Since 1987, nations have cut their consumption of ozone-depleting substances by 70%, although the phaseout is not complete and there is a significant black market in CFCs. **Free chlorine** in the stratosphere peaked around 1999 and is projected to decline for more than a century. Ozone loss is projected to diminish gradually until around 2050 when the polar ozone holes will return to 1975 levels. It will take another 100-200 years for full recovery to pre-1950 levels.

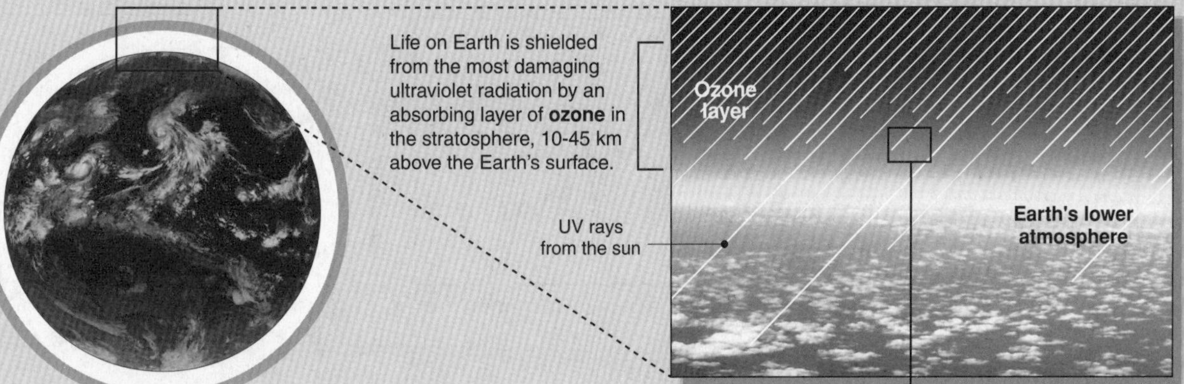

Sources of ozone depleting chemicals

The chemicals below drift up to the stratosphere, where ultraviolet radiation causes release of free chlorine, a highly reactive chemical.

Chloro-fluoro-carbons (CFCs)
- Propellants for aerosol cans
- Coolants in air-conditioners
- Coolants (freon) in refrigerators
- Styrofoam insulation/packaging
- Medical sterilisers

Halons
- Used in many fire extinguishers

Methyl bromide
- Used as a fumigant in agriculture

Methyl chloroform
- Used to degrease metals

Carbon tetrachloride
- Used in many industrial processes

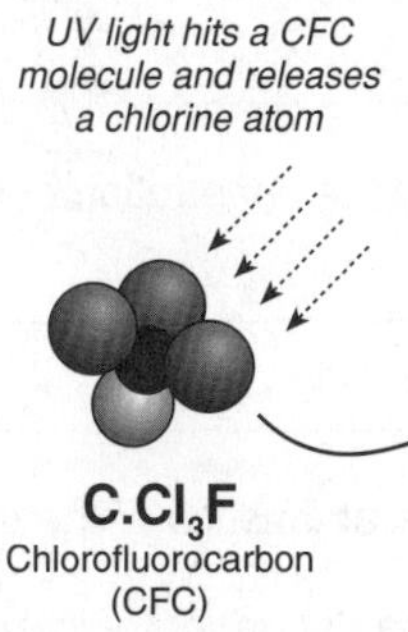

The destruction of ozone by free chlorine

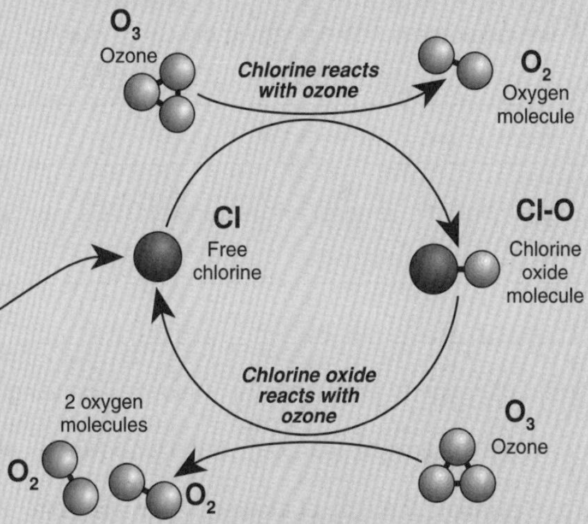

A large 'hole' in the ozone layer develops over Antarctica each summer, dropping the ozone well below its normal level. The size and intensity of the hole is growing each year, as can be seen in the satellite photos on the right. In recent years, a similar hole has developed over the Arctic.

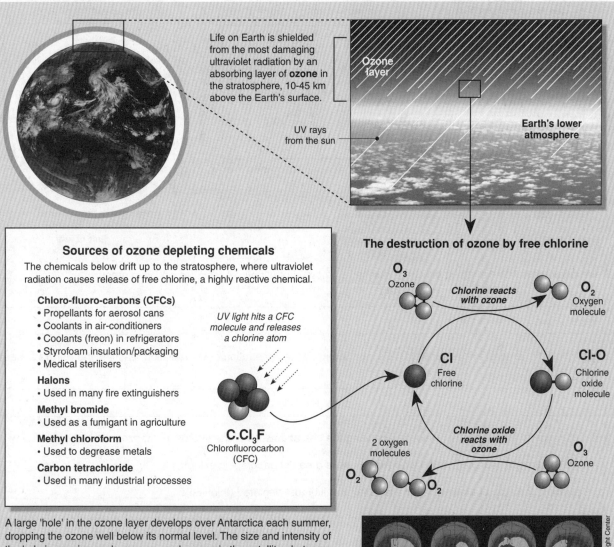

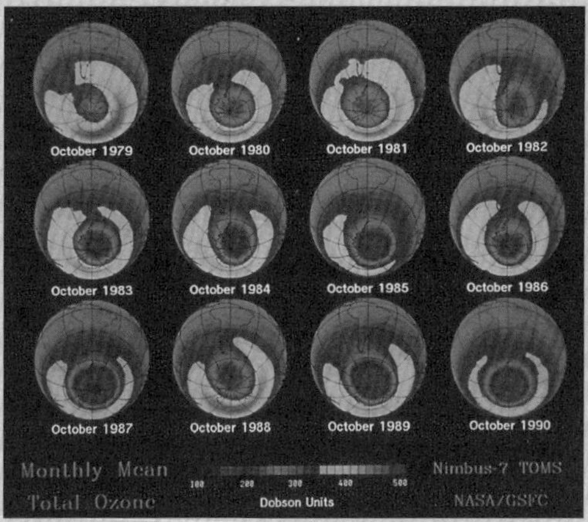

Photos: NASA/Goddard Space Flight Center

Dobson Unit (DU): A measurement of **column ozone** levels (the ozone between the Earth's surface and outer space). In the tropics, ozone levels are typically between 250 and 300 DU year-round. In temperate regions, seasonal variations can produce large swings in ozone levels. These variations occur even in the absence of ozone depletion. **Ozone depletion** refers to reductions in ozone below normal levels after accounting for seasonal cycles and other natural effects. For a graphical explanation, see NASA's TOMS site: ***http://toms.gsfc.nasa.gov/teacher/basics/dobson.html***

Characteristics of the ozone 'hole'

The ozone 'hole' (stratospheric ozone depletion) can be characterised using several measures. The five graphs on this page show how the size and intensity of the hole varies through the course of a year, as well as how the phenomenon has progressed over the last two decades. An explanation of the unit used to measure ozone concentration (Dobson units) is given on the opposite page. Graphs 2 and 5 illustrate readings taken between the South Pole (90° south) and 40° latitude.

Data supplied by NASA's Goddard Space Flight Center and the National Oceanic and Atmospheric Administration (NOAA) in the USA.

Graph 1: Ozone hole altitude profile

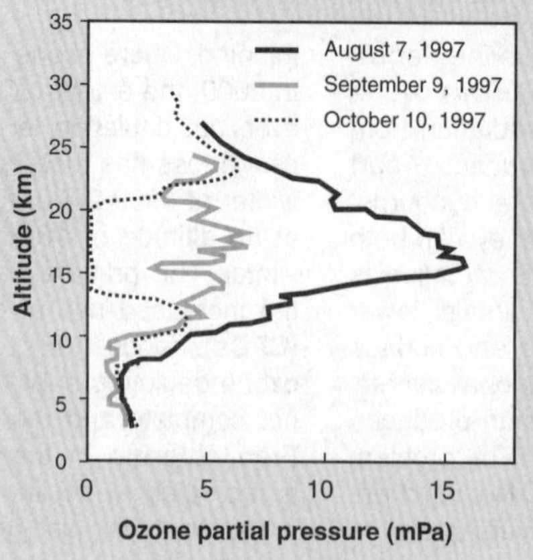

Graph 2: Antarctic ozone hole area (<220 DU, 40° – 90° South)

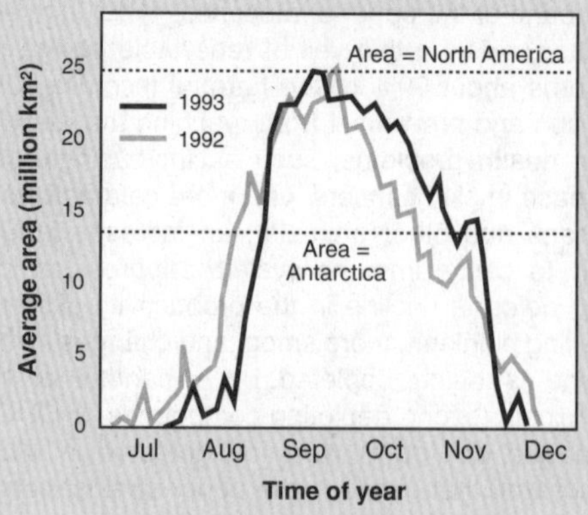

Graph 3: Change in area of the Antarctic ozone hole*

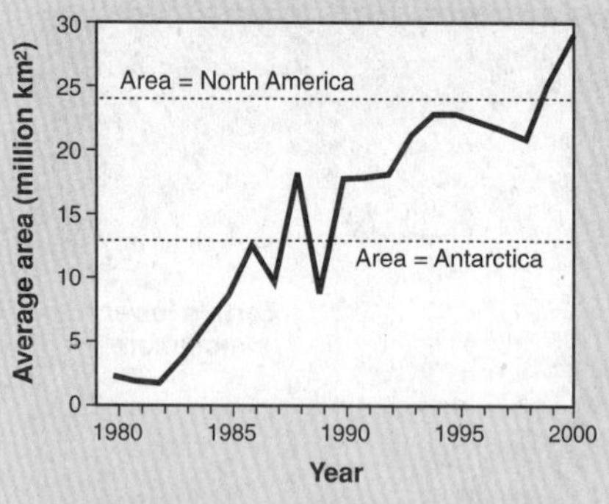

Graph 4: Antarctic ozone hole minimum values* (60° – 90° S)

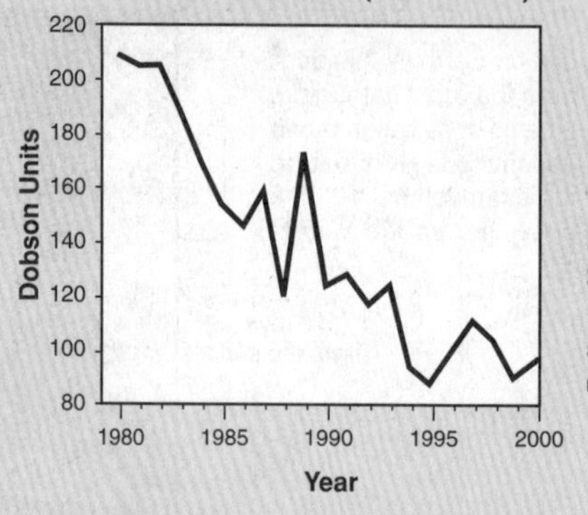

Graph 5: Antarctic ozone hole minimum values (40° – 90° South)

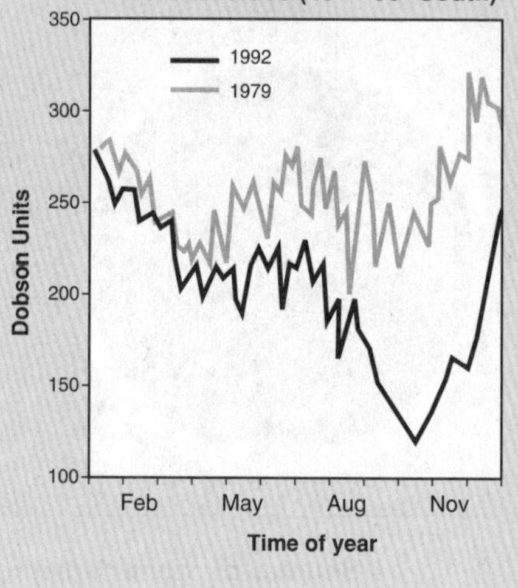

* Date range in which samples were collected in each year: 7 Sep–13 Oct
The ozone 'hole' is defined as region with less than 220 Dobson units

Sources: NASA Goddard Space Flight Centre; NOAA / CMDL

1. Describe some of the damaging effects of excessive amounts of ultraviolet radiation on living organisms:

2. Explain how the atmospheric release of CFCs has increased the penetration of UV radiation reaching the Earth's surface:

3. With reference to the graphs (1-5 above) illustrating the characteristics of the stratospheric ozone depletion problem:

   (a) State the time of year when the ozone 'hole' is at its greatest geographic extent:

   (b) Determine the time of the year when the 'hole' is at its most depleted (thinnest):

   (c) Describe the trend over the last two decades of changes to the abundance of stratospheric ozone over Antarctica:

   (d) Describe the changes in stratospheric ozone with altitude between August and October 1997 in Graph 1 (above):

4. Discuss some of the political and commercial problems associated with reducing the use of ozone depleting chemicals:

Acid Rain

Acid rain is not a new phenomenon. It was first noticed last century in regions where the industrial revolution began. Buildings in areas with heavy industrial activity were being worn away by rain. Acid rain, more correctly termed **acid deposition**, can fall to the Earth as rain, snow or sleet, as well as dry, sulfate-containing particles that settle out of the air. It is a problem that crosses international boundaries. Gases from coal-burning power stations in England fall as acid rain in Norway and Sweden, emissions from the United States produce acid deposition in Canada, while Japan receives acid rain from China. The effect of this fallout is to produce lakes that are so acid that they cannot support fish, and forests with sickly, stunted tree growth. Acid rain also causes the release of heavy metals (e.g. cadmium and mercury) into the food chain. Changes in species composition of aquatic communities may be used as **biological indicators** measuring the severity of acid deposition.

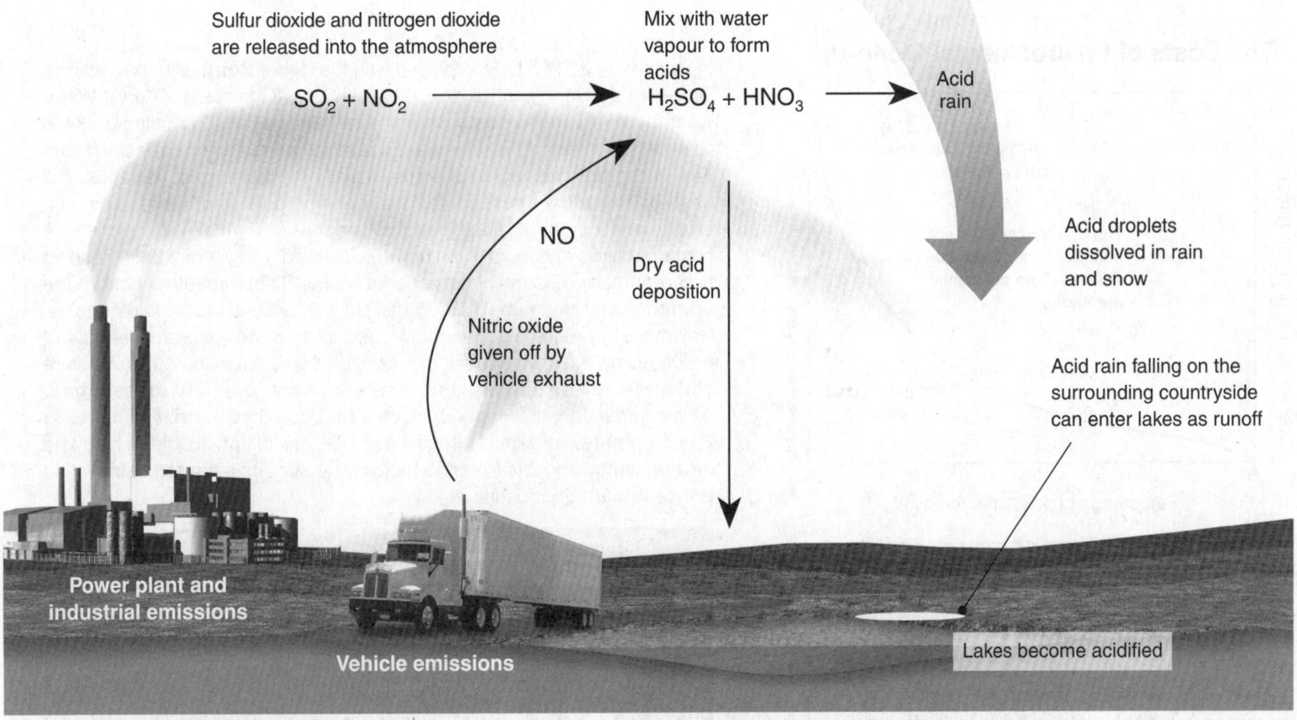

Acidity Tolerance in Lake Organisms

Different aquatic organisms have varying sensitivities to higher acidity (lower pH). The graph on the right shows how much acidity certain species can tolerate. The absence of certain indicator species from a waterway can provide evidence of pollution in the recent past as well as the present.

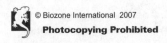

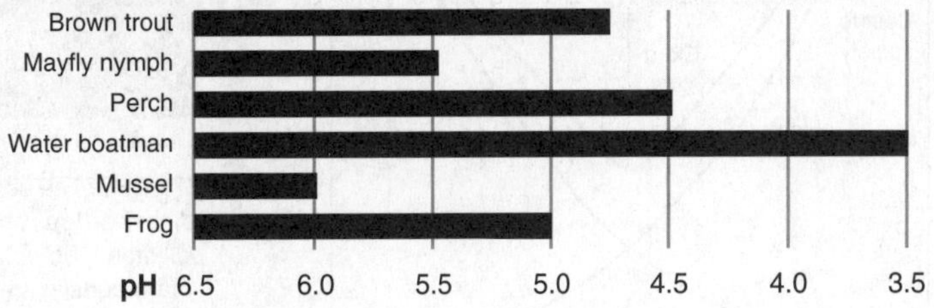

1. Describe the effect of acid deposition on communities of living organisms:

2. Study the graph illustrating the acidity tolerance of lake organisms (above).

   (a) State which species is the most **sensitive** to acid conditions:

   (b) State which species is the most **tolerant** of acid conditions:

   (c) Explain how you could use these kinds of measurements as an indicator of the ecological state of a lake:

3. Describe some of the measures that could be taken to reduce acid emissions:

4. Explain why these measures are slow to be implemented:

Related activities: Monitoring Change in an Ecosystem, Types of Pollution, Atmospheric Pollution

The Economic Impact of Pollution

Although there is an economic cost to pollution, placing a monetary value on it is difficult and controversial. The various sectors involved (health officials, economists, and industry) often disagree on how to estimate the cost of pollution, as it is difficult to assign monetary values to environment, health, and human life. Determining the economic impact of pollution can also be difficult because while a region as a whole may benefit from the economic activities of a polluter, groups within the region may suffer. Some economic costs associated with pollution are more easily determined than others. **Direct costs** (e.g. cleaning up an oil spill) are easily calculated, but **indirect costs** (e.g. estimating revenue losses) or **repercussion costs** (e.g. loss of public opinion) can be harder to quantify. **Cost-benefit analysis** is used to assess the cost of controlling pollution. The short-term and long-term costs and benefits for a variety of pollution control measures are compared and used to determine whether a control or regulation should be put in place. Environmental regulations, taxes and pollution quotas are commonly used to control levels of pollution and to promote **sustainable** use of resources and the environment.

The Costs of Environmental Clean-Up

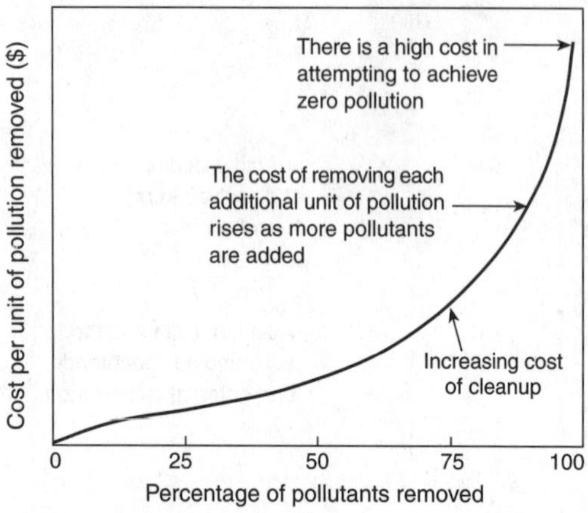

Graphs modified after Miller, 2002

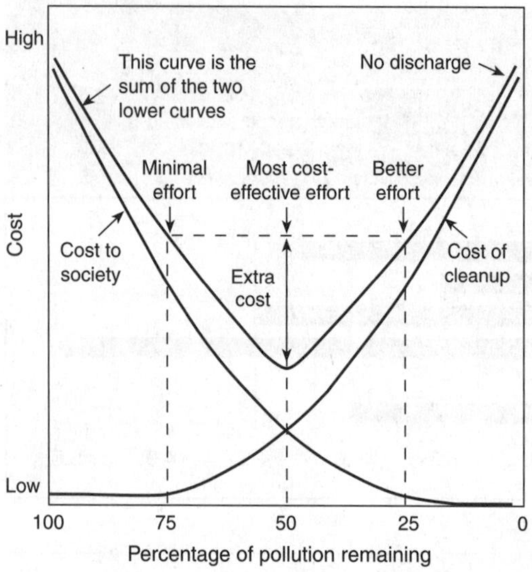

In December 2007 a barge collided with the **Hebei Spirit** oil tanker, spilling almost 11 000 tonnes of oil into the Yellow Sea off the coast of Korea. Warm weather, wind, and wave conditions dispersed the oil, producing a 33 km oil slick that affected 300 km of the coastline of the ecologically significant **Taean region**. This region is home to one of Asia's largest wetlands, it is used by migratory birds, contains a maritime park and hundreds of sea farms, and contains many beautiful beaches popular with tourists. A comprehensive and rapid clean up was achieved using over 200 000 people and a number of expert international teams. The immediate cost of the environmental clean-up (US$330 million) was relatively easy to determine. However, the full costs, including costs due to the long-term effects of pollutants on the environment, lost revenue from contaminated aquaculture activities, and reduced tourism are more difficult to estimate. As a comparison, the Hebei Spirit oil spill is one third the size of the 1989 **Exxon Valdez** oil spill, which cost US$2.5 billion to clean up. The environmental impacts from the Exxon Valdez oil spill are still being felt in Prince William Sound today.

EPA

Cleaning up of shoreline

Photo courtesy of Exxon Valdez Oil Spill Trustee Council

Dead seabirds after the Exxon-Valdex oil spill

GRAPHS: The cost of removing pollutants rises sharply as more pollutants are removed (top left), until the cleanup costs exceed the harmful costs of the pollution. The point at which the costs of the pollution and the costs of clean-up are equal marks the **break-even** point. This point is determined by separately plotting the clean-up cost and the cost of the pollution to society. The two curves are then added together to reveal the total costs (below left).

1. With reference to the **break-even point**, explain why total pollutant removal is often not cost effective:

2. Describe the following costs associated with a named major pollution incident:

(a) Direct costs:

(b) Indirect costs:

Loss of Biodiversity

The species is the basic unit by which we measure biological diversity or **biodiversity**. Biodiversity is not distributed evenly on Earth, being consistently richer in the tropics and concentrated more in some areas than in others. Conservation International recognises 25 **biodiversity hotspots**. These are biologically diverse and ecologically distinct regions under the greatest threat of destruction. They are identified on the basis of the number of species present, the amount of **endemism**, and the extent to which the species are threatened. More than a third of the planet's known terrestrial plant and animal species are found in these 25 regions, which cover only 1.4% of the Earth's land area. Unfortunately, biodiversity hotspots often occur near areas of dense human habitation and rapid human population growth. Most are located in the tropics and most are forests. Loss of biodiversity reduces the stability and resilience of natural ecosystems and decreases the ability of their communities to adapt to changing environmental conditions. With increasing pressure on natural areas from urbanisation, roading, and other human encroachment, maintaining species diversity is paramount and should concern us all today.

Biodiversity Hotspots

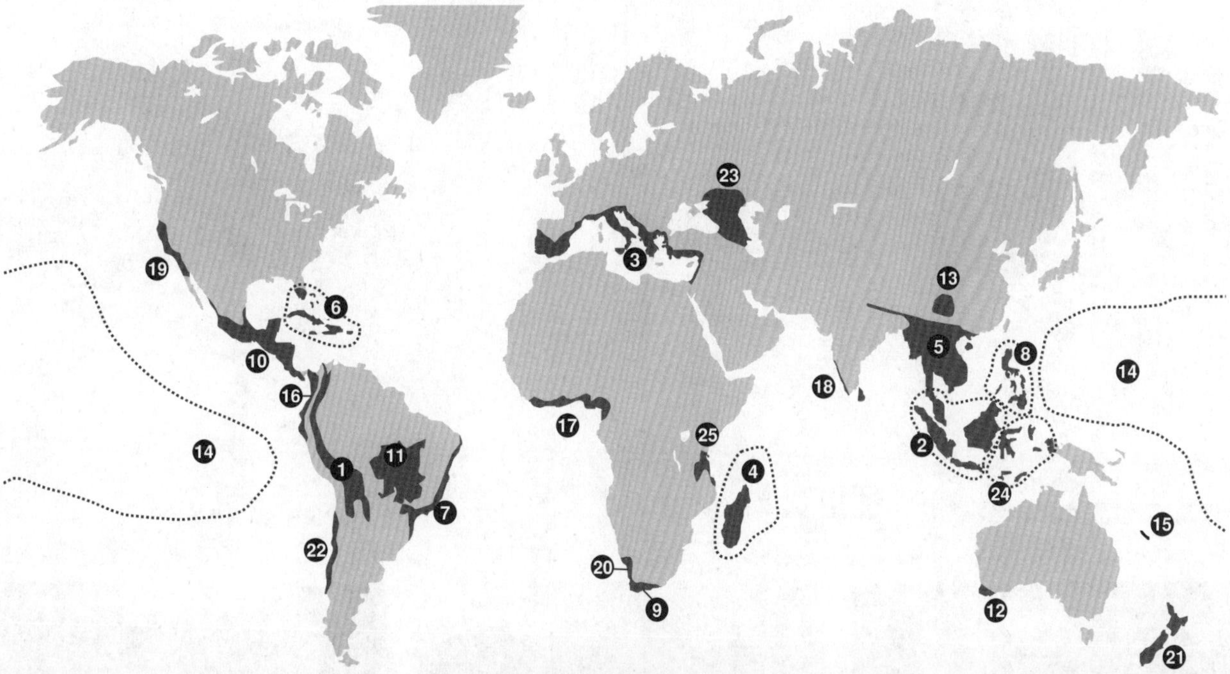

Threats to Biodiversity

Rainforests in some of the most species-rich regions of the world are being destroyed at an alarming rate as world demand for tropical hardwoods increases and land is cleared for the establishment of agriculture.

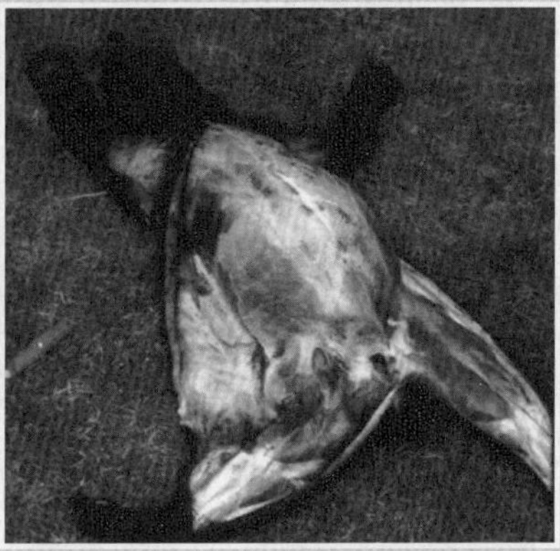

Illegal trade in species (for food, body parts, or for the exotic pet trade) is pushing some species to the brink of extinction. Despite international bans on trade, illegal trade in primates, parrots, reptiles, and big cats (among others) continues.

Pollution and the pressure of human populations on natural habitats threatens biodiversity in many regions. Environmental pollutants may accumulate through food chains or cause harm directly, as with this bird trapped in oil.

1. Use your research tools (e.g. textbook, internet, or encyclopaedia) to identify each of the 25 biodiversity hotspots illustrated in the diagram above. For each region, summarise the characteristics that have resulted in it being identified as a biodiversity hotspot. Present your summary as a short report and attach it to this page of your workbook.

2. Identify the threat to biodiversity that you perceive to be the most important and explain your choice:

Tropical Deforestation

Tropical rainforests prevail in places where the climate is very moist throughout the year (200 to 450 cm of rainfall per year). Almost half of the world's rainforests are in just three countries: **Indonesia** in Southeast Asia, **Brazil** in South America, and **Zaire** in Africa. Much of the world's biodiversity resides in rainforests. Destruction of the forests will contribute towards global warming through a large reduction in photosynthesis. In the Amazon, 75% of deforestation has occurred within 50 km of Brazil's roads. Many potential drugs could still be discovered in rainforest plants, and loss of species through deforestation may mean they will never be found. Rainforests can provide economically sustainable crops (rubber, coffee, nuts, fruits, and oils) for local people.

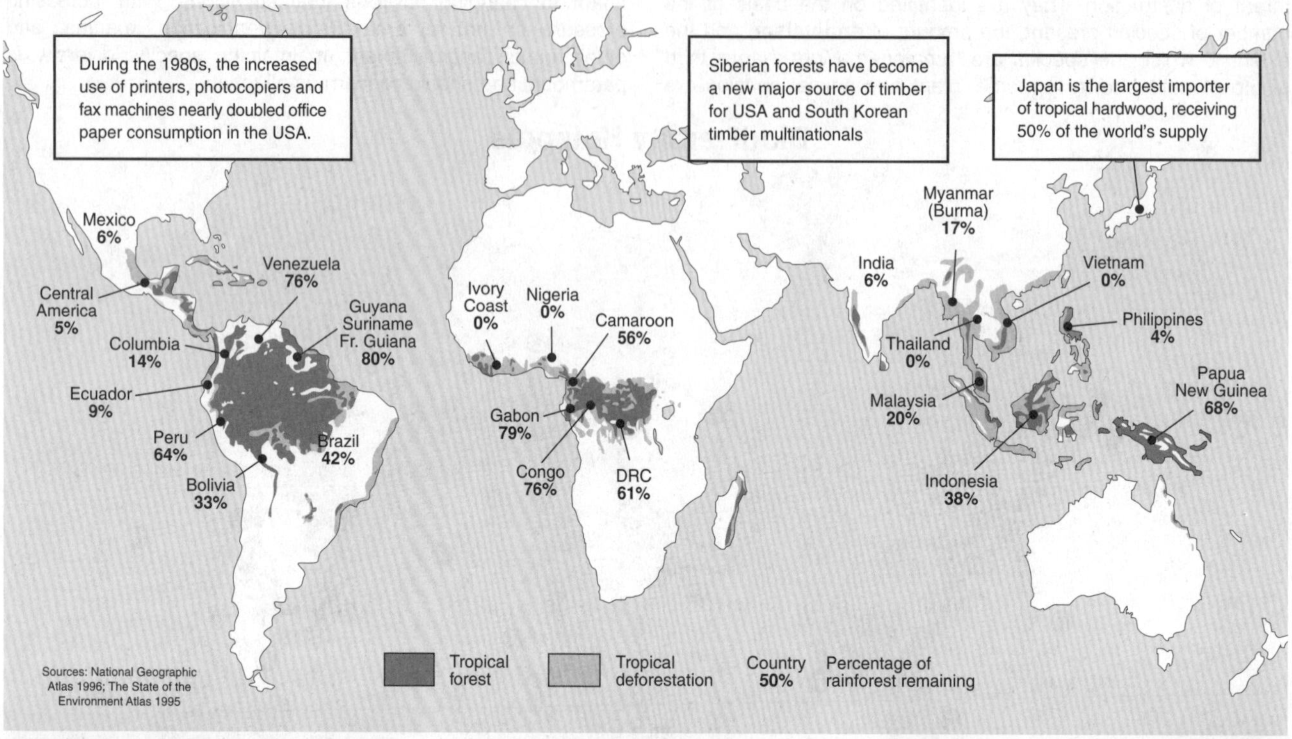

The felling of rainforest trees is taking place at an alarming rate as world demand for tropical hardwoods increases and land is cleared for the establishment of agriculture. The resulting farms and plantations often have shortlived productivity.

Huge forest fires have devastated large amounts of tropical rainforest in Indonesia and Brazil in 1997/98. The fires in Indonesia were started by people attempting to clear the forest areas for farming in a year of particularly low rainfall.

The building of new road networks into regions with tropical rainforests causes considerable environmental damage. In areas with very high rainfall there is an increased risk of erosion and loss of topsoil.

1. Describe three reasons why tropical rainforests should be conserved:

   (a) ______________________________

   (b) ______________________________

   (c) ______________________________

2. Identify the three main human activities that cause tropical deforestation and discuss their detrimental effects:

____

____

____

The Impact of Alien Species

Alien species is a term used to describe those organisms that have evolved at one place in the world and have been transported by humans, either intentionally or in advertently, to another region. Some of these alien species are beneficial, e.g. introduced agricultural plants and animals, and Japanese clams and oysters (the mainstays of global shellfish industries). **Invasive species** are those alien species that have a detrimental effect on the ecosystems into which they have been imported. They number in their hundreds with varying degrees of undesirability to humans. Humans have brought many exotic species into new environments for use as pets, food, ornamental specimens, or decoration, while others have hitched a ride with cargo shipments or in the ballast water of ships. Some have been deliberately introduced to control another pest species and have themselves become a problem. Some of the most destructive of all alien species are aggressive plants, e.g. mile-a-minute weed, a perennial vine from Central and South America, miconia, a South American tree invading Hawaii and Tahiti, and *Caulerpa* seaweed, the aquarium strain now found in the Mediterranean. Two animal aliens, one introduced unintentionally and other deliberately, are described below.

Brushtail Possum

A deliberate introduction

The brushtail possum (*Trichosurus vulpecula*) was deliberately introduced to New Zealand from its native Australia in the 1800s to supply the fur trade. In the absence of natural predators and with an abundance of palatable food, possums have devastated New Zealand's flora and fauna. They are voracious omnivores where they selectively feed on the most vulnerable plant parts and eat the eggs and nestlings of birds, and compete with native species for food. There are now more than 70 million of them and they are widespread throughout the country. Possums also carry bovine tuberculosis and pose a risk to livestock in regions bordering farmed lands.

Red Imported Fire Ant

An accidental invasion

Red fire ants (*Solenopsis invicta*) were accidentally introduced into the United States from South America in the 1920s and have spread north each year from their foothold in the Southeast. Red fire ants are now resident in 14 US states where they displace populations of native insects and ground-nesting wildlife. They also damage crops and are very aggressive, inflicting a nasty sting. The USDA estimates damage and control costs for red fire ants at more than $6 billion a year. Red fire ants lack natural control agents in North America and thrive in disturbed habitats such as agricultural lands, where they feed on cereal crops and build large mounded nests.

The Spread of Alien Species

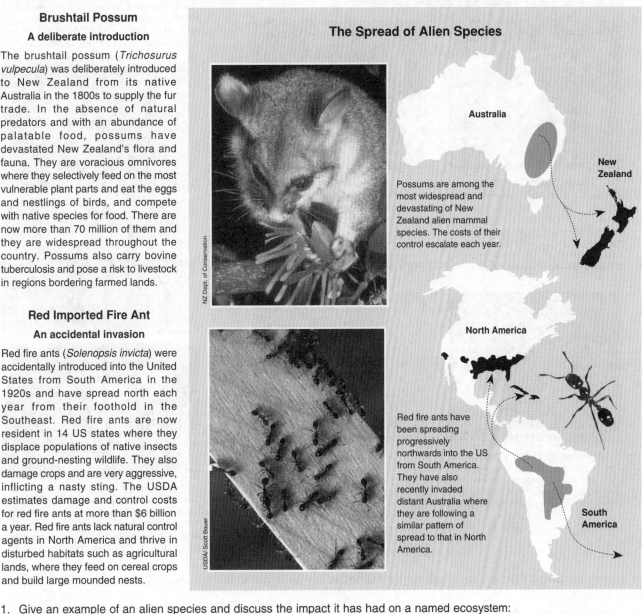

NZ Dept. of Conservation

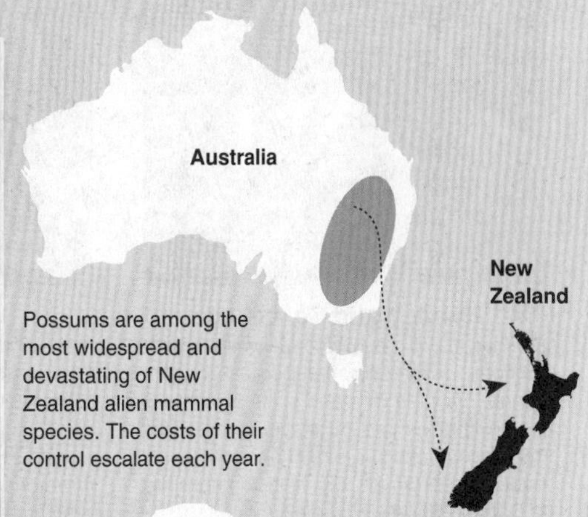
Australia
New Zealand
Possums are among the most widespread and devastating of New Zealand alien mammal species. The costs of their control escalate each year.

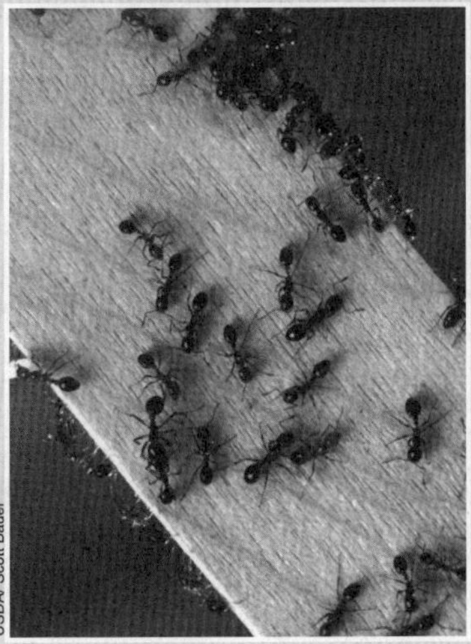
USDA/ Scott Bauer

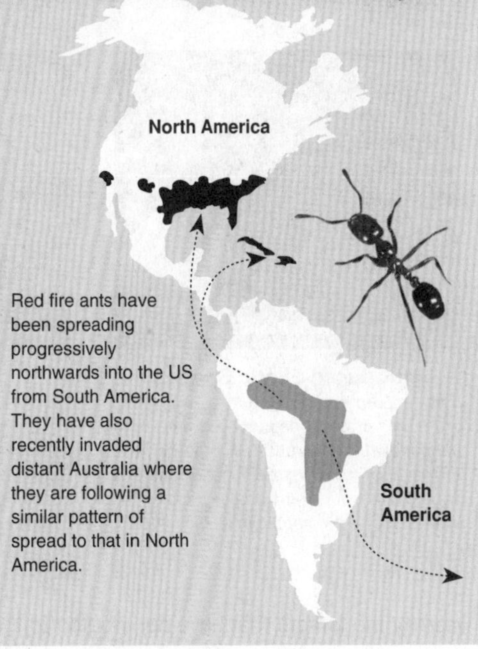
North America
South America
Red fire ants have been spreading progressively northwards into the US from South America. They have also recently invaded distant Australia where they are following a similar pattern of spread to that in North America.

1. Give an example of an alien species and discuss the impact it has had on a named ecosystem:

____

____

____

____

2. Describe an example of biological control of an invasive species: ______________________

____

____

____

Pollution and Global Change

Related activities: Interspecific Competition, Pest Control
Web links: Fire Ants Invade and Evolve

Endangered Species

Species under threat of severe population loss or extinction are classified as either **endangered** or threatened. An endangered species is one with so few individuals that it is at high risk of local extinction, while a threatened (or vulnerable) species is likely to become endangered in the near future. While **extinctions** are a natural phenomenon, the rapid increase in the rates of species extinction in recent decades is of major concern. It is estimated that every day up to 200 species become extinct as a result of human activity. Even if a species is preserved from extinction, remaining populations may be too small to be genetically viable. Human population growth, rising non-sustainable resource use, poverty, and lack of environmental accountability are the underlying causes of premature extinction of organisms. The two biggest direct causes are habitat loss, fragmentation, or degradation and the accidental or deliberate introduction of non-native species into ecosystems.

Causes of Species Decline

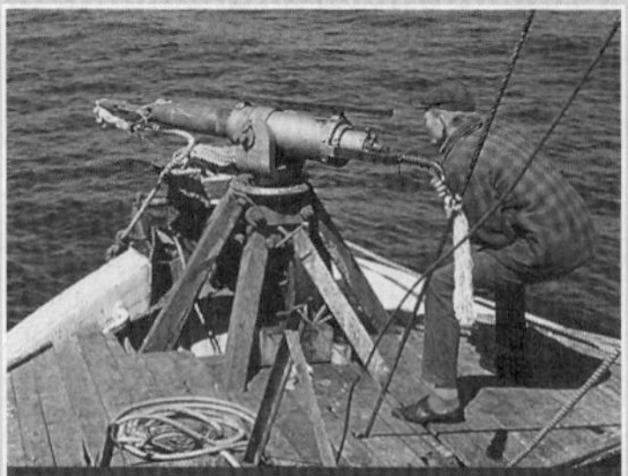

Commercial and "scientific" whaling

Hunting and Collecting

Species may be hunted or collected legally for commercial gain often because of poor control over the rate and scale of hunting. Some species are hunted because they interfere with human use of an area. Illegal trade and specimen collection threatens the population viability of some species.

Clear felling of native rainforest

Habitat Destruction

Natural habitat can be lost through clearance for agriculture, urban development and land reclamation, or trampling and vegetation destruction by introduced pest plants and animals. Habitats potentially suitable for a threatened species may be too small and isolated to support a viable population.

Weasel with stolen egg

DoC

Introduced Exotic Species

Introduced predators (e.g. rats, mustelids, and cats) prey on endangered birds and invertebrates. Introduced grazing and browsing animals (e.g. deer, goats) damage sensitive plants and trample vegetation. Weeds may out-compete endemic species.

Polluted discharge into waterway

Pollution

Toxic substances released by humans into the environment, e.g. from industry, cause harm directly or accumulate in food chains. Estuaries, wetlands, river systems and coastal ecosystems near urban areas are particularly vulnerable.

Case Study: Black Rhinoceros

Black rhinoceros *(Diceros bicornis)* were once plentiful throughout much of Africa. Now, only remnant populations remain. In Kenya, 98% of the population was lost in only 17 years.

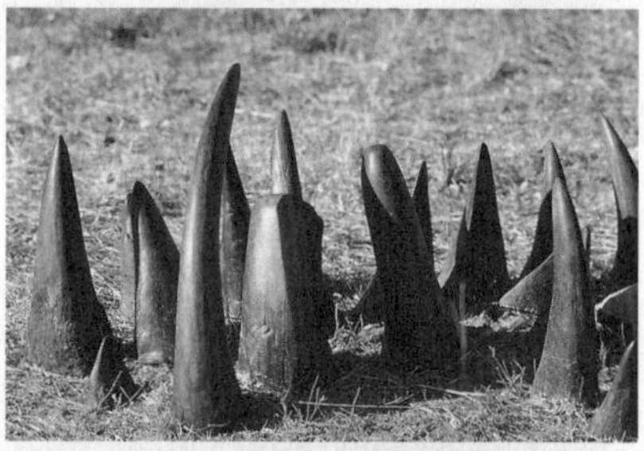

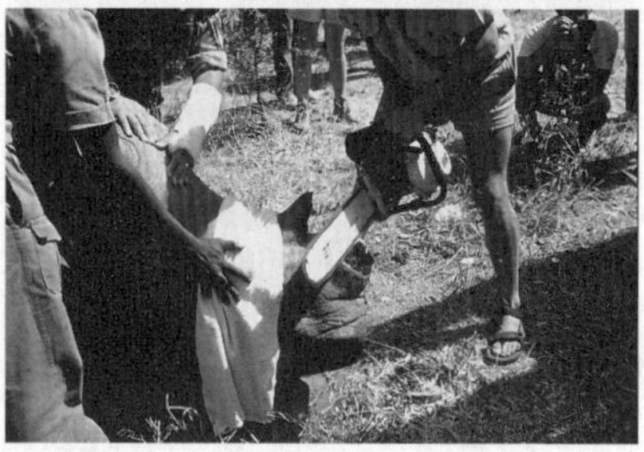

Dehorning programs (above) carried out in Zimbabwe in 1991 have not halted the slaughter. Large numbers of dehorned rhinos are still being shot; conservationists suspect that a trader with a large stockpile of horn is trying to cause rhinoceros extinction in order to increase the horn's value.

1. Identify the factors that have contributed to the **extinction** of one named animal species: ______________________

____

2. Describe two good reasons why any species should be preserved from extinction:

   (a) ____

   ____

   (b) ____

   ____

3. (a) Name an **endangered species** from your own country: ______________________

   (b) Describe the probable cause of its decline: ______________________

   ____

Nature Reserves

Conservation on a national scale generally involves setting up reserves or protected areas to slow the loss of biodiversity. **Nature reserves** may be designated by government institutions in some countries or by private landowners. The different types of nature reserves, e.g. wildlife, scenic and scientific reserves, and National Parks, all have varying levels of protection depending upon country and local laws. Various management strategies (below) are proving successful in protecting species already at risk, and helping those on the verge of extinction to return to sustainable population sizes. Internationally, there are a number of agencies concerned with monitoring and managing the loss of biodiversity. **The Nature Conservancy** is one such organisation. The mission of the Conservancy is to preserve the plants, animals, and natural communities that represent the diversity of life on Earth, by protecting the lands and waters they need to survive. With donations from over a million members, the Conservancy has purchased 12 621 000 acres in the USA and a further 96 386 000 acres outside the USA (an area greater than the combined size of Costa Rica, Honduras and Panama). Larger nature reserves usually promote conservation of biodiversity more effectively than smaller ones, with **habitat corridors** for wildlife and **edge effects** also playing a part.

Strategies for Managing Endangered Species

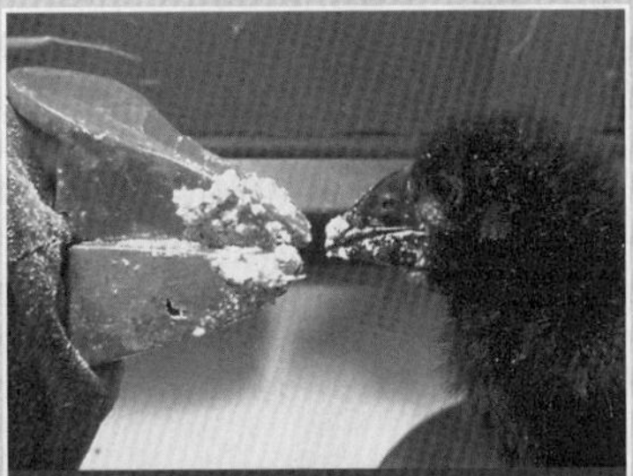
Puppet 'mother' feeds a takahe chick

Woodland-pond restoration (UK)

Captive Breeding and Relocation

Individuals are captured and bred under protected conditions. If breeding programs are successful and there is suitable habitat available, captive individuals may be relocated to the wild where they can establish natural populations. Zoos now have an active role in captive breeding programs.

Habitat Protection and Restoration

Most countries have a system of parks and reserves focused on whole ecosystem conservation. These areas aim to preserve habitats with special importance and they may be intensively managed through pest and weed control programs, revegetation, and reintroduction of threatened species.

Captive bred okapi (forest giraffe)

Orangutan (endangered species)

Zoos and Gene Banks

Many zoos specialize in captive breeding programs, while botanical gardens raise endangered plant species. They also have a role in public education. Universities and government agencies participate by providing practical help and expertise. **Gene banks** around the world have a role in preserving the genetic diversity of species.

CITES

The Convention on International Trade in Endangered Species (or CITES) is an international agreement between governments which aims to ensure that international trade in species of wild animals and plants does not threaten their survival. Unfortunately, even under CITES, species are not guaranteed safety from illegal trade.

Mainland Island Management

A new strategy in conservation involves intensive management of species within a well defined area. These programs have a goal of comprehensive ecosystem restoration, with species recovery being an important consideration. In New Zealand, this strategy has been used very successfully to restore populations of the endangered wattled crow, kokako.

kokako

IF

Kokako (above) are at risk through forest clearance and predation by introduced mammals, especially during the nesting season. Kokako recovery was implemented in a specified area of native forest which was large enough to sustain a viable population but small enough to implement long term pest control.

Chick survival is improved in a restored ecosystem

IF

These "mainland island" projects as they are called, involve very intensive pest control programs and continued monitoring of both pest populations and the species under threat. Such programs are costly but effective; through intensive ecosystem management, the kokako population decline has been reversed and chicks (above) now survive to breed.

1. Discuss how the following *ex situ* conservation measures are used in the restoration of endangered species:

   (a) Captive breeding of animals: ____________________

   (b) Botanic gardens and gene banks: ____________________

2. Identify the advantages of *in situ* (in place) conservation measures, such as island reserves, in conserving biodiversity:

____

Related activities: Endangered Species, Conservation of African Elephants, Loss of Biodiversity

National parks are usually located in places which have been largely undeveloped, and they often feature areas with exceptional ecosystems such as those with endangered species, high biodiversity, or unusual geological features. Canada's National Parks are a country-wide system of representative natural areas of Canadian significance. They are protected by law for public understanding, appreciation, and enjoyment, while being maintained for future generations. National parks have existed in Canada for well over a century. Some 83 million acres (11% of public lands) of the USA are in National Parks and Preserves, which protect natural resources, while allowing restricted activities. National wildlife refuges form a network across the USA, with at least one in every state. They provide habitat for endangered species, migratory birds, and big game.

Parks and Reserves in North America

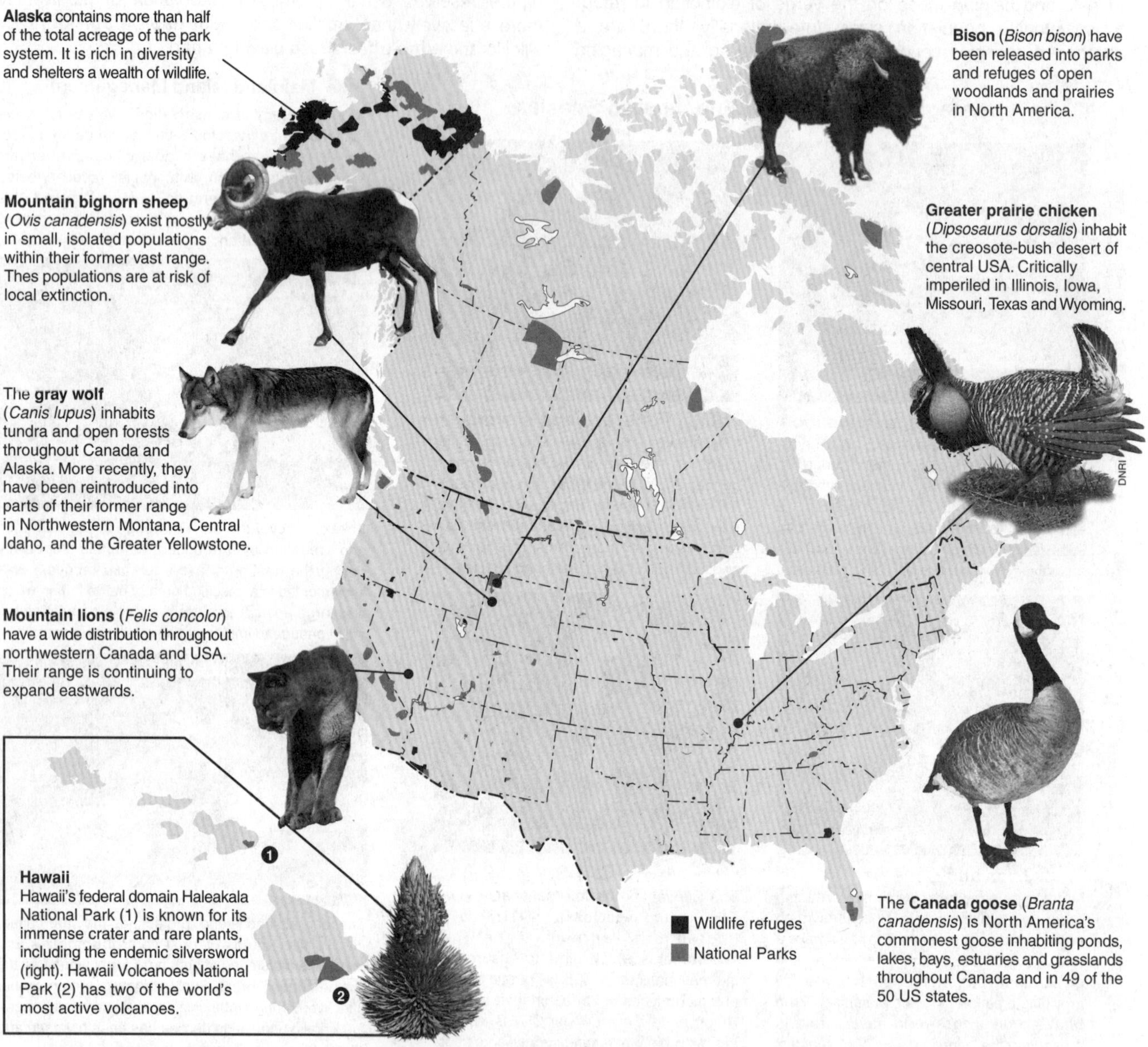
Alaska contains more than half of the total acreage of the park system. It is rich in diversity and shelters a wealth of wildlife.
Bison (*Bison bison*) have been released into parks and refuges of open woodlands and prairies in North America.
Mountain bighorn sheep (*Ovis canadensis*) exist mostly in small, isolated populations within their former vast range. Thes populations are at risk of local extinction.
Greater prairie chicken (*Dipsosaurus dorsalis*) inhabit the creosote-bush desert of central USA. Critically imperiled in Illinois, Iowa, Missouri, Texas and Wyoming.
DNRI
The **gray wolf** (*Canis lupus*) inhabits tundra and open forests throughout Canada and Alaska. More recently, they have been reintroduced into parts of their former range in Northwestern Montana, Central Idaho, and the Greater Yellowstone.
Mountain lions (*Felis concolor*) have a wide distribution throughout northwestern Canada and USA. Their range is continuing to expand eastwards.
1
2
Hawaii
Hawaii's federal domain Haleakala National Park (1) is known for its immense crater and rare plants, including the endemic silversword (right). Hawaii Volcanoes National Park (2) has two of the world's most active volcanoes.
Wildlife refuges
National Parks
The **Canada goose** (*Branta canadensis*) is North America's commonest goose inhabiting ponds, lakes, bays, estuaries and grasslands throughout Canada and in 49 of the 50 US states.

3. Using a local example, discuss the role of **active management** strategies in conservation of an endangered species:

4. Use your research tools (e.g. textbook, internet, or encyclopaedia) to identify a National park or wildlife refuge. Summarise the features that have resulted in it being identified as a protected area:

Conservation of African Elephants

Both African and Asian elephant species are under threat of extinction. The International Union for the Conservation of Nature (**IUCN**) has rated the Asian elephant as endangered and the African elephant as vulnerable. In India, the human pressure on wild habitat has increased by 40% in the last 20 years. Where elephants live in close proximity to agricultural areas they raid crops and come into conflict with humans. The ivory trade represents the greatest threat to the African elephant. Elephant tusks have been sought after for centuries as a material for jewellery and artworks. In Africa, elephant numbers declined from 1.3 million to 600 000 during the 1980s. At this time, as many as 2000 elephants were killed for their tusks every week. By the late 1980s, elephant populations continued to fall in many countries, despite the investment of large amounts of money in fighting poaching. From 1975 to 1989 the ivory trade was regulated under CITES, and permits were required for international trading. Additional protection came in 1989, when the African elephant was placed on *Appendix I* of CITES, which imposed a ban on trade in elephant produce. In 1997 Botswana, Namibia, and Zimbabwe, together with South Africa in 2000, were allowed to transfer their elephant populations from Appendix I to Appendix II, allowing limited commercial trade in raw ivory. In 2002, CITES then approved the sale, to Japan, of legally stockpiled ivory by Namibia, South Africa, and Botswana. African countries have welcomed this decision, although there is still great concern that such a move may trigger the reemergence of a fashion for ivory goods and illegal trade.

Two subspecies of African elephant *Loxodonta africana* are currently recognised: the **savannah elephant** *(L. a. africana)* and the less common **forest elephant** *(L. a. cyclotis)*. Recent evidence from mitochondrial DNA indicates that they may, in fact, be two distinct species.

In 1989 the Kenyan government publicly burned 12 tonnes of confiscated ivory. With the increased awareness, the United States and several European countries banned ivory imports. The photo above shows game wardens weighing confiscated ivory tusks and rhinoceros horns.

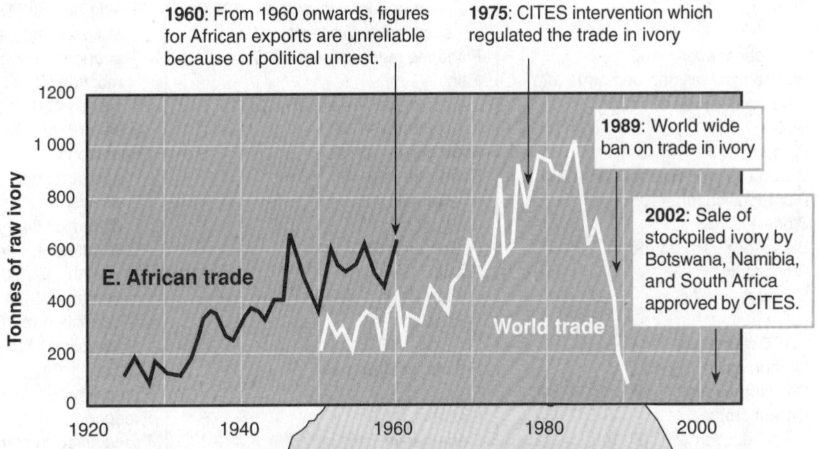

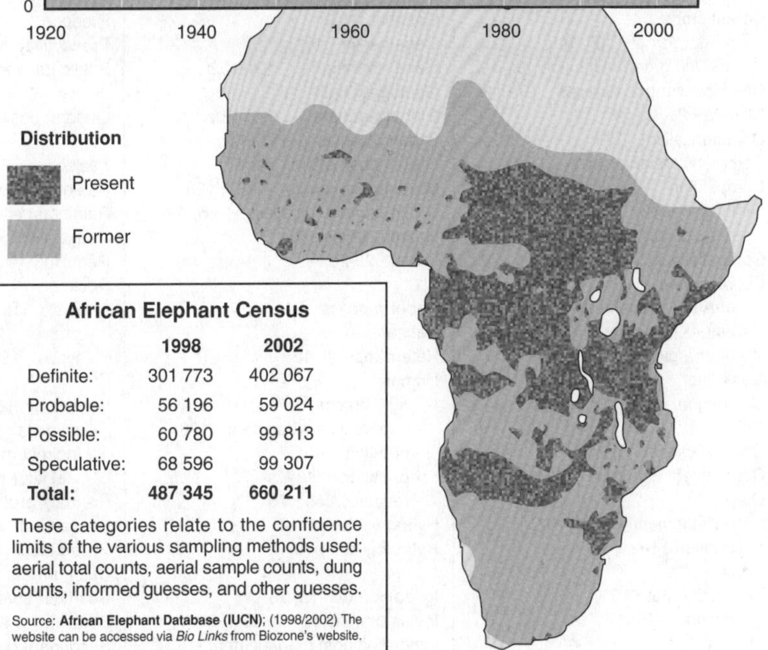

African Elephant Census

	1998	2002
Definite:	301 773	402 067
Probable:	56 196	59 024
Possible:	60 780	99 813
Speculative:	68 596	99 307
Total:	**487 345**	**660 211**

These categories relate to the confidence limits of the various sampling methods used: aerial total counts, aerial sample counts, dung counts, informed guesses, and other guesses.

Source: **African Elephant Database (IUCN)**; (1998/2002) The website can be accessed via *Bio Links* from Biozone's website.

1. Outline the action taken in 1989 to try and stop the decline of the elephant populations in Africa: ______________________

2. In early 1999, Zimbabwe, Botswana and Namibia were allowed a one-off, CITES-approved, experimental sale of ivory to Japan. This involved the sale of 5,446 tusks (50 tonnes) and earned the governments approximately US$5 million.

   (a) Suggest why these countries are keen to resume ivory exports: ______________________

   (b) Suggest two reasons why the legal trade in ivory is thought by some to put the remaining elephants at risk:

Pollution and Global Change

Index